Design and Applications of Positive Energy Districts

Design and Applications of Positive Energy Districts

Guest Editors

Tony Castillo-Calzadilla
Carlos Quesada-Granja

MDPI

Basel • Beijing • Wuhan • Barcelona • Belgrade • Novi Sad • Cluj • Manchester

Guest Editors
Tony Castillo-Calzadilla
Department of Industrial Engineering
University of La Laguna
San Cristóbal de La Laguna
Spain

Carlos Quesada-Granja
Department of Computing Science and Artificial Intelligence
University of Alicante
Sant Vicent del Raspeig
Spain

Editorial Office
MDPI AG
Grosspeteranlage 5
4052 Basel, Switzerland

This is a reprint of the Special Issue, published open access by the journal *Designs* (ISSN 2411-9660), freely accessible at: https://www.mdpi.com/journal/designs/special_issues/4351A1FDU0.

For citation purposes, cite each article independently as indicated on the article page online and as indicated below:

Lastname, A.A.; Lastname, B.B. Article Title. *Journal Name* **Year**, *Volume Number*, Page Range.

ISBN 978-3-7258-6372-3 (Hbk)
ISBN 978-3-7258-6373-0 (PDF)
https://doi.org/10.3390/books978-3-7258-6373-0

Contents

designs

MDPI

Editorial

Towards Net-Positive Energy Urban Districts: A Pillar of the Energy Transition

Tony Castillo-Calzadilla

Department of Industrial Engineering, School of Engineering and Technology, University of La Laguna, Camino San Francisco de Paula s/n, P.O. Box 456, 38200 San Cristóbal de La Laguna, Spain; tony.castillo@ull.edu.es

In the current energy transition landscape, characterised by the dual requirements of accelerated decarbonisation and increased resilience to disruptions in the electricity system, the concept of positive energy districts (PEDs) is becoming increasingly relevant [1,2]. A PED is defined as an urban area or neighbourhood (group of dwellings) composed of buildings of various types, such as commercial or residential, which simultaneously implements demand management strategies, on-site renewable energy generation, and operational flexibility. As a result, it is able to achieve an annual balance of net-imported non-renewable primary energy, lower than the equivalent of the primary energy used, due to local production from renewable energy sources [3,4].

More specifically, the most widely used definition describes PEDs as 'urban areas or groups of connected, energy-efficient and flexible buildings that produce near-zero net greenhouse gas emissions and actively manage surplus local or regional renewable energy production each year' [3]. In their design, these districts are not conceived as energy islands isolated from the system but as smart nodes in the urban–regional energy system, capable of interacting with networks, transport, electric mobility, urban heating/cooling systems, and energy management ICTs [5].

In other words, it is about making the leap from a "zero-energy building" to a "neighbourhood" or "district" that produces more renewable energy than it consumes annually increasing the scale, synergy among buildings, shared infrastructure, storage, and smart management key elements. This approach involves a three-pronged strategy:

(1) Intensive reduction in energy demand through increasing the efficiency of home appliances, enhancing building insulation, and optimising heating, ventilation, and air conditioning (HVAC) systems [5];
(2) On-site renewable energy generation via the integration of solar photovoltaic, solar thermal, geothermal, heat recovery, biomass, etc., to cover residual demand [1];
(3) Control, storage, and flexibility systems that allow for the adaptation of energy production, consumption, and export/import, optimising self-consumption and reducing the impact on central networks [4].

In this regard, the Joint Programming Initiative (JPI) Urban Europe programme (Positive Energy Districts and Neighbourhoods for Sustainable Urban Development) aimed to deploy and replicate 100 PEDs in Europe by 2025. A preliminary analysis of 61 European PED cases indicates that the concept is maturing, although it still faces technical, regulatory, and economic barriers [3].

The strategic role of PEDs is part of the logic of urban energy transformation: cities consume approximately two-thirds of the world's energy and generate around 70% of CO_2 emissions [6]. Therefore, the scalability of the PED model is essential for achieving climate neutrality targets (e.g., by 2050 in Europe) and responding to the urgent need to decarbonise not only the electricity supply but also the heating/cooling, urban transport, and existing building sectors [7].

From an economic and resource perspective, PEDs offer several advantages:

- By generating local renewable energy surpluses and promoting district self-consumption, transport losses are reduced, and the overall system efficiency is improved [1].
- The intrinsic flexibility of these districts (through managed demand, storage, and integrated electric mobility) mitigates the impact of renewable variability and reduces the dependence on peak imports of energy from fossil sources [7].
- In terms of replicability, a neighbourhood or district that achieves a positive energy balance can serve as a demonstrator and accelerator of integrated solutions for other urban areas, leading to economies of scale and reduced techno-systemic risks [8].

Particularly in the transition scenario marked by the need to activate regulatory, financial, and organisational "hindrances" (bottlenecks), PEDs are established as urban innovation laboratories, where new business models, interaction with citizens, energy communities, and local governance are tested [9]. The fact that scientific initiatives such as IEA EBC-Annex 83 are dedicated to systematising definitions, methodologies, and tools for PEDs reinforces their relevance in the field of research and public policy [10].

Looking ahead, several factors shape their critical relevance: the growing electrification of consumption (vehicles, heat pumps, and air conditioning), the demand for emissions reductions in existing buildings (it is estimated that a high percentage of the housing stock in the European Union does not meet optimal energy efficiency standards), and the need to deploy distributed energy infrastructure without compromising the quality of urban life [1,11].

For these reasons, we prepared this Special Issue to serve as a guide, both documentary and procedural/design, to the success stories of positive energy districts, beginning with two reviews. In line with recent analyses by [12,13], international research emphasises that the success of positive energy districts depends on both technological factors and integrated governance and planning processes. While the first study identifies a framework of critical factors from the ideation phase to valorisation that includes coordination between actors, flexible planning, and citizen participation as keys to long-term sustainability, the second highlights the role of dynamic and renewable energy building envelopes in energy efficiency and distributed generation within PEDs. These advances reinforce the vision of PEDs as smart urban ecosystems, where technological and social integration is essential to achieving positive energy balances and climate resilience.

Next, a variety of topics and case studies are explored in the articles published in this Special Issue. In [14], the combination of different energy vectors in the context of multi-energy systems is presented as a crucial opportunity to reach CO_2 reduction goals. According to [15], the transformation of multi-story residences into zero-net-energy buildings is a challenge that can be successfully addressed through a holistic energy efficiency improvement plan. [16] remarks that nature-based solutions (NBSs) offer a promising framework for addressing urban environmental challenges while also enhancing social and economic resilience. [17] develop a techno-economic study, which addresses the pressing energy constraints in nations such as Bangladesh by proposing the implementation of photovoltaic (PV) microgrids. Hashemi and [18] study the importance of DC–DC converters for energy management systems in positive energy districts (PEDs), because they enable efficient conversion between different voltage levels and ease the integration of various renewable energy sources, energy storage systems, and loads. In [19] it is proposed a novel assessment framework for PEDs based on climate neutrality, which is aimed at bridging the gap between research-grade high-resolution analysis and practical policy-aligned application. Meanwhile, [20] conduct a case study focusing on the environmental, social, and economic benefits that PEDs offer. And [21] shows that combining PEDs and urban-industrial symbiosis (UIS) enables self-sufficient systems with near-zero emissions, thus providing a model for sustainable urban development.

In summary, positive energy districts are not only a promising experiment but also a structural vector for operationalising the urban energy transition. Their widespread implementation has a multiplier effect on efficiency, renewable generation, grid flexibility, and citizen participation. Addressing them with scientific rigour, integrated planning, and institutional support is essential to meeting the climate, energy, and sustainable urban development goals that mark this century.

Conflicts of Interest: The author declare no conflict of interest.

References

1. Castillo-Calzadilla, T.; Garay-Martinez, R.; Andonegui, C.M. Holistic fuzzy logic methodology to assess positive energy district (PathPED). *Sustain. Cities Soc.* **2023**, *89*, 104375. [CrossRef]
2. Bottecchia, L.; Gabaldón, A.; Castillo-Calzadilla, T.; Soutullo, S.; Ranjbar, S.; Eicker, U. Fundamentals of Energy Modelling for Positive Energy Districts. In *Sustainability in Energy and Buildings 2021*; Littlewood, J.R., Howlett, R.J., Jain, L.C., Eds.; Springer Nature: Singapore, 2022; pp. 435–445.
3. Europe, J.P.U. Positive Energy Districts (PED)—Programme Page. 2020. Available online: https://jpi-urbaneurope.eu/ped/ (accessed on 1 November 2025).
4. Castillo-Calzadilla, T.; Oroya-Villalta, J.; Borges, C.E. Energy Management System for a Residential Positive Energy District Based on Fuzzy Logic Approach (RESTORATIVE). *Smart Cities* **2024**, *7*, 1802–1835. [CrossRef]
5. Moreno, A.G. How to Achieve Positive Energy Districts for Sustainable Cities: A Proposed Calculation Methodology. *Sustainability* **2021**, *13*, 710. [CrossRef]
6. Marquez-Ballesteros, M.J.; Mora-López, L.; Lloret-Gallego, P.; Sumper, A.; Sidrach-de-Cardona, M. Measuring urban energy sustainability and its application to two Spanish cities: Malaga and Barcelona. *Sustain. Cities Soc.* **2019**, *45*, 335–347. [CrossRef]
7. Castillo-Calzadilla, T.; Alonso-Vicario, A.; Borges, C.E.; Martin, C. E-Mobility in Positive Energy Districts. *Buildings* **2022**, *12*, 264. [CrossRef]
8. Kozlowska, A.; Guarino, F.; Volpe, R.; Bisello, A.; Gabaldón, A.; Rezaei, A.; Albert-Seifried, V.; Alpagut, B.; Vandevyvere, H.; Reda, F.; et al. Positive Energy Districts: Fundamentals, Assessment Methodologies, Modeling and Research Gaps. *Energies* **2024**, *17*, 4425. [CrossRef]
9. Krangsås, S.G.; Steemers, K.; Konstantinou, T.; Soutullo, S.; Liu, M.; Giancola, E.; Prebreza, B.; Ashrafian, T.; Murauskaitė, L.; Maas, N. Positive Energy Districts: Identifying Challenges and Opportunities. *Sustainability* **2021**, *13*, 10551. [CrossRef]
10. Europe, J.P.U. Europe Towards Positive Energy Districts (Booklet), 2020. Available online: https://jpi-urbaneurope.eu/wp-content/uploads/2020/06/PED-Booklet-Update-Feb-2020_2.pdf (accessed on 1 November 2025).
11. IEA EBC–Annex 83 Positive Energy Districts. n.d. Available online: https://annex83.iea-ebc.org/ (accessed on 1 November 2025).
12. Siakas, D.; Siakas, K.; Lampropoulos, G. Positive Energy District Success Factors: Learning from Global Challenges and Success Stories. *Designs* **2025**, *9*, 111. [CrossRef]
13. Almesbah, M.; Wang, J. Review of Dynamic Building Envelope Systems and Technologies Utilizing Renewable Energy Resources. *Designs* **2025**, *9*, 41. [CrossRef]
14. Capone, M.; Guelpa, E. Implementing Optimal Operation of Multi-Energy Districts with Thermal Demand Response. *Designs* **2023**, *7*, 11. [CrossRef]
15. Kitsopoulou, A.; Pallantzas, D.; Bellos, E.; Tzivanidis, C. Mapping the Potential of Zero-Energy Building in Greece Using Roof Photovoltaics. *Designs* **2024**, *8*, 68. [CrossRef]
16. Panori, A.; Komninos, N.; Latinopoulos, D.; Papadaki, I.; Gkitsa, E.; Tarani, P. Blending Nature with Technology: Integrating NBSs with RESs to Foster Carbon-Neutral Cities. *Designs* **2025**, *9*, 60. [CrossRef]
17. Ali, M.F.; Sarker, N.K.; Hossain, M.A.; Alam, M.S.; Sanvi, A.H.; Syam Sifat, S.I. Techno-Economic Feasibility Study of a 1.5 MW Grid-Connected Solar Power Plant in Bangladesh. *Designs* **2023**, *7*, 140. [CrossRef]
18. Hashemi, T.; Jafari Kaleybar, H. The Modeling and Simulation of Non-Isolated DC–DC Converters for Optimizing Photovoltaic Systems Applied in Positive Energy Districts. *Designs* **2024**, *8*, 130. [CrossRef]
19. Schneider, S.; Zelger, T.; Drexel, R.; Schindler, M.; Krainer, P.; Baptista, J. Declaration-Ready Climate-Neutral PEDs: Budget-Based, Hourly LCA Including Mobility and Flexibility. *Designs* **2025**, *9*, 123. [CrossRef]

20. Ntafalias, A.; Papadopoulos, P.; van Wees, M.; Šijačić, D.; Shafqat, O.; Hukkalainen, M.; Kantorovitch, J.; Lage, M. The Benefits of Positive Energy Districts: Introducing Additionality Assessment in Évora, Amsterdam and Espoo. *Designs* **2024**, *8*, 94. [CrossRef]
21. Shafiee Roudbari, E.; Menon, R.P.; Kantor, I.; Eicker, U. Toward Positive Energy Districts by Urban–Industrial Energy Exchange. *Designs* **2023**, *7*, 73. [CrossRef]

designs

MDPI

Review

Positive Energy District Success Factors: Learning from Global Challenges and Success Stories

Dimitrios Siakas [1], Kerstin Siakas [2,3,*] and Georgios Lampropoulos [4,5,*]

1 Sustainable Energy Systems Research Unit, HAMK Tech, Häme University of Applied Sciences, 13100 Hämeenlinna, Finland; dimi.siakas@hamk.fi
2 Department of Industrial Production, School of Technology and Innovations, University of Vaasa, 65200 Vaasa, Finland
3 Department of Information and Electronic Engineering, School of Engineering, International Hellenic University, 57400 Thessaloniki, Greece
4 Department of Applied Informatics, School of Information Sciences, University of Macedonia, 54636 Thessaloniki, Greece
5 Department of Education, School of Education, University of Nicosia, Nicosia 2417, Cyprus
* Correspondence: ksiakas@gmail.com (K.S.); lamprop.geo@gmail.com (G.L.)

Abstract

The aim of this study is to examine existing positive energy district (PED) initiatives by using an explanatory research approach for gaining insight, identifying patterns, clarifying underlying processes, exploring cause-and-effect relationships, and explaining phenomena in a greater depth. Specifically, studies from the existing literature that have explored multiple PEDs were analyzed. Current challenges, barriers, and obstacles, as well as success factors, good practices, and policy guidelines are thoroughly investigated, evaluated, categorized and compared to unveil lessons learnt from diverse existing international projects for turning urban areas into self-sustainable and greener urban neighborhoods. The proposed framework aims to reveal the required processes for successful PED creation and operation. It provides an overview of the current state of the art and enhances comprehension and know-how about the processes needed for the successful adoption and integration of PEDs based on lessons learnt from global challenges and success stories.

Keywords: positive energy district; renewable energy; urban energy systems; energy efficiency

1. Introduction

According to the United Nations (UN), urban cities are predicted to cover 70% of the worldwide population in 2050 and to contribute to approximately 78% of global energy use and consumption [1]. Furthermore, urban cities account for 60% of greenhouse gas (GHG) emissions. Global warming, a result of GHG emissions, continuously causes changes to the climate, which is why reducing GHG emissions globally is an important and urgent matter. This significantly concerns decision makers, leaders and energy providers. From a European Union (EU) point of view, the EU strives to become climate-neutral by 2050 [2], meaning an economy where anthropogenic GHG emissions and their removals are in balance over a given period.

The EU Directive 2024/1275 of the European Parliament and of the Council of 24 April 2024 on the energy performance of buildings (recast) [3] indicates that the EU, including all its member states, aims to accomplish net zero GHG emissions, entailing a balance

https://doi.org/10.3390/designs9050111

between emitted and absorbed atmospheric GHGs. Other legislative frameworks to support member states in reaching the EU targets are the Revised Renewable Energy Directive EU/2023/2413 [4], the Electricity Directive [5], Effort Sharing Regulation [6], the Energy Efficiency Directive [7], and the EU Emissions Trading System [8].

1.1. Creation and Operation of Positive Energy Districts

Sustainable energy refers to "energy resources that can fulfill existing energy requirements while preserving the capacity of future generations to meet their own energy needs" [9] and is highly interconnected with many of the sustainable development goals (SDG) of the UN [10]. In this context, the building energy sector plays a vital role in reducing global warming and in advancing intelligent buildings, intelligent cities, and intelligent regions to stay up-to-date and competitive [11,12] while capitalizing on renewable energy resources [13] and the context in general, e.g., geothermal energy.

The growth of renewable energy sources, distributed energy resources, smart grids, and prosumer engagement, has added complexity and created new challenges regarding the energy supply balance and stability of the overall energy system [14]. In regard to the EU 2050 aim, a joint initiative between the SET Plan and Joint Programming Initiative Urban Europe was created [15] and an Implementation Working Group handling Positive Energy Districts (PEDs) for sustainable urban development was established in Autumn 2018. A target of initiating 100 EU PEDs by 2025 was set [16]. PEDs, also known as "energy-efficient neighborhoods and energy-positive neighborhoods", are system-level concepts with local management of energy sources. The PED locally stores the surplus energy or exports it to the overlaid grid and buys energy from the grid when there is a shortage at the local level [14]. The definition of PEDs is that "Positive Energy Districts are energy-efficient and energy-flexible urban areas or groups of connected buildings which produce net zero greenhouse gas emissions and actively manage an annual local or regional surplus production of renewable energy. They require integration of different systems and infrastructures and interaction between buildings, the users and the regional energy, mobility and ICT systems, while securing the energy supply and a good life for all in line with social, economic and environmental sustainability" [17]. The PED concept is based on earlier ideas of single nearly/net/positive zero energy buildings, extended with energy mutuality between urban decarbonization and buildings [18]. The concept of PEDs aims to deliver environmental, economic, and social benefits through the optimization of local energy systems. PEDs constitute an innovative concept with challenging adoption and implementation. Data from existing PED projects aiming toward effective energy transition and sustainable urbanization a across Europe has been collected by JPI Urban Europe [19].

Through the "Clean energy for all Europeans package" [20], which was implemented in 2019, the EU introduced the concept of energy communities (EC) in its legislation. The aims were to empower citizens to drive the energy transition locally to benefit from advantages such as improved energy efficiency, reduced energy poverty, lower bills, and prospects for green local jobs [20]. The idea was that citizens would aid in improving the public acceptance of renewable energy projects as well as facilitating private investments and the restructuring of energy systems. In 2022, however, over 41 million EU citizens were unable to warm/cool their homes adequately [21]. Energy poverty is a multi-faceted phenomenon caused by low income, high energy expenses, and poor building energy efficiency. Hearn [22] interviewed 20 stakeholders who were directly involved in managing the development of PEDs, smart cities, or carbon-neutral districts, many of whom held multiple roles. They noticed that stakeholders recognize significant potential in PEDs for reducing energy vulnerability. Recognizing the potential contribution of ECs in realizing an affordable, secure, and cleaner energy system, the REPowerEU Plan [23] proposed the political objective of, by 2025, attaining one EC per municipality with a population of higher than 10,000. The EC goes hand in hand with PEDS.

PEDs are flexible and efficient energy areas and require the integration of different energy systems (electricity, heating, cooling, energy storage) and interaction among users, buildings, regional energy systems, transport, and information and communication technologies (ICT) systems. Energy-positive areas can be of many types and consist of many different actors. Most existing PED projects (66%) connect newly built neighborhoods to existing ones [24]. Technological limitations, complex ownership structures, building protections, and the preservation of cityscape may present challenges in implementing a PED, particularly in existing urban structures. However, approximately 75% of the building stock in the EU is energy inefficient [25], which make integrating and applying PED solutions to the existing urban structure a priority. Terés-Zubiaga et al. [26] consider that an efficient way to accelerate the reduction of energy consumption in the building sector is by increasing renovation rates at the district level, which involves upgrading multiple buildings within the same area.

An 8-step process of PED development was proposed by the Joint Research Centre report, that is, "the science and knowledge service of the European Commission" [27]. The report claims that the positive impacts of PEDs principally include a reduction in the EU carbon footprint and the empowerment of citizens and stakeholders to take the lead in their own communities. The report also states that the PED approach is complex due to the potentially high number of technological combinations at the district level. In order to facilitate PED, the local management system needs to utilize innovative ICT and autonomous cyber–physical systems (CPSs) for addressing the processes of controlling, managing, monitoring, and optimizing energy generation, consumption, and trading [28]. CPSs form part of the basic frameworks for the creation of smart and autonomous systems [29,30]. They enable, through multiple synchronous applications over diverse network structures, continuous communication and data analysis among physical devices, cyber platforms, and digital infrastructures [29,30]. Developments in computer science, electronics, and robotics drive the advancement of automated sensing systems, reducing the limitations of manual data collection. CPSs are being applied in the development of different smart applications, such as smart buildings, smart cities, smart transport, smart grids and utility systems, etc.

1.2. Achieving Positive Energy District Goals

In the 2010s, the European concept of smart cities put an emphasis on sustainable urban environments that utilized smart technologies integrating data and sensors collecting information from the surrounding environment. A minimum level of artificial intelligence is fundamentally incorporated into the smart innovative technologies [31]. Whilst smart home concepts aid end-users in comprehending their energy use, the concept of PEDs aims to reduce energy consumption and locally produce more energy than is consumed on an annual basis [32]. The PED concept originates in nearly zero-energy buildings and zero-energy buildings (ZEBs), which introduce an integrated approach to energy efficiency, emphasizing the utilization of renewable energy sources [27]. Urban development is transitioning from building solutions toward advanced PEDs with the aim of accelerating the transition of clean energy for achieving the EU's climate and energy targets to a district level [33].

The integration of different ICT and CPS tools is significant to develop smart cities and PEDs. Progress in computer science, electronics, and robotics impels the development of automated and autonomous sensing systems, which ensure consistency and enable the optimization of processes, minimization of human intervention, and facilitation of remote surveillance [34]. Digital predictive technologies that use analytical approaches to identify patterns and trends for predicting future events and outcomes are significant tools for adhering to the targets of ZEBs and PEDs. Together with sensor-based control systems, which activate diverse functions based on the detection of changes in the environment, ZEB policies for buildings and PEDs are supported [35]. Interactions between humans, infrastructures, and technologies are increasing; hence, there is an emphasis on human-centric PED designs [36]. Buildings, cities, and regions are utilizing technological improvements, aiming for smarter performances and improved sustainability. This phenomenon is called the urban intelligence concept [35]. Future cities will progressively be managed through smart city solutions that consider the efficiency of water and energy consumption, waste management, and reducing noise, pollution, and traffic congestion. Energy consumption is the main concern of PEDs.

The three Ds, namely decarbonization, decentralization, and digitalization, are the solutions proposed to meet the increased energy demand caused by urbanization, low-carbon policy requirements, and a transactive energy market [37].

Decarbonization: The EU goals of climate-neutrality by 2050 are in harmony with the decarbonization goal of developing smart CPS distribution systems that promote energy efficiency, local energy accommodation, and energy transmission flexibility [38].

Decentralization: An increasing number of distributed energy resources are nowadays integrated into the distribution system. Such DERs are building PV systems, electric vehicles, and distributed energy storage. Consumers represent a significant component of any smart energy system. In emerging distribution systems, electricity consumers are going to be more actively involved in decisions regarding energy, including home energy management and participation in peer-to-peer markets [39]. Since human behavior is important in PEDs, emphasis toward cyber–physical–social distribution systems should be considered [40].

Digitalization: To achieve real-time monitoring and control, smart devices are installed in a distributed manner. These devices include control mechanisms, communication, and a big collection of data. They are connected to each other or to the control center via a computer network by wired or wireless communication techniques. This type of digitalization through distribution systems describes CPSs [38].

1.3. Research Question, Aims of This Study

The research question of this study is as follows: What lessons can be applied from diverse existing international projects regarding the transformation of neighborhoods into new self-sustainable, positive and greener urban areas? To answer the research question, a critical examination of existing PED initiatives, their success stories, good practices, lessons learnt, and policy guidelines, together with challenges and obstacles, needs to be carried out. This study aims to expand the current understanding of PEDs by thoroughly examining, identifying, categorizing, evaluating, comparing, and presenting earlier research findings in a structured way enriched with reasoning regarding the results. The ultimate aim is to enhance comprehension and know-how about the creation and operation of PEDs and to contribute to further development and research in this important domain.

1.4. Structure of This Article

The structure of this article is as follows: Section 2 describes the research method. In Section 3, the analysis of experiences from existing PEDs is presented. This analysis forms the basis of the PED Success Factor framework. The analysis is based on insights from studies that examined multiple PED projects. In Section 4, the PED Success Factor frame-work is presented. In Section 5, the related outcomes are further discussed, and in Section 6 the conclusions of the study and future research directions are presented.

2. Method

The explanatory research method was used in this study for exploring cause-and-effect relationships, understanding underlying processes, and explaining phenomena in more depth. This method supports researchers in gaining valuable insights into a topic, clarifying concepts, and identifying patterns, before conducting more conclusive and structured research. Opposed to descriptive research, that only describes phenomena, explanatory research seeks to answer questions such as "Why does this happen?" and "What factors influence this outcome?" Skinner et al. [41] argue that explanatory studies are trying to answer two questions, namely "What are the causes of normative and differential patterns of development as they unfold in the actual contexts of daily life?" and "What are the mechanisms by which those causes exert their effects in shaping development?" The aims of this explanatory study are to gain insights into what factors are important for successful PED implementation and operation.

A systematic literature review following suitable approaches, such as the PRISMA guidelines, is often regarded as a valid method for identifying and collecting related documents, but a broad and wide-ranging topic, such as PED challenges and success stories, requires a broader perspective and more flexible approaches [42]. Hence, although this can be regarded as a limitation of the study, a systematic literature review approach was not adopted.

A typical explanatory research method inevitably includes a literature review for reviewing existing information in order to identify gaps or trends, seek new insights, and evaluate phenomena from a different point of view. The overview on PED experiences, such as drivers, success factors, challenges, and barriers, involves investigating various scientific articles published in journals and conferences. As this study covers a wide-ranging, complex, and interdisciplinary topic, a comprehensive critical literature review, following suitable guidelines [43], was performed by searching for suitable scientific articles in well-known library repositories, such as "Web of Science", "Scopus", "ACM", "IEEE", and "Google Scholar", using suitable keywords, such as "drivers", "experiences", "success factors", "challenges", "barriers", "positive energy districts", and "PED" to identify relevant documents that were then further processed and to critically analyze prior research without bias.

This study strives to present a comprehensive collection of fundamental PED success factors and to propose a suitable framework with the aim of improving the understanding of critical aspects in the creation of PED projects. To narrow the research scope, this study applies an explanatory research approach by exploring why something happens (the creation of PEDs), explaining the causes (minimizing global warming) and consequences (challenges and success factors).

Therefore, this study put emphasis on identifying studies within the literature that explored the processes and outcomes of multiple PED projects. Hence, this study focuses on expanding the existing understanding of PEDs by examining real-life PED experiences of multiple projects to identify key success factors. These factors were further examined, which led to the creation of the PED Success Factor framework. The framework highlights key aspects and processes that need to be created and nurtured throughout the different stages (e.g., initiation, planning, prototyping, execution/implementation, evaluation, etc.) of the PED lifecycle. The process of mapping is important to understand and improve operations by visualizing workflows, identifying inefficiencies, and promoting communication and collaboration. Specifically, it improves the understanding of how work is performed, locates areas for improvement, and standardizes processes for increased efficiency and compliance. It also ensures a more all-inclusive identification of stakeholders, capturing the dynamic nature of their emergence, changes in relative importance, and withdrawal [44].

3. Experiences from Existing PEDs

3.1. First Study—Analysis of 62 PEDs

Saheb et al. [45] revealed findings from 62 pioneering PEDs identified in EU project databases, such as Intelligent Energy Europe, CONCERTO, FP7, Covenant of Mayors. The drivers they distinguished for utilizing PED approaches (in addition to contributing to combating against global warming) were motivated by urban renewal and the conversion of industrial or brownfield areas into residential or mixed communities. They emphasize that "from a policy perspective, the concept of a zero-energy community has proven to be hard to implement and evaluate" despite the fact that there is evidence for "the feasibility of the zero-energy concept for individual buildings". They emphasized the lack of consistent terminology and targets used when implementing 'zero energy communities', which makes it difficult for policymakers and academics to carry out independent assessment. They identified a hierarchical approach that enabled decision makers to adapt to the specific challenges and opportunities of the local context. In our opinion, a democratic approach would be preferable since stakeholder commitment is required and in a hierarchical structure full commitment might be difficult. Moreover, they identified that cross-sectoral planning is beneficial because in addition to energy issues, it might include other fields like waste or water optimization. To improve citizen commitment, they identified capacity building and training of municipalities and citizens as a way to enable their complete and effective participation. They also emphasize that the adequate documentation and monitoring of projects aims to offer a solid evidence base regarding what went well/did not go well and mistakes and is significant for future PED development.

By analyzing the outcomes of Saheb et al. [45], some key aspects that lead to successful PED development, adoption, and integration were identified, as presented in Table 1 and further discussed and analyzed.

Table 1. Key aspects in developing effective PEDs.

Key Aspects in Developing Effective PEDs Based on Saheb et al. [45]	Analysis and Comments
"Clearly defined long-term targets, community boundaries and values to support citizen engagement"	The deep-rooted values and interests of individual societies display divergence of values, having an impact on business models, governance structures, decision making, knowledge sharing, social acceptance, etc., [46]. Therefore, it is of utmost importance that targets, community boundaries, and values are discussed, aligned and clearly defined by stakeholder collaboration from the very beginning. The value that each strategic initiative is expected to create for the different stakeholders also needs to be discussed and elaborated on. In particular, citizen engagement is considered important in PEDs [47], and to make an impact it needs to be embedded in every stage of the decision-making process. Involving citizens is considered important in PEDs, as they undertake an integral role in defining themes, finding solutions, and identifying priorities for action [48]. Citizen involvement is, however, voluntary, and it requires deep commitment, proper allocated time, and an interest in co-creation.

Table 1. *Cont.*

Key Aspects in Developing Effective PEDs Based on Saheb et al. [45]	Analysis and Comments
"Linking targets to community priorities such as economic development, urban renewal, energy poverty, energy security (especially for isolated communities)"	In order to obtain political-level and municipal commitment and financial support, it is significant to align PED targets with municipal priorities. After identifying municipal priorities and having decided on strategic initiatives, effective communication is needed so that all involved stakeholders understand and agree on the initiatives. Strategic project management improves decision making, project performance, and competitiveness [22]. Strategic planning forms the path to follow, but smooth alignment with project execution may be difficult if all stakeholders are not onboard, giving rise to inadequate alignment that causes inefficiency, and may not bring expected results. Hearn [22] emphasizes that stakeholders perceive the reduction of energy poverty as having significant potential in PEDs and highlights that PED replication should, in addition to decarbonization, also address energy poverty mitigation.
"Transposing long-term goals into milestones and short-term objectives to avoid discouraging the community"	PED implementation necessitates a profound understanding and consideration of policies, priorities, strategies, resources, solutions, and contextual conditions [11]. In strategic management, the vision refers to mental images of the future (zero-energy communities), which become tangible in the form of mission statements (100 PEDs by 2025), a definition of the primary purpose that articulates the responsibilities the stakeholders (distributed decision-making and stakeholder commitment). The goals are attempts to improve performance by making mission statements more concrete. Objectives represent the operational definitions of goals in more precise terms and describe what needs to be accomplished in order to reach the goals. A significant factor that was identified in the four projects was to divide long-term goals into short-term objectives and to use milestones. Balancing a long-term vision with short-term milestones is critical for stakeholder commitment and effective productivity.
"Measurable targets and a clear implementation timeline"	An effective implementation plan with measurable targets and a clear implementation timeline provides a feasible roadmap for project execution. It outlines key elements to guide the PED project toward successful project completion and goal achievement. The foundation of the implementation plan relies on a well-defined project scope and clear goals. Cai and Gou [49] proposed a set of performance indicators (KPIs) based on the KPIs of existing positive energy buildings (PEB), including the geometric data of the building and energy-related indicators, such as the power of the PVs and the area of the PVs deployed. Similarly, Barrutieta et al. [50] looked at building geometry, location, energy consumption, and building-integrated photovoltaics and their interrelations to create PED KPIs. Both studies emphasize that balance assessment KPIs can include a renewable energy supply compared to energy demand or annual energy exchanges with the grid. The Smart Readiness Indicator (SRI) has recently also emerged as a promising KPI aimed at energy savings and assessment of the ability of buildings to respond to user needs and energy flexibility. SRI evaluates the capability of a building to integrate future emerging technologies and adapt to occupant requirements regarding the functionality levels of various smart services [51].

Table 1. *Cont.*

Key Aspects in Developing Effective PEDs Based on Saheb et al. [45]	Analysis and Comments
"Transparency about progress toward these targets and any revisions or trade-offs made during project implementation"	Transparency and inclusiveness toward KPI targets are essential for tracking progress, making well-informed decisions, and driving success. A communication strategy considering KPI transparency creates alignment, trust, and engagement. Innovation is underpinned by the concepts of invention (creation of a new idea or concept) and creativity (the act of turning new and imaginative ideas into reality). Stakeholder engagement is a key driver of PED value, enabling, inclusive, effective, and sustainable energy solutions. Engaging stakeholders, in particular citizens, enhances co-creation, aligns interests, and supports the successful planning, implementation, and long-term sustainability of PEDs. It maintains that PEDs are not only technically effective but also socially accepted. The emphasis is on optimizing the value of the project and innovation for diverse stakeholders (society, community, institutions, and individuals) and boosting its impact [52].
"The supply and demand of urban services (energy, waste, water and transport) needs to achieve integration at a local level"	A strong awareness exists regarding increased well-being of people as a result of accessibility to services and amenities provided in urban areas and, as a consequence, the management of these services is a prior concern for both citizens and PED planners. A PED performance-based planning approach is built around urban ecosystem services, such as energy, waste, water and transport [53]. The type of performance, for example, what kind of ecosystem services should be targeted, depends on service demand of the ecosystem, and the level of performance depends on the impacts on ecosystem service supply [54].
"The local residents and community need to be meaningfully engaged"	Local residents' active participation is not only beneficial but essential for successful PED development. It ensures that the development process is inclusive, sustainable, and reflective of the needs and aspirations of the PED. PEDs need citizens who are users, producers, consumers, and owners, in addition to political actors. A combined effort from these actors may have a significant impact on PEDs, driving the climate transition, preserving the environment, and advancing the economy [47].

Furthermore, some key challenges were also identified based on the analysis of Saheb et al. [45]. These challenges are presented and further analyzed in Table 2.

Table 2. Key challenges in developing effective PEDs.

Key Challenges in Developing Effective PEDs Based on Saheb et al. [45]	Analysis and Comments
"Lack of agreement regarding the definition of a zero-energy community, its boundaries and which urban services are included in it"	Kozlowska et al. [18] highlighted the need for a clear and comprehensive PED definition, and the agreement of a coherent PED design approach. In order to find an adequate balance between different energy sources in a PED, it is imperative to identify which renewable energy resources are suitable in the climate zone of the PED, which specific needs and ambitions they represent, and what different functions and guiding principles need to be optimized against each other [55]. When developing PEDs, the specific situation of the urban context should be considered, such as density, type of buildings, available local renewable energy resources, etc. Most projects take different pathways due to local circumstances regarding energy sufficiency measures, use of renewable energy produced locally or nearby, and use of technologies for minimizing energy needs. Also, the level and kind of local integration of supply and demand of urban services play an important role.

Table 2. *Cont.*

Key Challenges in Developing Effective PEDs Based on Saheb et al. [45]	Analysis and Comments
"Replication difficult because the geographic, political, economic, historical, and social context of urban areas differ significantly"	Across different countries, cities, or districts, standardized PED solutions are difficult to apply because they are content-specific both regarding existing values [55,56] and also because of their spatial and morphologic differences [18,57]. The identification of common PED characteristics was proposed for the creation of generic PED archetypes [18]. Key PED characteristics were grouped into the following: (1) facts and figures (built form, climate, density, energy demand, land use, renewable energy potential, physical geographical location, sizes/population size); (2) technologies (energy distribution (e.g., co-generation, district network), energy-efficiency measures, energy storage, mobility solutions, renewable energy supplies); (3) quality of life (accessibility to green space, accessibility to services (e.g., bike lane, public transportation)), health impacts (e.g., air pollution, noise pollution), local value/sense of community, social-economic conditions, user comfort; and (4) other (impacts of PEDs, local challenges, local targets and ambitions, regulations/policies, stakeholder involvement.
"Governance and citizen engagement"	Energy citizenship denotes active participation in energy systems and engagement in energy-related discourse through conscious decisions related to energy [58,59]. The analysis approach, a combination of citizenship critical social psychology and governmentality (a society where members play an active role in their own self-government), reveals neoliberal energy citizenship (free-market capitalism, reduction in government spending and deregulation) as the dominant social representation. It identifies unrestricted representations of energy citizenship grounded in entrepreneurial energy engagement at a local level. The image of an ethical PED prosumer has been argued to exclude information for illiterate people to become prosumers [60] because of their lack of energy awareness and energy action. Focusing on local and human rights claims, the analysis of Nguyen and Batel [58] also revealed the following two types of energy citizenship, namely the "active vs. vulnerable consumer" (at national policy level tackling existing and future energy poverty) and the "local–global citizen" (tackling global climate matters by local energy actions).

3.2. *Second Study—Insights from 61 PEDs*

The insights from 61 PED projects were presented by Bossi et al. [24] following the analysis of the JPI Urban Europe booklet [19] for identifying common characteristics to advise and guide PED stakeholders. Further, 29 cases of the total 61 cases outlined in the booklet reported a PED ambition. Bossi et al. [24] found that there is a geographic imbalance between PED/toward-PED projects, most of them being in Norway (eight) and only two in Eastern and Southeastern Europe, which they attribute to the existence of particular matters or national programs that focus on PEDs.

In 2020, the majority of the PED projects were in the stage of implementation (69%); 24% were in the stage of planning and only a few in the stage of realization (3%) or in operation (3%) [24]. Toward-PED projects were in their implementation stage (44%); 13% were in the planning stage and realized (9%) or in operation (31%). By analyzing the building structures, they were grouped into "Newly Built", "Existing Neighborhood", and "Mixed".

In total, 66% of the projects with PED ambition were a mixed type of neighborhood with both newly built and existing buildings and infrastructures, 28% were newly built districts, and 7% were existing neighborhoods. Totally, 41% of the toward-PED projects were based on existing neighborhoods, 38% on mixed type, and 16% were newly built.

A common PED approach seemed to be to combine a high number of energies. Totally, 29% of the PED projects used "moderate mixed" (2–3) energy sources, 50% used "highly mixed" (4–5) energy sources, and 13% "very highly mixed". Only 8% of the PEDs used a single source of energy. Totally, 60% of the toward-PED projects were "moderate mixed", 30% were "highly mixed", 7% used only a single source, and only 3% used "very highly mixed" energy typologies. Hence, we see that PEDs use more energy sources (4–5) than toward-PEDs (2–3) for reaching the required energy surplus from renewable energy sources. The preferred types of energies were district and local heating, photovoltaic, heat pump systems, and geothermal energy [24].

Bossi et al. [24] summarize that the PED main success factors are as follows: "Stakeholders; citizen involvement strategies, integrated technology and political support". Siakas et al. [46] also studied the JPI Urban Europe booklet and carried out a thematic analysis for categorizing, extracting, and synthesizing texts. The data were examined to identify common themes and topics, ideas and patterns of meaning that were expressed repeatedly. The success factors they identified included early and broad stakeholder involvement, interdisciplinary collaboration, and context-sensitive planning for addressing local needs and avoiding unintentional consequences (i.e., gentrification, inequality). They also recognized that enabling policies and financing instruments are significant for supporting innovation, and scaling and replicating PEDs in diverse urban settings. However, for replicating successful PED models, culture is a decisive factor that needs particular attention. A socio-cultural, technological, economic, environmental, political, legal, ethical, and demographic (STEEPLED) factor analysis [46] was proposed for a more complete assessment including multiple aspects. Interdependencies between technology, governance, and market dynamics characterize PEDs. Emerging contemporary technologies considered important for the creation and operation of PEDs, such as cyber–physical systems, CPS, Digital Twins (DT), artificial intelligence, Internet of Things, edge computing, and Blockchain technology, are integrated and intertwined in different ways. A holistic, cross-disciplinary, and systematic understanding is crucial for managing complexities and urban challenges.

The primary PED barriers are "access to adequate funding and business models". In comparison to traditional projects, PED projects are more expensive and complicated. Hence, there is a need to adopt and apply advanced business models and obtain multiple financial sources. Flexible regulations and a devoted legal framework are required for meeting the new challenges. Based on the work of Bossi et al. [24], key areas to be examined were identified, which are presented and further discussed in Table 3.

Table 3. Key areas to be examined.

Key Areas to Be Examined Based on Bossi et al. [24]	Analysis and Comments
"Using PED Labs for testing, experimenting with different approaches and strategies, elaborating with guidelines and tools and monitoring of existing approaches."	A living lab is an open innovation ecosystem where new ideas and solutions are developed and tested in a real-world context. Open innovation is a tool for integrating customers/end-users in the innovation process, particularly in the ideation stage of innovation, where the voice of the customer is crucial for later customer acceptance due to a feeling of participation and influence on the process/product [61]. A living lab is an excellent place for developing and testing innovative ideas in real life. In a living lab, end-users provide feedback at each stage of the iterative process. The continuous feedback enables adaptation and improvement of the innovations to meet end-user needs and requirements. PED projects usually create living labs across all countries of the project aiming to pilot real-world solutions for more inclusive, affordable, and sustainable energy systems.

Table 3. *Cont.*

Key Areas to Be Examined Based on Bossi et al. [24]	Analysis and Comments
"Developing appropriate governance structures and adaption of the legal framework to fit PEDs consideration of functional urban areas (regional perspective); Collaboration within and between stakeholders, such as city administrations, civic society, energy providers, real estate industry etc."	Governance structures enabling policy and planning integration at both the vertical and horizontal level are crucial for allowing multi-level coordination at the EU level, national level, regional level, and local level. Public and private partnerships involving local authorities, energy providers, real estate developers, and citizen groups develop shared ownership and investment constructs that foster accountability. In particular, citizen support is needed for urban transformation solutions [47]. Smart integration and cross-sectoral collaboration between energy, transport, construction, water, and waste sectors are desirable features that will add value to the creation and operation of a PED.
"Incorporating of PED strategies into a comprehensive urban planning: mainstreaming of energy planning in urban planning strategies; connecting energy aspects with climate action; adapting high-quality regarding functions and design."	Due to changing meteorological conditions imposed by climate change, the modeling of energy systems faces a transformation. As a result, a community of practice in energy–climate modelling, aiming to increase the integration of energy system models with weather and climate models, has developed [62]. Connecting energy aspects through coordinated action across technology, policy, and society to climate action is essential for effective climate change mitigation, a reduction in GHG emissions, and the achievement of climate goals. A shift toward renewable energy, increased efficiency, and inclusive, context-sensitive approaches requires coordinated action across sectors and disciplines to attain a sustainable, low-carbon future [63].
"Developing achievable business models and identifying funding opportunities: raising awareness and aspiring for political support regarding national programs; consideration of alternative renewable energy sources and technological solutions; exploring job creation and boosting local/regional economy."	A basic requirement for a PEB business model is the inclusion of a set-up of renewable energy and energy storage. The development of common business models and protocols to manage complexity and interdependencies is required and expected, but practice has shown that PEB business models are different according to geographical conditions, which impact both technology choices and energy needs. In addition, every country has its own building traditions, socioeconomic conditions, legislation, and building regulations. A practical roadmap, based on the business model, effectively guides the implementation and operation process [46].
"Investigating strategies suitable for existing urban structures: merging renewal and greening strategies; including stakeholders with a particular focus on landowners and citizens."	Aging infrastructure, heritage preservation, and dense populations are challenges that PEDs meet in existing urban structures. However, several retrofitting and integration strategies can facilitate existing districts to meet or exceed energy-positive targets [13]. Merging renewal and greening strategies in PEDs while meaningfully including stakeholders, particularly landowners and citizens, is both an opportunity and a challenge. Greening strategies concentrate on supporting well-being and livability by focusing on vegetation and green spaces, improved air quality, biodiversity, and urban cooling. Synergies between different approaches, such as circular economy, greening strategies, building renovation, and energy retrofits emphasize solutions such as PV-integrated shading, permeable surfaces, and green mobility. Circular economy is an evolving business model that is considered restorative. It is increasingly considered as an appropriate solution to achieving prosperity whilst acknowledging ecological and social boundaries.

The "Covenant of Mayors for Climate and Energy" is a voluntary movement of local authorities in the EU that develops and implements sustainable energy and climate policies that created a guidance framework for potential PED developers by focusing on leveraging the Sustainable Energy and Climate Action Plan (SECAP) methodology [64]. Additionally, common features between PED and SECAP processes were identified to effectively plan and monitor sustainable energy and climate action while also being aligned with specific KPIs. The framework underlines the importance of interoperability and synergies, concentrating on "shared principles, compatible frameworks, common data management, and specific integration points", and focuses on four main stages, that is, initiation, planning, implementation, and monitoring stages.

The project claims that the proposed framework supports PED replicability through its "comprehensive energy simulations, collaborative planning, and adaptive measures". PEDs are considered to be a significant solution toward sustainable urban areas, aiming to reduce GHG emissions and to enhance local resilience by integrating energy planning with broader urban development and stakeholder engagement. However, a good understanding and consideration of strategies, project features, and experiences form the basis for designing and developing a PED project. The SECAP highlights a strategic systemic approach that integrates climate, territorial, and energy planning. This entails analysis of the data regarding morphoclimatic zones (landforms associated with a particular climate), environmental variables and energy consumption, and the identification of targets and selection of site-specific actions, supported by robust indicators for progress monitoring, developing adaptation/mitigation actions, and increasing the environmental awareness of the involved actors. The four-stage methodology is not a new approach, and it was already defined as the project management lifecycle/project lifecycle in 1996 by the Project Management Body of Knowledge [65] and Prince2 [66] and is a continuously evolving process. The typical phases are initiation (project definition at a broad level), planning (establishing the scope, objectives, and procedures), implementation/execution (carrying out the project plan), monitoring/control (tracking progress and making necessary adjustments). In addition, a closure (finalizing and closing the project) follows. However, the SECAP introduces a new EU municipality integrating an urban planning paradigm related to energy efficiency and climate challenges. Citizen engagement, important in PED projects, is nevertheless a gray area [67] that needs a multi-faceted approach that requires particular caution, including application of combinations of factors, such as suitable incentives, adequate education/training, inclusive governance, effective communication, and active co-creation for boosting understanding, trust, and ownership regarding the measures adopted in the district of the citizen [47].

3.3. *Third Study—The Annex 75 Project*

The Annex 75 project of the "International Energy Agency" and "Energy in Building and Community", with 25 collaborating partner from 13 countries, investigated similarities and differences between them, emphasizing the balance between measures regarding energy efficiency and renewable energy [26]. According to the related outcomes, the technologies used in PEDs can be broadly categorized into (i) demand reduction/energy demand reduction technologies, (ii) energy distribution and supply systems energy, and (iii) storages. The readiness of the technologies should be assessed together with their cost efficiency. Similarly, the required stakeholder engagement, policy implications, and replicability should also be assessed, and prosumer models based on new energy community legislation should be considered.

The Annex 75 project developed a methodology for examining the renovation of buildings at the district level and its balance between cost-effective energy efficiency measures and renewable energy sources. Three KPI were identified as the most noteworthy in the methodology [26], namely,

1. Primary energy use (kWh/m^2 year);
2. Annualized total costs (EUR/m^2 year);
3. GHG emissions (CO_2 eq/m^2 year).

The sustainability and cost-effectiveness level of the renovation projects was defined with reference to these KPIs. Dynamic simulations (e.g., with an hourly time interval), were used for the evaluation of whole-system performance. The energy demands for heating and cooling are based on the building dimensions and thermal properties buildings. Electricity demand is calculated according to standard PED country profiles. Regarding the supply side, centralized and decentralized energy systems and specific interactions amongst the different technologies are considered in this methodology. The energy use for space heating and cooling, ventilation, lighting and building supplementary electricity consumption-integrated technical systems (pumps, fans, electric valves, pumps, etc.) was considered in the assessment. The characterization of energy systems involves four factors, namely "cost (as a function of capacity), service lifetime, conversion efficiency, and associated energy carrier". GHG emissions and consumption of primary energy are taken into consideration by using corresponding county emission factors and primary energy factors. This methodology was employed in a Portuguese real-life case study and the results were analyzed in detail. The study was considered significant for demonstrating the applicability of the methodology despite the fact that the results are from a certain Portuguese context.

Based on the Annex 75 framework, fifteen success stories were collected from seven European countries [26]. The success stories disclosed that renovation on a PED level is very complicated due to the high number of stakeholders involved, and the wide knowledge and considerable financial resources needed. A main challenge (bottleneck) identified was "the balance between energy efficiency measures and renewable energy use is based on expert estimations and not on calculations". District renovation was primarily initiated by the readiness and support of "the municipality, a housing association, a residents' association, or the wish of the inhabitants (tenants/residents)". Early-stage communication with tenants and citizens was significant for their engagement, which led to a significant overall satisfaction rate. Also, a suitable coordinator/facilitator is essential. Main drivers for PED utilization and its success were as follows:

- Raising the building to new modern standards regarding comfort and energy use: New modern standards for comfort and energy use in buildings focus on energy efficiency, indoor environmental quality, sustainability, and smart technology integration. These standards evolve regularly to reflect advances in climate science, design practices, materials, and occupant expectations. It should be noted that all new buildings in the EU should show zero-emission by 2030, and existing buildings by 2050.
- Improving open space attractiveness: PEDs should not only be sustainable but also livable, functional, live, energetic and vibrant in order to improve social acceptance and the accomplishment of sustainable initiatives, and encourage citizen and community participation in energy-saving behaviors. Open space attractiveness in PEDs is raised by (i) architectural esthetics design of urban areas, including green spaces, pedestrian-friendly streets, public spaces for interaction, events, community use, smart infrastructure and interaction, interactive energy displays and real-time feedback on consumption; (ii) microclimate control, such as shading, wind buffers, cooling via vegetation; (iii) noise reduction, adequate air quality, natural light; (iv) comfort and well-being; (v) integration of energy infrastructure, such as solar panels, into visual appearances; (vi) bike lanes and public transport access; (vii) accessibility (for all ages and abilities) and functionality; (viii) mixed-use zoning to safeguard vicinity to work, services, and leisure; (ix) social inclusiveness; (x) preservation of heritage and local identity; and (xi) spaces that evoke pride, belonging, and enjoyment.
- Developing the image of the district: Involving citizens from the very beginning in the PED project is identified as an effective measure for user acceptance and successful PEDs [46]. Improving open space attractiveness will also develop the image of the PED and increase economic value by attracting businesses and residents.

Funding is the main challenge regarding district renovation, but there also exist challenges such as coordination of stakeholders, creation of optimal renovation measures, and relocation of tenants during the renovation. Some of these challenges can be overcome with an appropriate business model, whilst other challenges, such as the legal framework, are more difficult to overcome.

Based on the related outcomes [26], the following implications can be highlighted:

- Combining energy efficiency upgrades in district renovations with renewable energy initiatives and wider urban or infrastructure improvements and social enhancement can create greater value and result in more efficient use of financial investments.
- Technologically, there are opportunities that would not be possible through individual solutions. However, there is a lack of technical know-how, knowledge, and protocols to simplify the complex process, as well as a lack of resources for coordination work depending on PED heterogeneity and complexity.
- The availability of financial resources is significant. The majority of PED projects are partially funded by European funding or other public funding.
- Good coordination and flexibility are required. Project phases should overlap to shorten long project timelines. The municipality has a critical role in integrated planning and municipal action.
- Public support is central for enabling stakeholder dialog, stakeholder commitment, and coordination work to successfully realize district renovation projects. The neighborhood association or similar social stakeholders in the district are proposed to be developed to aid this time- and resource-consuming process.

3.4. Fourth Study—Examination of Four PED Projects (Austria, Italy, the Netherlands, and Romania)

Bruckner et al. [57] reported their experiences and lessons learnt regarding four PEDs situated in Austria, Italy, Netherlands, and Romania, each of them with different climatic conditions and unique energy challenges, such as outdated infrastructure or heritage protection. They emphasized the significance of innovative technologies like photovoltaic-thermal systems, application of demand-side actions, and flexible grid usage. They emphasized that implementing PEDs in existing urban areas presents several challenges, including the absence of widely accepted methods for assessing energy balances, the complexity of integrating renewable energy systems into dense and concentrated urban infrastructures, and the need for multi-stakeholder awareness and collaboration to ensure smooth progress and foster local acceptance.

System boundaries were determined from several perspectives, namely, (i) Spatial: the geographical limits enclosing energy services and supplies; (ii) Temporal: the balancing period (one operational year); and (iii) Functional: Use of energy, functions, requirements (included or excluded), which can be grouped into the following:

- The innermost level—PED operational energy and user electricity;
- PED mobility aspects—everyday motorized private mobility;
- Climate-neutral PED—the outermost layer, PED-embodied energy and emissions associated with district construction, mobility, maintenance, and repair.

A simulation and assessment framework entitled "PEExcel", which is a decision support system, was developed for the "climate neutral Positive-Energy-District" in Austria [68]. The aims of the framework were to allow practitioners and researchers to speedily model and assess the PED primary energy and emission balance in an early planning phase.

4. PED Framework

The most significant success factors and lessons learnt from diverse existing international PED projects that can be applied to turn new PEDs into self-sustainable and greener urban neighborhoods are summarized in the PED Success Factor framework presented in Table 4. Specifically, the success factors and lessons learnt are categorized according to PED factors and presented following the order in which the different processes occur in the PED lifecycle. The framework is a process-oriented description of a PED that considers how such a district is planned, developed, and operated over its lifecycle. Process orientation entails by focusing on business processes rather than highlighting functional structure or hierarchy [69].

Table 4. PED Success Factor framework.

Stage	PED Factors	Success Factors
1. Initiation	Ideation Feasibility	Defining key objectives and high-level business case and rationale. What is the project trying to achieve? Defining broad scope and constraints. What are the initial risks, resources, and assumptions? Investigating funding potential, job creation and boosting of local/regional economy.
	Stakeholders	Identifying key stakeholders that will be affected and who should be involved.
	Technology	Identifying and creating an overview of diverse PED technology options by considering alternative renewable energy sources and technological solutions.
	Funding	Raising awareness and aspiring for political support regarding national programs.
	Policy	Securing political commitment.
	Aims	Identifying and aligning existing policies to complement PED aims and objectives
2. Planning	Business Goal	Developing achievable business models and identifying funding opportunities.
	Stakeholder Engagement	Motivating, encouraging, and engaging local stakeholders (community, citizens, residents, businesses, and civic organizations) and promoting collaboration among all stakeholders.
	Strategy	Developing a detailed strategy with the involvement of all relevant stakeholders.
	Measures	Defining specific, measurable, achievable, realistic, and timely (SMART) long-term goals and targets linked to community priorities.
	Action Plan	Transferring long-term goals into an action plan consisting of short-term objectives connected to specific milestones for greater clarity, direction, and focus.
3. Prototyping	Prototype	Potential implementation of a PED Lab for elaboration of guidelines and tools, experimentation and testing of different approaches and strategies, and monitoring of results (this step is not mandatory but provides insights before real-life application).
	Data	Ensuring early-stage interoperability through effective and robust data management frameworks and standardized protocols for successful interaction between energy generation, consumption, and storage systems at the district level.
4. Implementation	Monitoring	Implementing monitoring processes to track progress toward goals and objectives.
	Transparency	Ensuring progress transparency regarding the implementation of the action plan.
	Follow-up	Ensuring that the strategy is successfully implemented.
5. Evaluation	Assessment	Evaluating the effectiveness of the implemented actions.
6. Valorization	Dissemination Exploitation	Dissemination and exploitation of best practices for potential replication of identified PED success factors.

The success factors derive from the key aspects identified in the studies examined. The related studies focused on exploring multiple PEDs across different settings; hence, the related outcomes can be applied in various sectors. Figure 1 presents the key aspects identified within the studies examined. Specifically, Saheb et al. [45] displayed conclusions from a study of 62 projects regarding factors that are decisive for the proposed framework, namely success factors and challenges. The insights from 61 PED projects presented by Bossi et al. [24] focused on success factors and action areas, while the Annex 75 project with 25 collaborating partners by Terés-Zubiaga et al. [26] focused on lessons learnt and main drivers for the adoption, integration, and utilization of PEDs. Through the study of Bruckner et al. [57], which examined four PED projects, success factors and challenges were identified.

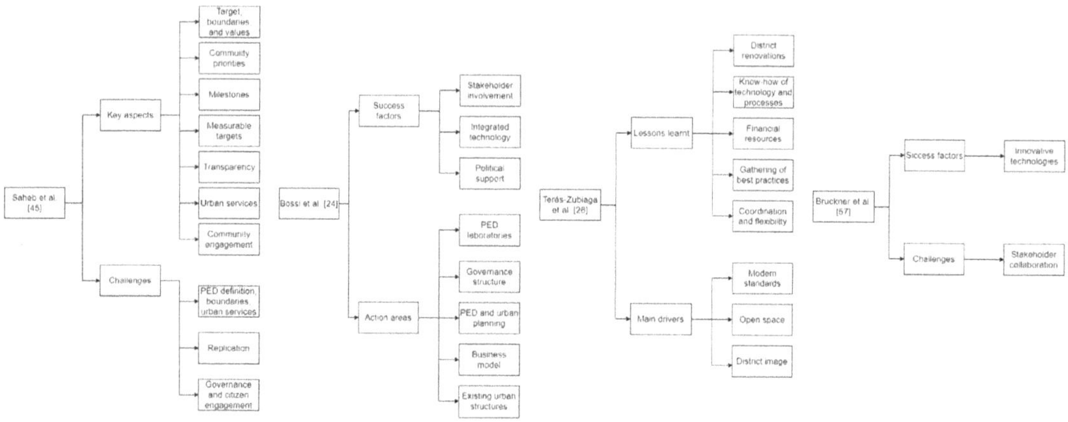

Figure 1. Key aspects identified from the related studies.

The PED framework can be used in business environments to make organizations more efficient and productive. It enables processes to be reviewed, optimized, and standardized and help in dealing with problems and challenges that might arise. The PMBOK [65] is a reference guide on how to complete a project in the best way. It is a set of processes, guidelines, best practices, and tools for managing projects. The principles of the PMBOK are used for the creation of the PED Success Factor framework.

Both PEDs and organizations are systems with interconnected components striving toward a common goal, but with different focuses. A PED aims to reach energy self-sufficiency and sustainability within an urban area, while an organization focuses on achieving specific business objectives. Both involve strategic planning, resource management, and stakeholder engagement to achieve their respective goals.

The overarching PED objective involves the creation of an urban district that produces more renewable energy than the energy it consumes over a year. The physical scope (neighborhood, business district, or mixed-use area) needs to be decided at an early stage through stakeholder engagement, such as city planners, residents, building owners, energy providers, and mobility operators. An important factor to take into consideration concerns whether the planned PED will be created in a new area or if building retrofits will be needed. New areas are easier to create, while in existing areas there may be challenges regarding aging infrastructure, heritage preservation, and ownership. In addition to the energy factor, PEDs also promise holistic sustainability, sharing of resources, and integrating environmental, economic, and social well-being. This entails addressing social challenges such as energy poverty and gentrification, and promoting inclusivity, citizen engagement, and improved well-being.

In the planning and design stage, the urban design integration and the energy demand need to be assessed together. Urban design integration includes high-efficiency buildings, mixed-use layouts, optimized spatial orientation, and mobility planning, while the energy demand concentrates on current and projected building energy use, mobility patterns, and infrastructure requirements. Energy balance scenarios are usually tested in a lab or with simulation to examine whether the positive energy goals set are feasible. Moreover, the consideration of contemporary, smart autonomous technology selection is required for optimizing the use of energy via smart applications, storage, and flexible consumption patterns.

In the operation stage, energy production and consumption across the district continuously need to be tracked to follow-up the collective self-consumption and peer-to-peer trading of energy, including mobility-energy integration of electric vehicles, change demand, and potential storage. Performance is evaluated by comparing actual performance to targets based on real data energy sharing rules and adjustable operations.

The PED Success Factor framework consists of success factors that are categorized in five stages according to the processes in the PED lifecycle, namely,

1. **Initiation:** This stage is the first phase in the PED project lifecycle. It is a high-level phase that plays a vital role in laying the foundation for the project's success. During this stage, the idea for the project is discussed, defined, evaluated, and authorized. Without an adequate initiation, the PED project may lack clear direction or stakeholder buy-in, increasing the risk of failure.
2. **Planning:** This stage is a critical phase where the project's roadmap is developed. It ensures clarity among stakeholders, provides direction for the implementation phase, and sets a baseline for tracking progress and for performance evaluation. It involves defining in detail the objectives, scope, timelines, resources, risks, and strategies necessary to complete the project successfully.
3. **Prototyping:** This stage in a project lifecycle is a phase where a preliminary version of the PED is developed to test key ideas or concepts, receive feedback, identify design and usability issues, and validate feasibility. The living labs in PEDs are prototypes that have been identified as particularly important for stakeholders and developers to understand how the final PED might be. As it involves stakeholders early, it reduces the risks of not meeting their needs. It also has the potential to detect design flaws or technical challenges before full-scale development begins.
4. **Implementation:** This stage in a project lifecycle is the phase where plans and strategies developed in earlier stages are put into action to create the actual deliverables of the project. It is the main phase of the PED project, transforming plans into tangible outcomes where resources, time, and effort are spent.
5. **Evaluation:** In this phase, the project is reviewed and assessed to determine how well it met its objectives, what did/did not work well, and what lessons can be learned for future projects. This stage is important because it aims to support accountability and transparency, improve future project planning and execution, and encourage continuous process improvement.
6. **Valorization**: Dissemination and exploitation of lessons learnt for making best practices and success factors available to other differing contexts.

Figure 2 provides an overview of the PED Success Factor framework in which the main stages of the PED lifecycle are shown in the sequence they occur. Below every stage, within the dotted box, the PED factors that need attention in that particular stage are highlighted. These factors have been identified to be pivotal for the successful creation and operation of PEDs.

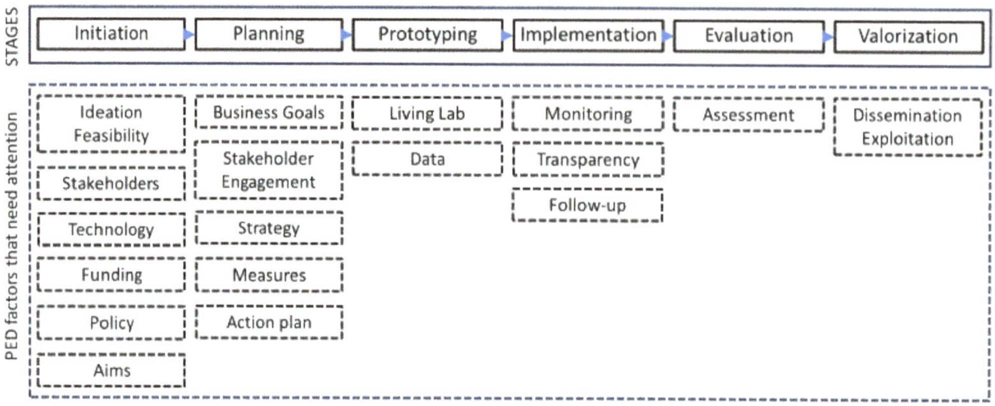

Figure 2. The PED Success Factor framework.

5. Discussion

In the strive for the EU to become climate-neutral by 2050 and a global role model regarding energy transition, open sharing of knowledge and sustainable energy solutions through collaboration and innovation is promoted. The aims of a PED ecosystem include the development and maintenance of attractive, self-sustainable, and economically viable EU urban areas, in which citizens, communities and their vicinity can thrive. The pioneering concept of PEDs is winning ground in the EU. The EU's SET-Plan, with its goal of establishing 100 PEDs in Europe by 2025, focuses on resource pooling, knowledge sharing, and collaborative solution development among stakeholders. The emphasis is on the planning, implementation, and replication of energy-efficient self-sustainable PEDs. Replication must be adapted to local needs and constraints. The main aim is to create a surplus of renewable energy in the 100 PEDs and to gradually expand the concept to wider areas. However, the results from the studies have shown that every PED context is different. Each district has unique climate, cultural, political, financial, and technical circumstances, making "one-size-fits-all" solutions unachievable. Recent studies have also highlighted the impact of autonomous and hydrogen-based hybrid energy systems, which can further facilitate the generation and distribution of energy even in rural areas and aid in achieving a sustainable future [70–72]. Additionally, it is evident that PEDs are interdependent [11] and have significant environmental impact [73], techno-economic implications [74], and a highly regarded role in achieving SDGs [75]. As a result, it is important to consider the social, environmental, and economic aspects of PEDs when assessing their development, adoption, and integration [76]. Focusing on approaches that encourage social innovation [61] and using suitable assessment methodologies [18] is vital for developing human-centric PEDs that emphasize well-being, inclusion, and justice [36].

Our proposed framework, which focuses on processes, is in line with the outcomes of other related studies which have also highlighted the significance of processes within PEDs. Processes orientation in PEDs refers to the structured, stepwise, and context-sensitive approaches needed to plan, implement, and scale PEDs effectively. Very few studies can be found in the literature that emphasize processes orientation in PEDs. Sareen et al. [77] for example, argued that a well-designed processes according to the various development phases is required in PEDs. The PED Success Factor framework brings forward the processes in all phases of the PED lifecycle. Other studies mention processes as a major component of PEDs. Turci et al. [78], for example, proposed the use of the Theory of Change (ToC) to facilitate PED implementation. The ToC is a structured process-oriented roadmap that outlines consecutive actions regarding how a PED project is expected to achieve its planned goals. Hence, processes have a central role in the ToC, which enables a shared understanding of a problem and supports decision making toward a solution. Similarly, Krangsås et al. [11] stated that the main challenges for developing PEDs are governance, incentive, social, process, market, technology, and context. To address the challenges, which are both interrelated and interdependent, the processes behind them need to be understood. Natanian et al. [79] also focused on a process-oriented perspective and emphasized that a holistic framework is needed to address the multi-dimensional challenges of the PED design process. Similarly, Kozlowska et al. [18] highlighted that to make coherent decisions in the PED design and planning process, social, ecological, and cultural factors need to be considered. Therefore, the proposed PED framework is process-oriented and focuses on key processes that lead to the successful adoption and implementation of PEDs.

The outcomes of this study contribute to widespread PED implementation through learning from challenges and success stories. The explanatory research approach used in this study enabled a deeper understanding of the PED phenomenon, clarification of diverse PED concepts, and the identification of patterns related to outcomes and lessons learned from existing PED projects. Reported success factors, good practices, policy guidelines, challenges, barriers, and obstacles were thoroughly investigated, evaluated, categorized, and compared to unveil patterns of lessons learnt from diverse existing international PED projects. The outcome of our study resulted in the PED Success Factor framework, which illustrates a description of the success factors that most frequently occur in the literature, organized according to the PED lifecycle processes.

Nevertheless, the two most frequently emphasized PED characteristics for success are stakeholder commitment and technological solutions. In particular, citizen commitment has already been recognized by the EU already through the creation of ECs. Stakeholder dynamics, which include the changing roles and perceptions of stakeholders, have been studied and found to be of utmost importance for successful PED implementation. In PEDs, stakeholders have shifting roles and diverse incentives among actors. Stakeholders need to be organized through interdisciplinary collaboration and supported by well-defined policies, stakeholder management frameworks, and effective governance to align interests and address multifaceted challenges. The main concerns for successful PED implementation are stakeholder co-operation (commitment and cross-disciplinary collaboration) and new measures for achieving high energy efficiency, local renewable energy generation, and energy storage and flexibility, as well as energy sufficiency.

A PED ecosystem depends on a wide range of technological solutions, including smart meters, smart technologies, intelligent software agents, Internet of Things, peer-to-peer networks, and blockchain-based systems for secure and transparent transactions. Standardized protocols and robust data management frameworks are needed to integrate the different technologies and maximize interoperability, which is a critical aspect for microgrid operation and interconnected distributed solar home systems, known as mesh grids or swarm electrification. Data exchange platforms enable interoperability within the ecosystems, enabling seamless interaction and integration among various energy systems, technologies, and stakeholders. Security and integration standards are evolving. The clean energy market, which utilizes renewable energy sources with minimal impact on the environment, is increasingly dependent on digital technologies and interconnected infrastructures, which in turn creates increased potential cyber threats and cybersecurity concerns regarding how to secure reliable operations of interconnected PED systems.

6. Conclusions

A main incentive for piloting a number of PED cases in the EU until 2025 has been the creation of a foundation for understanding the benefits, values, scalability, and replicability. The results from diverse projects have shown that a holistic cross-disciplinary approach for mainstreaming PEDs or their fundamental components across a larger number of settings is crucial, requiring the integration of STEEPLED aspects for effective planning and design.

PEDs are fundamental for achieving sustainable urban development through their enhancement of energy self-sufficiency and reduction in GHG emissions. They have steadily gained significance and recognition on the policy agenda of the European Commission regarding energy transition. PEDs aim at achieving a net positive energy balance and reduced GHG emissions through high energy efficiency, renewable energy generation, and energy flexibility at the district level, as well as extensive stakeholder involvement and collaboration. PEDs face challenges associated with governance, context-specific factors and integration of STEEPLED factors.

The emphasis of this study was to gather insight into the factors used by the EU municipalities that have been the frontrunner in setting up PEDs. It was noticed that buildings that are substantially energy-efficient are at the cornerstone of low-carbon communities. The aim of this study was to examine existing PED initiatives by looking at PED eco-systems and documented success factors to unveil what lessons learnt from various existing international projects could be applied to turn new PEDs into greener and self-sustainable urban neighborhoods. The main contribution of this study is the proposed PED Success Factor framework, which is a process-oriented framework that presents key processes that are required for the successful creation, adoption, and operation of PEDs based on the experiences and outcomes of PEDs.

Future work will concentrate on analyzing the influences of STEEPLED factors on PEDs. Specifically, emphasis will be placed on the role of each STEEPLED factor in the context of PEDs and how they may be interrelated to each other. This will help us extend the PED Success Factor framework processes with a more detailed description of each process, including technology integrations, time and cost, tasks, roles, and risk factors, such as cybersecurity threats, data privacy and safety, digital literacy, and energy poverty. Additionally, given the exploratory nature of this study, the proposed framework not yet being validated through a case study and the study not adopting a systematic approach can be regarded as the main limitations. Hence, future studies are encouraged to further examine the key aspects of PEDs through a systematic literature approach to identify key enabling technologies and issues that need to be addressed. Moreover, there is a need for case studies to be conducted to explore the implications and impact of PEDs and the role of stakeholders in their successful realization. It is also important to avoid gentrification when greening districts and ensuring equal access to the required resources and infrastructure. Hence, it is vital to identify the key mechanisms that ensure that low-income and underprivileged residents as well as vulnerable groups are not excluded from the merits that PEDs can bring. However, given their social influence, it is important to further examine the social KPIs of PEDs, such as energy affordability and resident satisfaction. As PEDs are highly reliant on technology, future studies should put emphasis on identifying key cybersecurity threats in interconnected grids, on ethical concerns, and on data security and privacy issues.

Author Contributions: Conceptualization, D.S. and K.S.; methodology, D.S. and K.S.; formal analysis, D.S., K.S. and G.L.; investigation, D.S. and K.S.; data curation, D.S. and K.S.; writing—original draft preparation, D.S., K.S. and G.L.; writing—review and editing, D.S., K.S. and G.L.; visualization, D.S.; supervision, K.S. All authors have read and agreed to the published version of the manuscript.

Funding: This work was partly carried out as part of the "Positive Energy Districts in Kanta-Häme (EKA+)" project by Häme University of Applied Sciences HAMK and Sykli Oy. The project was partly funded by the European Union and the Regional Council of Kanta-Häme.

Institutional Review Board Statement: Not applicable.

Informed Consent Statement: Not applicable.

Data Availability Statement: All of the data are contained within the article.

Conflicts of Interest: The authors declare no conflicts of interest.

References

1. United Nations. *United Nations Generating Power.* Available online: https://www.un.org/en/climatechange/climate-solutions/cities-pollution (accessed on 30 March 2025).
2. European Commission. 2050 Long-Term Strategy—European Commission. Available online: https://climate.ec.europa.eu/eu-action/climate-strategies-targets/2050-long-term-strategy_en (accessed on 15 May 2025).
3. European Commission. Directive—EU-2024/1275—EN—EUR-Lex. Available online: https://eur-lex.europa.eu/eli/dir/2024/1275/oj/eng (accessed on 5 March 2025).
4. European Commission. Directive (EU) 2023/2413 of the European Parliament and of the Council of 18 October 2023 Amending Directive (EU) 2018/2001, Regulation (EU) 2018/1999 and Directive 98/70/EC as Regards the Promotion of Energy from Renewable Sources, and Repealing Council Directive (EU) 2015/652. Available online: https://eur-lex.europa.eu/eli/dir/2023/2413/oj/eng (accessed on 20 May 2025).
5. European Commission. Directive (EU) 2019/944 of the European Parliament and of the Counceil of 5 June 2019—On Common Rules for the Internal Market for Electricity and Amending Directive 2012/27/EU. Available online: https://eur-lex.europa.eu/eli/dir/2019/944/oj/eng (accessed on 20 May 2025).
6. European Commission. Effort Sharing Regulation (ESR) | EESC. 2021. Available online: https://www.eesc.europa.eu/en/our-work/opinions-information-reports/opinions/effort-sharing-regulation-esr (accessed on 25 April 2025).
7. European Commission. Energy Efficiency Directive. Available online: https://energy.ec.europa.eu/topics/energy-efficiency/energy-efficiency-targets-directive-and-rules/energy-efficiency-directive_en (accessed on 25 April 2025).
8. European Commission. EU Emissions Trading System (EU ETS)—European Commission. Available online: https://climate.ec.europa.eu/eu-action/eu-emissions-trading-system-eu-ets_en (accessed on 25 April 2025).
9. Lv, Y. Transitioning to sustainable energy: Opportunities, challenges, and the potential of blockchain technology. *Front. Energy Res.* **2023**, *11*, 1258044. [CrossRef]
10. United Nations. The 17 Goals. Available online: https://sdgs.un.org/goals (accessed on 3 March 2025).
11. Krangsås, S.G.; Steemers, K.; Konstantinou, T.; Soutullo, S.; Liu, M.; Giancola, E.; Prebreza, B.; Ashrafian, T.; Murauskaitė, L.; Maas, N. Positive Energy Districts: Identifying Challenges and Interdependencies. *Sustainability* **2021**, *13*, 10551. [CrossRef]
12. Wu, Y. Decentralized transactive energy community in edge grid with positive buildings and interactive electric vehicles. *Int. J. Electr. Power Energy Syst.* **2022**, *135*, 107510. [CrossRef]
13. Soutullo, S.; Giancola, E.; Sánchez, M.N.; Ferrer, J.A.; García, D.; Súarez, M.J.; Prieto, J.I.; Antuña-Yudego, E.; Carús, J.L.; Fernández, M.Á.; et al. Methodology for Quantifying the Energy Saving Potentials Combining Building Retrofitting, Solar Thermal Energy and Geothermal Resources. *Energies* **2020**, *2020*, 5970. [CrossRef]
14. Monti, A.; Pesch, D.; Ellis, K.A.; Mancarella, P. Chapter One—Introduction. In *Energy Positive Neighborhoods and Smart Energy Districts*; Monti, A., Pesch, D., Ellis, K.A., Mancarella, P., Eds.; Academic Press: Cambridge, MA, USA, 2017; pp. 1–5. [CrossRef]
15. European Commission. SET Plan Progress Report 2024. Available online: https://setis.ec.europa.eu/set-plan-progress-report-2024_en (accessed on 26 April 2025).
16. Positive Energy Districts—European Commission. Available online: https://setis.ec.europa.eu/working-groups/positive-energy-districts_en (accessed on 15 May 2025).
17. JPI Urban Europe. Positive Energy Districts (PED). Available online: https://jpi-urbaneurope.eu/ped/ (accessed on 15 May 2025).
18. Kozlowska, A.; Guarino, F.; Volpe, R.; Bisello, A.; Gabaldòn, A.; Rezaei, A.; Albert-Seifried, V.; Alpagut, B.; Vandevyvere, H.; Reda, F.; et al. Positive Energy Districts: Fundamentals, Assessment Methodologies, Modeling and Research Gaps. *Energies* **2024**, *17*, 4425. [CrossRef]
19. Gollner, C.; Hinterberger, R.; Noll, M.; Meyer, S.; Schwarz, H.-G. Booklet of Positive Energy District in Europe. Urban Europe. 2019. Available online: https://jpi-urbaneurope.eu/wp-content/uploads/2019/04/Booklet-of-PEDs_JPI-UE_v5_NO-ADD.pdf (accessed on 20 May 2025).

20. European Commission. Clean Energy for all Europeans Package Completed: Good for Consumers, Good for Growth and Jobs, and Good for the Planet—European Commission. Available online: https://commission.europa.eu/news/clean-energy-all-europeans-package-completed-good-consumers-good-growth-and-jobs-and-good-planet-2019-05-22_en (accessed on 26 April 2025).
21. European Commision. Energy Poverty. 2025. Available online: https://energy.ec.europa.eu/topics/markets-and-consumers/energy-consumers-and-prosumers/energy-poverty_en (accessed on 30 March 2025).
22. Hearn, A.X. Positive energy district stakeholder perceptions and measures for energy vulnerability mitigation. *Appl. Energy* **2022**, *322*, 119477. [CrossRef]
23. European Commision. REPowerEU 2022. Available online: https://commission.europa.eu/strategy-and-policy/priorities-2019-2024/european-green-deal/repowereu-affordable-secure-and-sustainable-energy-europe_en (accessed on 26 April 2025).
24. Bossi, S.; Gollner, C.; Theierling, S. Toward 100 Positive Energy Districts in Europe: Preliminary Data Analysis of 61 European Cases. *Energies* **2020**, *13*, 6083. [CrossRef]
25. European Commision. In Focus: Energy Efficiency in Buildings—European Commission. 2020. Available online: https://commission.europa.eu/news/focus-energy-efficiency-buildings-2020-02-17_en (accessed on 2 May 2025).
26. Terés-Zubiaga, J. Cost-effective building renovation at district level combining energy efficiency & renewables—Methodology assessment proposed in IEA EBC Annex 75 and a demonstration case study. *Energy Build.* **2020**, *224*, 110280. [CrossRef]
27. Shnapp, S.; Paci, D.; Bertoldi, P. *Enabling Positive Energy Districts Across Europe: Energy Efficiency Couples Renewable Energy*; Publication Office of the European Union: Luxembourg, 2020. [CrossRef]
28. Habib, M.K.; Chimsom, I.C. CPS: Role, Characteristics, Architectures and Future Potentials. *Procedia Comput. Sci.* **2022**, *200*, 1347–1358. [CrossRef]
29. Cicceri, G.; Tricomi, G.; D'Agati, L.; Longo, F.; Merlino, G.; Puliafito, A. A Deep Learning-Driven Self-Conscious Distributed Cyber-Physical System for Renewable Energy Communities. *Sensors* **2023**, *23*, 4549. [CrossRef]
30. Siakas, D.; Lampropoulos, G.; Siakas, K. Autonomous Cyber-Physical Systems Enabling Smart Positive Energy Districts. *Appl. Sci.* **2025**, *15*, 7502. [CrossRef]
31. Vinuesa, R.; Azizpour, H.; Leite, I.; Balaam, M.; Dignum, V.; Domisch, S.; Fuso Nerini, F. The Role of Artificial Intelligence in Achieving the Sustainable Development Goals. *Nat. Commun.* **2020**, *11*, 233. [CrossRef]
32. Clerici Maestosi, P. Harmonizing Urban Innovation: Exploring the Nexus between Smart Cities and Positive Energy Districts. *Energies* **2024**, *17*, 3422. [CrossRef]
33. Zhang, X.; Shen, J.; Saini, P.K.; Lovati, M.; Han, M.; Huang, P.; Huang, Z. Digital Twin for Accelerating Sustainability in Positive Energy District: A Review of Simulation Tools and Applications. *Front. Sustain. Cities* **2021**, *3*, 663269. [CrossRef]
34. Borah, S.S.; Khanal, A.; Sundaravadivel, P. Emerging Technologies for Automation in Environmental Sensing: Review. *Appl. Sci.* **2024**, *14*, 3531. [CrossRef]
35. Agostinelli, S.; Cumo, F.; Guidi, G.; Tomazzoli, C. Cyber-Physical Systems Improving Building Energy Management: Digital Twin and Artificial Intelligence. *Energies* **2021**, *14*, 2338. [CrossRef]
36. Nguyen, M.T.; Batel, S. A Critical Framework to Develop Human-Centric Positive Energy Districts: Toward Justice, Inclusion, and Well-Being. Front. *Sustain. Cities* **2021**, *3*, 691236. [CrossRef]
37. Lee, K.; Man, K.L. Edge Computing for Internet of Things. *Electronics* **2022**, *11*, 1239. [CrossRef]
38. Wang, Y.; Chen, C.-F.; Kong, P.-Y.; Li, H.; Wen, Q. A Cyber–Physical–Social Perspective on Future Smart Distribution Systems. *Proc. IEEE* **2023**, *111*, 694–724. [CrossRef]
39. Delicato, F.C.; Al-Anbuky, A.; Wang, K.I.-K. Editorial: Smart Cyber–Physical Systems: Toward Pervasive Intelligence systems. *Future Gener. Comput. Syst.* **2020**, *107*, 1134–1139. [CrossRef]
40. Zhang, J.J.; Wang, F.-Y.; Wang, X.; Xiong, G.; Zhu, F.; Lv, Y.; Hou, J.; Han, S.; Yuan, Y.; Lu, Q.; et al. Cyber-Physical-Social Systems: The State of the Art and Perspectives. *IEEE Trans. Comput. Soc. Syst.* **2018**, *5*, 829–840. [CrossRef]
41. Skinner, E.; Dancis, J.; Descriptive and Explanatory Designs. Human Development. 2020. Available online: https://pdx.pressbooks.pub/humandevelopment/chapter/descriptive-and-explanatory-designs/ (accessed on 20 May 2025).
42. Snyder, H. Literature review as a research methodology: An overview and guidelines. *J. Bus. Res.* **2019**, *104*, 333–339. [CrossRef]
43. Jesson, J.; Lacey, F. How to do (or not to do) a critical literature review. *Pharm. Educ.* **2006**, *6*, 139–148. [CrossRef]
44. Rouse, P.; Sherkat, O.; O'Grady, W. Dynamic Stakeholder Analysis Through Process Mapping. *Financ. Account. Manag.* **2025**, *41*, 398–418. [CrossRef]
45. Saheb, Y.; Shnapp, S.; Johnson, C. The Zero Energy concept: Making the whole greater than the sum of the parts to meet the Paris Climate Agreement's objectives. *Curr. Opin. Environ. Sustain.* **2018**, *30*, 138–150. [CrossRef]
46. Siakas, D.; Rahanu, H.; Georgiadou, E.; Siakas, K.; Lampropoulos, G. Positive Energy Districts Enabling Smart Energy Communities. *Energies* **2025**, *18*, 3131. [CrossRef]
47. Fatima, Z.; Pollmer, U.; Santala, S.-S.; Kontu, K.; Ticklen, M. Citizens and Positive Energy Districts: Are Espoo and Leipzig Ready for PEDs? *Buildings* **2021**, *11*, 102. [CrossRef]

48. Mamidi, A.P. Strategic Project Management: Bridging the Gap Between Strategy and Execution. Master's Thesis, University of Vaasa, Vaasa, Finland, 2025.
49. Cai, S.; Gou, Z. Transitioning Positive Energy Buildings toward Positive Energy Communities: Leveraging Performance Indicators for Site Planning Assessments. *Energy Build.* **2024**, *325*, 114976. [CrossRef]
50. Barrutieta, X.; Kolbasnikova, A.; Irulegi, O.; Hernández, R. Decision-making framework for positive energy building design through key performance indicators relating geometry, localization, energy and PV system integration. *Energy Build.* **2023**, *297*, 113442. [CrossRef]
51. Chatzikonstantinidis, K.; Giama, E.; Fokaides, P.A.; Papadopoulos, A.M. Smart Readiness Indicator (SRI) as a Decision-Making Tool for Low Carbon Buildings. *Energies* **2024**, *17*, 1406. [CrossRef]
52. Lampropoulos, G.; Garzón, J.; Misra, S.; Siakas, K. The Role of Artificial Intelligence of Things in Achieving Sustainable Development Goals: State of the Art. *Sensors* **2024**, *24*, 1091. [CrossRef]
53. Cortinovis, C.; Geneletti, D. A performance-based planning approach integrating supply and demand of urban ecosystem services. *Landsc. Urban Plan.* **2020**, *201*, 103842. [CrossRef]
54. JPI Urban Europe/SET Plan Action 3.2. White Paper on PED Reference Framework for Positive Energy Districts and Neighbourhoods. 2020. Available online: https://jpi-urbaneurope.eu/wp-content/uploads/2021/10/setplan_smartcities_implementationplan-2.pdf (accessed on 20 May 2025).
55. Derkenbaeva, E.; Vega, S.H.; Hofstede, G.J.; Van Leeuwen, E. Positive Energy Districts: Mainstreaming Energy Transition in Urban Areas. *Renew. Sustain. Energy Rev.* **2022**, *153*, 111782. [CrossRef]
56. Boccalatte, A.; Fossa, M.; Gaillard, L.; Ménézo, C. Microclimate and Urban Morphology Effects on Building Energy Demand in Different European Cities. *Energy Build.* **2020**, *224*, 110129. [CrossRef]
57. Bruckner, H.; Alyokhina, S.; Schneider, S.; Binder, M.; Abdin, Z.U.; Santbergen, R.; Verkou, M.; Zeman, M.; Isabella, O.; Pagliarini, M.; et al. Lessons Learned from Four Real-Life Case Studies: Energy Balance Calculations for Implementing Positive Energy Districts. *Energies* **2025**, *18*, 560. [CrossRef]
58. Nguyen, M.T.; Batel, S. Which energy citizenship in positive energy districts? A governmentality social psychological analysis of participatory governance. *J. Environ. Policy Plan.* **2024**, *26*, 117–130. [CrossRef]
59. Silvast, A.; Valkenburg, G. Energy citizenship: A critical perspective. *Energy Res. Soc. Sci.* **2023**, *98*, 102995. [CrossRef]
60. Ryghaug, M.; Skjølsvold, T.M.; Heidenreich, S. Creating energy citizenship through material participation. *Soc. Stud. Sci.* **2018**, *48*, 283–303. [CrossRef]
61. Baer, D.; Loewen, B.; Cheng, C.; Thomsen, J.; Wyckmans, A.; Temeljotov-Salaj, A.; Ahlers, D. Approaches to Social Innovation in Positive Energy Districts (PEDs)—A Comparison of Norwegian Projects. *Sustainability* **2021**, *13*, 7362. [CrossRef]
62. Craig, M.T.; Wohland, J.; Stoop, L.P.; Kies, A.; Pickering, B.; Bloomfield, H.C.; Browell, J.; De Felice, M.; Dent, C.J.; Deroubaix, A.; et al. Overcoming the disconnect between energy system and climate modeling. *Joule* **2022**, *6*, 1405–1417. [CrossRef]
63. Kılkış, Ş.; Krajačić, G.; Duić, N.; Rosen, M.; Al-Nimr, M. Effective mitigation of climate change with sustainable development of energy, water and environment systems. *Energy Convers. Manag.* **2022**, *269*, 116146. [CrossRef]
64. Davide, M.; Bastos, J.; Bezerra, P.; Hernandez, M.G.; Palermo, V.; Pittalis, M.; Todeschi, V.; Treville, A.; Barbosa, P.; Melica, G. *How to Develop a Sustainable Energy and Climate Action Plan (SECAP)—Covenant of Mayors Guidebook—Main Document*; Publications Office of the European Union: Luxembourg, 2025. [CrossRef]
65. PMI. PMBOK Guide | Project Management Institute. Available online: https://www.pmi.org/standards/pmbok (accessed on 1 July 2025).
66. Axelos. PRINCE2 Certification | Qualifications and Exams | Axelos. Available online: https://www.axelos.com/certifications/propath/prince2-project-management (accessed on 1 July 2025).
67. Scorza, F.; Santopietro, L. A systemic perspective for the Sustainable Energy and Climate Action Plan (SECAP). *Eur. Plan. Stud.* **2024**, *32*, 281–301. [CrossRef]
68. Schneider, S.; Drexel, R.; Zelger, T.; Baptista, J. PEExcel: A Fast One-Stop-Shop Assessment and Simulation Framework for Positive Energy Districts. In Proceedings of the BauSim Conference 2024, IBPSA-Germany and Austria, Vienna, Austria, 10–12 September 2024; Volume 10, pp. 80–88. [CrossRef]
69. Kohlbacher, M.; Gruenwald, S. Process orientation: Conceptualization and measurement. *Bus. Process Manag. J.* **2011**, *17*, 267–283. [CrossRef]
70. Khan, S.A.; Tao, Z.; Agyekum, E.B.; Fahad, S.; Tahir, M.; Salman, M. Sustainable rural electrification: Energy-economic feasibility analysis of autonomous hydrogen-based hybrid energy system. *Int. J. Hydrog. Energy* **2025**, *141*, 460–473. [CrossRef]
71. Rosen, M.A.; Koohi-Fayegh, S. The Prospects for Hydrogen as an Energy Carrier: An Overview of Hydrogen Energy and Hydrogen Energy Systems. *Energy Ecol. Environ.* **2016**, *1*, 10–29. [CrossRef]
72. Yue, M.; Lambert, H.; Pahon, E.; Roche, R.; Jemei, S.; Hissel, D. Hydrogen Energy Systems: A Critical Review of Technologies, Applications, Trends and Challenges. *Renew. Sustain. Energy Rev.* **2021**, *146*, 111180. [CrossRef]

73. Volpe, R.; Bisello, A.; Tuerk, A.; Guarino, F.; Giancola, E.; Sanchez, M.N.; Reda, F. Linking Environmental Impact Assessment and Positive Energy Districts: A Literature Review. *Clean. Environ. Syst.* **2025**, *16*, 100264. [CrossRef]
74. Wang, G.; Gilmont, O.; Blondeau, J. Pathways to Positive Energy Districts: A Comprehensive Techno-Economic and Environmental Analysis Using Multi-Objective Optimization. *Energies* **2025**, *18*, 1134. [CrossRef]
75. Cellura, M.; Fichera, A.; Guarino, F.; Volpe, R. Leveraging Positive Energy Districts Surplus for the Achievement of the Sustainable Development Goals. *Energies* **2025**, *18*, 506. [CrossRef]
76. Casamassima, L.; Bottecchia, L.; Bruck, A.; Kranzl, L.; Haas, R. Economic, social, and environmental aspects of Positive Energy Districts—A review. *Wiley Interdiscip. Rev. Energy Environ.* **2022**, *11*, e452. [CrossRef]
77. Sareen, S.; Albert-Seifried, V.; Aelenei, L.; Reda, F.; Etminan, G.; Andreucci, M.-B.; Kuzmic, M.; Maas, N.; Seco, O.; Civiero, P.; et al. Ten questions concerning positive energy districts. *Build. Environ.* **2022**, *216*, 109017. [CrossRef]
78. Turci, G.; Civiero, P.; Aparisi-Cerdá, I.; Marotta, I.; Massa, G. Transition Approaches towards Positive Energy Districts: A Systematic Review. *Buildings* **2024**, *14*, 3039. [CrossRef]
79. Natanian, J.; Guarino, F.; Manapragada, N.; Magyari, A.; Naboni, E.; De Luca, F.; Cellura, S.; Brunetti, A.; Reith, A. Ten questions on tools and methods for positive energy districts. *Build. Environ.* **2024**, *255*, 111429. [CrossRef]

designs

MDPI

Article

The Benefits of Positive Energy Districts: Introducing Additionality Assessment in Évora, Amsterdam and Espoo

Aristotelis Ntafalias [1,*], Panagiotis Papadopoulos [1], Mark van Wees [2], Danijela Šijačić [2], Omar Shafqat [2], Mari Hukkalainen [3], Julia Kantorovitch [3] and Magui Lage [4]

1 Motor Oil Hellas S.A, 15124 Marousi, Greece; ppapadopoulos@moh.gr
2 Department of Technology, Amsterdam University of Applied Sciences, Amstelcampus, Rhijnspoorplein 2, 1091 GC Amsterdam, The Netherlands; m.t.van.wees@hva.nl (M.v.W.); d.sijacic@hva.nl (D.Š.); o.shafqat@hva.nl (O.S.)
3 VTT Technical Research Centre of Finland Ltd., 02150 Espoo, Finland; mari.hukkalainen@vtt.fi (M.H.); julia.kantorovitch@vtt.fi (J.K.)
4 CNET—Centre for New Energy Technologies, S.A, Rua Particular À Rua Cidade de Goa, N° 2, 2685-038 Sacavém, Portugal; magui.lage@edp.com
* Correspondence: antafalias@moh.gr

Abstract: Positive Energy Districts (PEDs) are a promising approach to urban energy transformation, aiming to optimize local energy systems and deliver environmental, social and economic benefits. However, their effectiveness and justification for investment rely on understanding the additional value they provide (additionality) in comparison to current policies and planning methods. The additionality perspective is not used yet in current evaluations of PED demonstrations and pilots. Therefore, this paper introduces the concept of additionality in the evaluation of PEDs, focusing on the additional benefits they bring and the circumstances under which they are most effective. We discuss the additionality of PEDs in addressing the challenges of climate neutrality and energy system transformation in three European cities that are funded by the European Commission's H2020 Programme. It should be noted that given the ongoing status of these projects, the assessment is mainly based on preliminary results, as monitoring is still ongoing and quantitative results are not yet available. The paper discusses the drivers and barriers specific to PEDs, and highlights the challenges posed by technical complexities, financing aspects and social and legal restrictions. Conclusions are drawn regarding the concept of additionality and its implications for the wider development of PEDs as a response to the challenges of climate neutrality and energy system transformation in cities. We conclude that the additionality perspective provides valuable insights into the impact and potential of PEDs for societal goals and recommend this approach for use in the final evaluation of R&I projects involving PEDs using actual monitored data on PEDs.

Keywords: positive energy districts; citizen engagement; additionality; urban energy systems; sustainability

1. Introduction

Many cities across Europe aim towards climate neutrality, in order to drastically reduce greenhouse gas (GHG) emissions to mitigate climate change [1]. Since energy production and consumption are responsible for major GHG emissions, green and digital transition efforts aim to transform the energy system to achieve significant reductions in environmental impact [2]. At the city level, this transformation requires a redesign of districts' energy systems, encompassing energy production, distribution and utilization. For this purpose, the concept of Positive Energy Districts (PEDs) has been introduced as one possible response for integrating energy systems locally and as an initiative to maximize the energy-performance-buildings sector on a larger scale [3]. PEDs are designed to generate more energy than they consume, with adaptability to market fluctuations and reduced reliance on centralized grids. This is achieved through decentralized solutions, renewable energy, storage technologies and enhanced management practices. The EU has proposed regulations to support the implementation of PEDs by 2050, with a target to plan 100 PEDs by 2025 [4,5]. Although still in the early stages, this project could significantly influence the global adoption of PEDs. The state of the art of its implementation and effective penetration into the current energy systems of European cities, according to the evidence provided by the literature [6], shows that fewer than half of the documents analyzed refer to actual case studies, 80% of which were funded as part of research projects.

A review of spatiotemporal urban energy system modeling for urban decarbonization strategy formulation [7] examines the complexity and effectiveness of various tools and concepts for supporting urban energy system transformation toward sustainable, low-carbon cities. The study presents a systematic screening and evaluation method, identifying 10 key modeling dimensions based on a literature review. These dimensions are used to assess 20 different urban energy system modeling frameworks, considering the needs of both the building sector and the mobility sector and the unique spatiotemporal characteristics of urban energy systems.

PEDs offer a range of potential benefits, including the optimization of energy systems beyond individual buildings, benefitting from solutions that have more impact or are easier to implement at a wider scale (district or neighborhood) [3]. The optimization of energy systems includes increased energy efficiency, higher density of renewable energy, flexibility and storage (considering congestion), as well as the integration of building energy systems with other city energy related infrastructures, such as electric vehicle (EV) charging points [8]. Additionally, these dynamic district-level energy systems are controlled in a smart way to optimize their positive impact withing the wider city-level energy systems and energy markets [9].

However, PEDs move beyond the traditional focus on buildings' energy consumption and climate impact. They are meant to achieve a broader range of environmental, social and economic benefits, aiming to foster inclusivity, citizen engagement, improved well-being and opportunities for local businesses [10]. Additionally, PEDs can be used to combat energy poverty in historic urban districts [11]. However, preserving historic structures while simultaneously promoting sustainable practices and addressing community needs presents a significant challenge [12,13]; therefore, many requirements must be considered to ensure that these structures can be securely adapted for new functions and uses [14].

The development of PEDs in European cities is the subject of many EU-funded research and innovation projects (R&Is), including POCITYF [15], ATELIER [16] and SPARCS [17]. These projects exploit the potential of PEDs by utilizing local energy systems, which typically combine multiple buildings, integrating external renewable energy sources and, in many cases, are designed to be governed by citizen-driven energy communities. Given the complex nature of PEDs, it is unsurprising that the potential drivers of and barriers to their design, implementation and operation are multifaceted.

The aim of this paper is to introduce the concept of additionality in the evaluation of PEDs, addressing the following fundamental question: "Do PEDs offer tangible additional benefits, and if so, under what circumstances and through which mechanisms?". To answer this question, we examine three PED developments in three different cities: Amsterdam, Évora and Espoo. These are lighthouse cities in the three aforementioned EU-funded projects, representing a range of city characteristics and contexts, as well as different PED designs.

2. PEDs and Additionality Concept

2.1. What Are Positive Energy Districts?

The PED concept is an emerging approach to achieving sustainable urban development and addressing the challenges of climate change and energy security; however, its definition is not unambiguous [8]. In broad terms, a PED can be considered as an urban area that aims to achieve a positive or net-zero energy balance by producing more energy from renewable sources than it consumes [18]. This is achieved by the installation and integration of renewable energy systems (RESs) (such as photovoltaics (PVs), battery energy storage systems (BESSs), etc.) and by optimizing energy consumption in buildings in the district. PEDs can be of any size, from a small neighborhood to a large city district, physical or digital, and address different aspects of the energy spectrum, such as use in buildings and transportation and energy-grid stability [3,19].

From an environmental standpoint, PEDs aim to reduce GHG emissions, playing a crucial role in mitigating climate change [11]. Moreover, they contribute to improved air quality, fostering healthier living environments and enhancing the overall well-being of communities [20]. Additionally, from an economic perspective, PEDs aim to serve as catalysts for local growth, increasing job creation and driving innovation in sustainable technologies [21]. It is worth noting that PEDs are expected to foster a sense of community unity by actively engaging residents in energy conservation practices and encouraging their participation in renewable energy production initiatives [22].

While PEDs are gaining traction, there are various factors that influence their progress, both positively (drivers) and negatively (barriers). Advancements in technology and RES penetration, political will and policy development, economic incentives, supportive governance, active community engagement and ongoing research and development efforts are key drivers of the wider development of PEDS as they promote and make them more viable and accepted [23–25].

On the other hand, barriers pose challenges that need to be addressed for the successful implementation and extensive adoption of PEDs. These obstacles include technical complexities, financing aspects, governance and legal restrictions, social impacts and difficulties in replicating successful models [26–29]. Understanding these major —and, in some cases, conflicting—categories of drivers and barriers is crucial for stakeholders aiming to promote the sustainable development of PEDs. For instance, RESs play a crucial role in achieving PED targets [30]; however, there are cases in which promising legislative frameworks aimed at promoting RESs have not been translated into comprehensive policy implementations [31].

Barriers to and drivers of PEDs have been studied and addressed in many research and innovation projects, such as, for example, in [9,32]. However, we observed that these projects often include aspects unrelated to the PED concept, meaning that they could also be applied to individual and unconnected buildings. For example, the barriers to energy efficiency in buildings are very relevant for policies and measures, but they are already being applied extensively without any associations with PEDs. This shows that addressing these barriers and enhancing drivers, including energy efficiency measures in PED projects, are not enough to accelerate energy efficiency.

2.2. *Assessing the Additionality of PEDs*

Additionality refers to the concept of causing a benefit that is additional to what would have occurred without the implementation of a specific project or initiative [33]. In the context of PEDs, additionality encompasses broader considerations of district-scale energy planning. It often relates to the additional energy savings, renewable energy generation and utilization, or GHG emission reductions that result from the implementation of sustainable energy projects within a district.

In this paper, our focus is on identifying the drivers and barriers specific to PEDs that can potentially be influenced by a PED approach.

Through this assessment, we aim to draw conclusions regarding the concept of additionality. Furthermore, ensuring that a specific project, initiative, or service genuinely contributes to the goals of PEDs is crucial. Therefore, establishing a baseline is necessary to determine whether the PED approach has resulted in any additional impact and is worth pursuing. It should be noted that additionality serves as a complementary, albeit distinct approach, for assessing barriers and drivers based on monitoring key impact indicators, a practice commonly employed by most R&I projects. This paper focusses on assessing the three main impact areas of PEDs [34], as presented in Figure 1. The PED areas selected for assessment (Figure 1) are analyzed below.

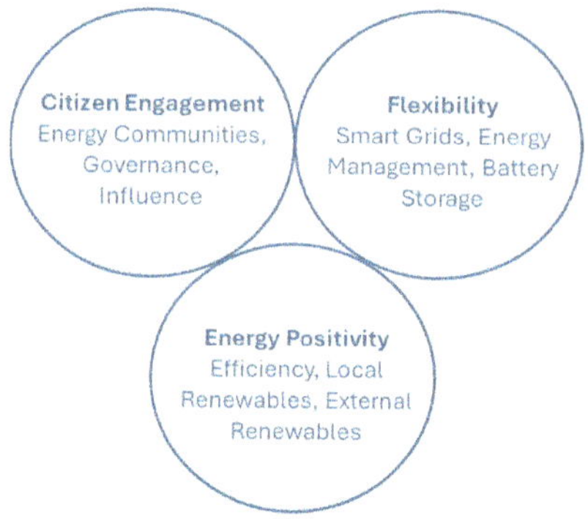

Figure 1. PED areas selected for assessment.

Impact area 1: High energy efficiency and local renewable energy production to achieve energy positivity

Energy positivity in PEDs can be achieved through increasing energy efficiency, local renewable energy production and integrating external renewable energy production into the PED energy system. This represents a demanding benchmark, even for new constructions, despite current regulations and practices already leading to near-zero-energy buildings at baseline. However, the PED energy balance may also include energy use from appliances and other sources not directly connected to the buildings. Achieving this benchmark is even more challenging in existing buildings, particularly through renovation.

Impact area 2: Increasing the flexibility of the local energy system

PEDs aim to increase energy flexibility by maximizing self-consumption and minimizing peak power demand through the installation of smart (micro) grids, energy management and battery storage. An energy system is flexible if it can meet (and reduce) peak load demand cost-effectively, reliably, and across all time scales. This means that the system must be able to balance supply and demand, even when there are fluctuations in variable renewable energy output. To achieve this, the system requires sufficient storage capacity, including both electricity storage and, through sector coupling, renewable heat and gas storage. This allows the system to balance periods of high variable renewable energy generation with periods of high demand, but low generation. In locations where the existing grid infrastructure is becoming a constraint on new connections and the growth of renewable energy production, increasing energy flexibility is particularly important. For example, in the city of Amsterdam, energy flexibility is crucial due to net congestion issues. Similarly, in the city of Espoo, the district heating and cooling system, as well as heating peak demands, are also critical considerations in achieving energy flexibility.

Impact area 3: Citizen engagement in the governance of PEDs

PEDs aim to involve citizens and other local stakeholders in governance and provide opportunities for residents to influence and interact with energy systems, particularly through energy communities. This is based on the hypothesis that increased involvement would enhance the benefits for citizens and increase the impact of PED's societal goals.

The framework for assessing additionality

In the context of PED developments in Amsterdam, Évora and Espoo, the question of additionality arises as follows: What is the realized impact of a demonstration PED project compared to the baseline if the R&I project had not been implemented? It is worth noting that impact is also correlated with cost-effectiveness; if a PED achieves the same impacts as the baseline but at a lower cost, this represents an added benefit. Figure 2 illustrates the systematic evaluation of the additional benefits derived from the PED approach. The workflow begins with identifying the key barriers and drivers within the three main impact areas relevant to PEDs. It then assesses whether these factors are influenced by the PED approach. Depending on this assessment, the analysis either considers the typical benefits derived from standard interventions or identifies the additional benefits that can be directly attributed to the PED approach. Finally, the overall benefit is determined by combining both typical and additional benefits.

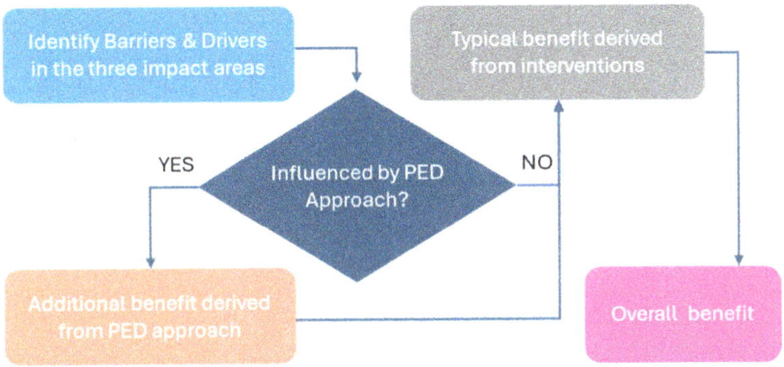

Figure 2. Flow diagram reflecting PED additionality.

The logical framework for addressing the additionality of PED cases is explained in Figure 3. For instance, the GHG emissions in an area after the implementation of the PED concept are calculated, partly based on measured energy production and demand data, and then compared to a baseline scenario without a PED. If these emissions are lower for the PED scenario, the PED can be considered as providing additional benefits for this specific impact.

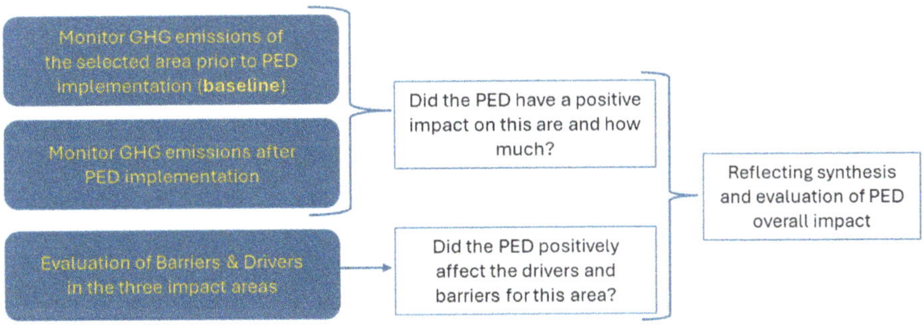

Figure 3. Example of addressing additionality framework.

Objectives of the R&I projects and the role of PED demonstration cases

The objectives of the three R&I projects, as presented in Figure 4, share a common goal: to explore and investigate the feasibility and design of PEDs and their impact, resulting in scientific insights and recommendations for policymakers, urban planners and other stakeholders. The primary R&I activity is related to the implementation and operation of actual PEDs, which undergo both ex ante and ex post monitoring and evaluation of their impacts. These demonstrations also contribute valuable input to research on business models, social aspects, governance structures, energy system optimization and replication.

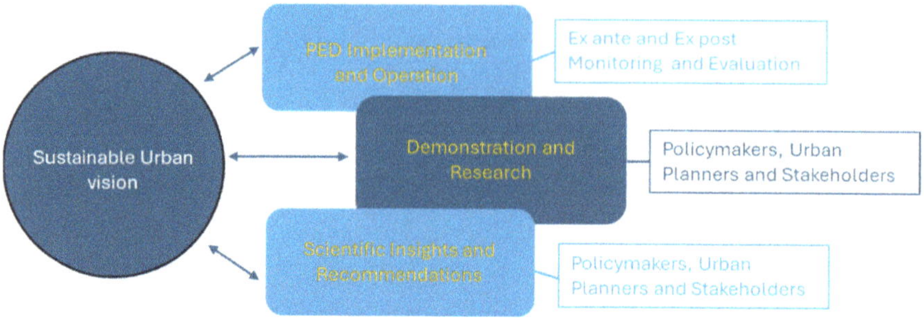

Figure 4. Objectives of the three R&I EU-funded projects.

At a broader level, cities are shaping their visions of a more sustainable future and decarbonizing urban environments. The expected impacts of these demonstrations are based on central hypotheses regarding the added value of the PED concept for relevant societal goals, compared to a baseline scenario without the PED approach and without the R&I project (as the baseline). The scope of impacts to be considered was specified in the terms of reference for the Horizon 2020 call for proposals [35], aligning with the R&I questions on PEDs stemming from emerging European policies aimed at the ambitious climate targets set by the European Union.

As PEDs vary in size and shape, the three PEDs studied have served as demonstration areas in three European research and innovation (R&I) projects. The assessment of the case studies is presented in the following sections.

3. Évora Case Study

Description of the PED in Évora

Évora holds national and international recognition for its recovery policies and heritage preservation. Since 2012, Évora has been implementing its Sustainable Energy Action Plan (SEAP), aiming at reducing its energy consumption, reducing carbon dioxide (CO_2) emissions by 20% in comparison with 1990 levels and promoting a sustainable use of energy. Energy efficiency has become a policy priority for Évora municipality due to its ability to address challenges such as reducing dependence on imported energy, reducing GHG emissions and improving economic competitiveness. The city is divided into three regions: the city center, a rural area and a technological and industrial park, where three PEDs are implemented. Different conditions enable testing different solutions in different environments, resulting in multiple conclusions and lessons learned, however, in this paper, only the city center will be analyzed in detail. The Évora PED is a pioneering project located in the historic city center of Évora, a World Heritage Site. The project incorporates eight municipal buildings, two parking lots and eleven residential and commercial clients, with the goal of achieving energy positivity while preserving the city's cultural heritage.

The PED's innovative solutions address the unique challenges of this location, where heritage preservation legislation mandates the careful consideration of energy efficiency and production. Key solutions include the installation of building-integrated photovoltaics (BIPVs) and building management systems (BMSs) in municipal buildings. The BIPV solutions that will be installed in the municipal buildings in the historical city center, namely PV glasses and PV shingles, will contribute to the positivity with 380 kWp of installed capacity estimated in 440 MWh of energy production yearly. This is aligned with the EU 2030 Climate and Energy Framework, which aims to cut GHG emissions by 40% compared with 1990 levels. A 5 MW community solar farm (CSF) has been established outside the city center, enabling citizens to invest in renewable energy production, as they are not permitted to install conventional PV panels on their rooftops. This solution allows city center residents to virtually participate in solar energy production, overcoming heritage-related barriers and promoting the goal of energy positivity. At the end of the project, this PED will have 117% of the total energy need covered by local RES, which mean not only that all the energy need will be met by renewable energy production, but also that it will have a surplus of 17%.

The PED project also features a range of cutting-edge technologies, including energy routers, smart inverters, EV chargers integrated with flexibility algorithms, smart lamp posts and a pay-as-you-throw (PAYT) system. These solutions demonstrate a multifaceted approach to enhancing energy efficiency and management, showcasing the potential for sustainable energy solutions in historic urban environments. With regard to residential clients, the introduction of a home energy management system (HEMS) offers a way for households to engage in energy conservation and optimization actively. These systems empower residents to manage their energy consumption better while contributing to the district's broader energy sustainability goals.

Impact area 1: Barriers and drivers related to energy efficiency and renewable energy production in the PED

When developing a PED, it is important to acknowledge that historical cities are under heavy European and national legislation, and so, to install solutions that can result in a positive energy balance, it is essential to obtain necessary authorizations. For this purpose, it is crucial to have the local municipality engaged, as they are the authorities that will have to interact with the local cultural agency. The municipality of Évora was required to prepare a comprehensive document to explain and guarantee that the solutions that would be installed would not interfere with the aesthetic appearance of the city. To achieve this, selecting technologies that can be discreetly installed, such as BIPVs, is important. In the case of POCITYF [15], manufacturing companies were involved in the project consortium. Alternatively, if companies are not part of the consortium, it is essential to research and identify companies that offer these technologies and ensure that their availability aligns with the project timeline. The municipality of Évora prepared a comprehensive document, which was submitted to the Cultural Regional Agency (in Portuguese, Direção Regional da Cultura); however, the authorization was only given after a second round of interaction, almost two years after the beginning of the process.

Given the time-consuming nature of these processes, it is crucial to initiate the authorization process at a very early stage of the PED project. Failure to do so may jeopardize the project timeline, particularly during the lengthy monitoring phase. Following the authorization phase, if the solutions are to be installed in municipal buildings, public tenders must be prepared. Similarly, the tender process is also time-consuming. However, if the municipality recognizes the benefits of these solutions for the city, it is easier to move through that process.

The development of this PED has faced two major challenges: the legislative obstacles surrounding BIPV installation and the limited availability of injection points. The PED's objectives are multifaceted, aiming to achieve energy positivity, navigate regulatory challenges related to its World Heritage Site status, boost local energy generation, deploy EV chargers with flexible algorithms, integrate smart waste management within the PAYT system and promote citizen engagement through mobile apps and HEMS. This holistic approach tackles the pressing issue of urban energy management while encouraging community participation and engagement in sustainable practices.

Table 1 presents the main drivers of, and Table 2 presents the barriers that may be faced by, the PED approach in the city center of Évora.

Table 1. Drivers of energy efficiency and RES, and observations on the impact of the PED approach in Évora.

Drivers	Relevance to PED Approach	Has the PED Strengthened This Driver?
EU Funding and Legislation: EU-funded projects	Yes. For historical cities, it is necessary to have innovative solutions, such as BIPV, which are still very expensive, to reach positivity (or at least, contribute to local production).	Yes. The grants made additional solutions installation possible.
Sustainability Goals and Urban Transformation: Urban environment and climate change goals	Yes. Possessing a proven track record of efforts to meet the European Union's goals for building a resilient energy system and implementing a bold climate policy is crucial for ensuring security, sustainability, competitiveness and affordability to achieve a PED.	No. These goals were already set before inclusion of the buildings in the PED.
EU Funding and Legislation: Changes to existing legislation to realize the full potential of sustainable technologies	Yes. The objective is to have more production than consumption within the city center, taking advantage of current technologies to reduce GHG emissions and increase citizens wellbeing.	Yes. The project created solutions to overcome this type of inequality.

Table 2. Barriers to energy efficiency and RES and observations on the impact of the PED approach in Évora.

Barriers	Relevance for PED Approach	Has the PED Reduced This Barrier?
Regulatory and Contractual Barriers: European Legislation	High. The installation of the RES in the city center is necessary to reach positivity.	No. The PED approach did not contribute to the reduction of this barrier.
Technical Barriers and Grid Integration: Injection-point unavailability	High. To have more production in a certain area, it is important to have availability of the grid to install these technologies.	No. The PED approach did not contribute to the reduction of this barrier.
Resident and Community Engagement Challenges: Citizens' engagement	High. Contributes to achieving energetic positivity for the area, and fosters a sense of community ownership and participation in sustainable practices.	Yes. The PED approach contributed to the reduction of this barrier. The CSF offered a solution, enabling citizens residing in the city center to participate in solar energy production virtually, as if the panels were installed on their own rooftops.

Significant drivers are EU-funded projects, which align with the EU's 2030 targets by utilizing funds to accelerate the energy transition, particularly through the adoption of expensive technologies. Urban environment and climate change goals present opportunities for historical cities. However, these cities struggle to meet climate change and energy transition goals due to existing legislation, which hinders the achievement of EU targets. Changes to existing legislation to realize the full potential of sustainable technologies are also drivers as revising current laws could allow for the installation of RESs in historical cities, creating opportunities for technologies that respect cultural heritage.

Several challenges hinder the energy transition in historical cities. One major obstacle is European legislation, which restricts the installation of RESs in historical city centers, such as Évora, which is listed on the World Heritage List. This protective legislation, while essential for preserving cultural heritage, makes it difficult to deploy sustainable technologies in areas that need energy upgrades the most.

Another significant barrier is the injection-point unavailability in the electrical grid. The grid in Portugal was originally designed for centralized energy production, and with the increased penetration of renewables, grid congestion is becoming a problem. This lack of capacity to integrate more renewable energy, like solar PV plants, creates bottlenecks, slowing down the expansion of clean energy infrastructure.

Citizens' engagement is also crucial, but challenging. Involving the public in the energy transition is essential to ensure they are at the center of the process. Building trust and maintaining open communication with citizens are necessary to encourage their active participation in and acceptance of sustainable projects. Without this engagement, local support may falter, affecting project outcomes.

Impact area 2: Barriers and drivers related to flexibility of the energy system

Within the PED are installed three EV chargers, each with a capacity of 7.2 kW, alongside three batteries, each with a storage capacity of 30 kWh. As part of the initiative for decarbonization, one EV charger and battery will be deployed within the Living Laboratory for Decarbonization (LVpDÉ), with an additional two EV chargers and two batteries to be installed at the City Hall.

These assets are planned for integration into the EV Management Platform, where flexibility algorithms will be employed to optimize self-consumption and enhance cost savings. Moreover, the implementation of BMS across all eight municipal buildings offers users the ability to monitor energy consumption and production in real-time. This data-driven approach facilitates optimization strategies, empowering active user participation in energy management initiatives. The main drivers of energy flexibility are EU project financing to cover equipment costs, such as batteries and EV chargers, and installation expenses. Another positive factor is the limited availability of available solutions to increase savings due to legal restrictions on installing RESs. However, since these solutions were installed in municipal buildings, it was necessary to launch public procedures for the installations, which proved to be very time-consuming.

The following tables present the main drivers (Table 3) and barriers. One important driver is financing from EU projects, which plays a crucial role in supporting the development of sustainable technologies. Since many of these technologies remain costly, EU funds are essential in boosting their advancement. Another key factor is the limited range of options for increasing self-consumption. Technologies that are compatible with the preservation of cultural heritage are both limited and expensive, making it difficult for historical cities to fully participate in the energy transition.

Table 3. Drivers of energy system flexibility and observations on the impact of the PED approach in Évora.

Drivers	Relevance to PED Approach	Has the PED Strengthened This Driver?
EU Funding and Legislation: Financing from EU projects	High.	Yes, PED development is a critical component in the context of sustainable cities, as it increases the possibilities of funding for such initiatives. PEDs are being developed as key components for energy-efficient environments, recognizing the significant role of cities in the energy transition.
Local Renewable Energy and Efficiency: Limited options to increase self-consumption	High. Storage and RES equipment increases self-consumption, helping to compensate for the low capacity installed in the buildings.	Yes, increasing self-consumption is critical to decrease dependency on the grid and reach positivity.

Table 4 shows how energy system flexibility may be impacted by the PED approach in the city center of Évora.

Table 4. Barriers to energy system flexibility and observations on the impact of the PED approach in Évora.

Barriers	Relevance to PED Approach	Has the PED Reduced This Barrier?
Regulatory and Contractual Barriers: Public tenders	High. Without the public tenders, the systems could not be installed.	No, this procedure is mandatory given that the municipalities were responsible for installing the solutions, and the PED approach did not influence it.

One important driver is financing from EU projects, which plays a crucial role in supporting the development of sustainable technologies. Since many of these technologies remain costly, EU funds are essential in boosting their advancement. Another key factor is the limited options for increasing self-consumption. Technologies that are compatible with the preservation of cultural heritage are both limited and expensive, making it difficult for historical cities to fully participate in the energy transition.

Public tenders in Portugal present a barrier due to lengthy bureaucratic processes. The internal inertia within municipalities causes significant delays in decision-making and implementation. These time-consuming processes slow down the development of critical energy transition projects and add layers of complexity to an already challenging process.

Impact area 3: Barriers and drivers related to citizen engagement in PED governance

The citizen engagement aspect emerged as a key factor in realizing the objectives of Évora's PED project, which was primarily aimed at placing citizens at the forefront of the energy transition. Central to this approach were three distinct engagement campaigns:

1. Targeting city center residents equipped with PV panels—installed prior to the legislation forbiting their installation—where the installation of HEMS was essential in enhancing energy efficiency through behavioral changes. This initiative was crucial in enhancing energy efficiency through behavioral changes, fostering a culture of sustainability among residents.
2. Engaging clients to participate in testing the PAYT solution underscored our commitment to innovative waste-management practices. This approach not only promoted efficient waste management but also encouraged citizens to take an active role in reducing waste and promoting a cleaner environment.
3. The utilization of digital applications developed within the project framework constituted a third avenue of engagement. These applications included a tourist app incentivizing exploration through point accrual, a city information platform offering access to pertinent urban data such as weather, traffic and energy community presence, as well as a cultural experience marketplace facilitating access to events while rewarding citizen participation.

These engagement campaigns were critical in realizing the objectives of Évora's PED project, as they placed citizens at the forefront of the energy transition, empowering them to make behavioral changes that enhance energy efficiency and contribute to the societal objectives of PEDs (Table 5).

Table 5. Drivers of citizen engagement and observations on the impact of the PED approach in Évora.

Drivers	Relevance to PED Approach	Has the PED Strengthened This Driver?
Sustainability Goals and Urban Transformation: Include citizens in the transformation of the city	High. Needed for the wide penetration of PEDs.	Yes, through the changes in citizens' behavior, it is possible to increase self-consumption and decrease the amount of energy purchased from the grid.

A key driver is the need to include citizens in the transformation of the city. In Portugal, public tenders often face significant delays due to internal inertia within municipalities, where bureaucratic processes and administrative inefficiencies slow down decision-making and implementation. This hampers the ability to involve citizens effectively in urban transformation efforts, which is essential for achieving energy transition and sustainability goals.

On the other hand, another hurdle is the interest of citizens to be active throughout the project. While initial enthusiasm may be high, maintaining this engagement over the duration of the project is difficult. Building trust can take time, and the lack of technical knowledge among citizens often hinders their full understanding of the project's benefits, which in turn diminishes their motivation and involvement over time.

Table 6 presents the drivers of citizen engagement that may be impacted by the PED approach in Évora.

Table 6. Barriers to citizen engagement and observations on the impact of the PED approach in Évora.

Barriers	Relevance for the PED Approach	Has the PED Reduced This Barrier?
Resident and Community Engagement Challenges: Interest of the citizens in being active throughout the project	High. Needed for energy positivity.	Yes, it is essential to develop strategies for realizing participatory processes that involve citizens from the beginning phase. Building trustful and reliable relationships between citizens and local authorities is critical to the success of PEDs.

In Évora, the concept of additionality is clearly demonstrated through the PED project's tangible impacts, which would not have occurred without its implementation. The PED has driven technological innovation, policy alignment and stakeholder engagement, while overcoming barriers like regulatory constraints and financing challenges. It has also fostered energy communities that reduce energy costs, increase local production and promote social cohesion. The project's financial benefits are clear, with BIPV systems installed at no cost to the municipality, leading to long-term energy savings. These outcomes showcase the PED's role in accelerating Évora's sustainable energy transition.

4. Amsterdam Case Study

Description of the PED in Amsterdam

Amsterdam is undertaking the transformation of a former industrial neighborhood, Buiksloterham, into a sustainable area featuring a blend of residential and commercial spaces. Within this development, two new building blocks in separate locations totaling 22,000 m^2 are being constructed to become part of a PED. The first block, named República, consists of 55 apartments alongside various utility buildings, such as a hotel, restaurants and a shared office space, totaling six structures [36]. The second block, "Poppies", comprises 100 apartments.

It is envisaged that these will be complemented by an external PV plant outside the actual PED area. This additional PV production is necessary, because the two blocks cannot be made energy-positive, as a result of the limitation on integrating more PVs or other forms of renewable energy production (lack of roof space, grid congestion and unfavorable business case). Additionally, a biogas plant using the waste streams from the PEDs is planned.

All three PED components will be integrated into a local energy market, with an energy storage battery installed at the República site.

The construction phase of the PED in Amsterdam has now been successfully completed, marking the beginning of the official monitoring period. This monitoring phase, which is critical for gathering data on the district's performance and impacts, will continue until the end of 2025. After the conclusion of the monitoring, a comprehensive evaluation will be conducted to assess the actual impacts of the PED. As the monitoring has just started, no concrete results are available at this time. Therefore, the current assessment of additionality presented in this document is based on preliminary estimates and is intended to serve as an illustrative example of how the additionality concept will be applied in future evaluation processes.

Impact area 1: Barriers and drivers related to energy efficiency and renewable energy production

A wide range of energy efficiency measures and RES integrations have been implemented in the two building blocks of the PEDs, incorporating a range of cutting-edge technologies. These measures include the following:

- Greater-than-building-standards insulation to minimize heat loss and enhance thermal efficiency.
- A centralized heat pump plant for heating, complemented by distributed booster heat pumps. These are combined with subsoil Aquifer Thermal Energy Storage (ATES) to increase the coefficient of performance for both heating and cooling.
- Extensive coverage of the building roofs with the integration of 191 high-performance PV panels.
- Heat recovery systems to capture and recycle waste heat.
- Additional thermal storage buffers to afford flexibility in controlling heat pumps.

As a result, the level of energy efficiency and RES production in the PED building blocks was higher than the baseline, as reflected by the relevant regulation. However, this was not enough to achieve energy positivity, although this technically would have been possible [37,38].

The primary drivers behind the project's sustainability standards were the initial developer-set standards and additional financing from the R&I project. The PED approach's potential drivers of energy efficiency and RES did not have a verifiable impact. In contrast, the second building block of the Amsterdam PEDs, where the municipality owned the land and initiated a tender with high energy performance demands, showed a different outcome.

The additionality of external RES production is debatable. For the Amsterdam PED demonstration, the project requirement for energy-positive buildings was the main driver. The PED approach enabled the integration of buildings and infrastructure, including a largely RES-based heating system. The connection to local PED grids and potential financial benefits could improve the business case for external RES production, but this will not be demonstrated in the Amsterdam project due to construction and sourcing delays (Table 7).

Table 7. Drivers of energy efficiency and RES and observations on the impact of the PED approach in Amsterdam.

Drivers	Relevance to PED Approach	Has the PED Strengthened This Driver?
EU Funding and Legislation: Changes to existing legislation to realize the full potential of sustainable technologies	Current building regulations in the Netherlands require new buildings to become close to zero-energy.	Regulation is part of the baseline and is not influenced by a PED approach. Furthermore, the baseline is dynamic, as regulation of building performance is becoming strict, reducing the difference between baseline and PED.
Sustainable Building Projects: Quality standards for buildings set by project developers and/or the municipality	High energy efficiency and RES standards are needed to achieve energy positivity. In some cases, the standards are set at a higher level by the developers and/or the municipality (selling the land).	These standards were already set before inclusion of the buildings in the PED. However, by demonstrating the viability of the concept, they could move the baseline forward in the long term.
EU Funding and Legislation: EU-funded projects	Additional energy efficiency and RESs are needed to achieve energy positivity.	The grants made additional EE and RES measures possible. Some of the EE and RES measures in the PED were covered by the EU grant.

A range of factors propel the energy transition in historical cities. EU funding and legislation remain pivotal, with EU funding projects aligned with the European targets for 2030 aimed at boosting the energy transition through the adoption of expensive, yet essential technologies. Additionally, changes to existing legislation to realize the full potential of sustainable technologies are critical; by updating laws, the EU can remove barriers and facilitate the integration of renewable energy systems in historical contexts. Quality standards for buildings set by projects ensure that new developments meet stringent sustainability criteria, promoting energy-efficient construction while preserving cultural heritage. Table 8 presents a summary of the barriers related to energy efficiency and renewable energy production.

Table 8. Barriers to energy efficiency and RES and observations on the impact of the PED approach in Amsterdam.

Barriers	Relevance to PED Approach	Has the PED Reduced This Barrier?
Market and Financial Barriers: Integrating energy efficiency and RES in the building design	Both are crucial to achieve energy positivity, beyond current (regulated) practice in new buildings in the Netherlands.	In the Amsterdam PED, there is no indication that the PED approach has directly contributed to the building design, as a result of (1) a mismatch in time (the design was almost finished before the PED project started) and (2) the lack of a business case or other arguments for subsequent modifications to the design.
Regulatory and Contractual Barriers: Additional costs of energy efficiency and RES	Needed to achieve energy positivity. Currently, in most cases, the business case for EE and RES beyond regulation is not positive.	The business model of the developers did not allow for enough energy efficiency and RES to achieve energy positivity. The PED approach did not provide additional drivers for an improved business case for EE and RES.
Regulatory and Contractual Barriers: Barriers to the implementation of energy efficiency and RES measures	In the Netherlands, new buildings, in almost all cases, comply with regulation. Investments in additional energy efficiency and RES is rare.	The experiences with implementation of energy efficiency and RES measures in the PED has provided insight into ways to address these barriers.
Technical Barriers and Grid Integration: Difficulties in the building development and design process and the PED project	High. The PED approach should be able to influence the design of buildings from the start.	Partly. The development and project design had started before a PED approach was adopted.

Various barriers complicate the implementation of energy efficiency and RES cities. Integrating energy efficiency and RES in the building design can be challenging, as aligning sustainability goals with architectural standards may be complex. Moreover, the additional costs of energy efficiency and RES measures increase the financial burden on projects, making them less attractive for investors and building owners. Furthermore, the difficulties in the building development and design process and the PED project add another layer of complexity, as balancing sustainability with architectural preservation often creates design challenges that slow down project progress.

Impact area 2: Barriers and drivers related to flexibility of the energy system

The PED in Amsterdam features a centralized battery system, smart EV chargers and distributed smart-grid-ready heat pumps, managed by an advanced EMS to optimize self-consumption and minimize peak power demand. However, the project faced challenges in legal and organizational aspects, including contract establishment and asset integration. Delays were caused by issues with billing, tax application and the integration of the battery into the management system. An unforeseen barrier emerged due to congestion in the electricity grid, limiting energy system usage. The local flexible energy market is crucial, but current congestion issues pose challenges.

The main drivers of energy flexibility are the strategic and commercial interests of private sector parties developing flexibility technologies for local grids, the urgency of congestion management in the city and the project financing available for measures such as the battery and development costs (Table 9). In the city, a growing number of R&I projects address energy flexibility; however, there is no evidence yet that the integration of energy flexibility within a PED has strengthened these drivers or reduced the existing barriers (Table 10), beyond the availability of project resources.

Table 9. Drivers of energy system flexibility and observations on the impact of the PED approach in Amsterdam.

Drivers	Relevance to PED Approach	Has the PED Strengthened This Driver?
Local Renewable Energy and Efficiency: Increasing congestion bottlenecks in the city	PED aims for increased flexibility to allow higher densification of RES in the PED and to become a source of income (selling services to the grid). This strongly depends on the location.	In the Amsterdam PED, the flexibility system is not yet operational, so its impacts have not been demonstrated yet. The project faces both technical and financial obstacles, so that both technical feasibility and a positive business case, crucial for upscaling and replication, cannot yet be demonstrated.
Sustainable Building Projects: Increasing markets for products and services in energy flexibility	R&I projects on PEDs aim to bring these products and services closer to the market to speed up implementation.	The Amsterdam PED was only one of many R&I projects over recent years that addressed flexibility in energy system. None of these projects were part of an overall PED approach.
EU Funding and Legislation: Grant financing for investments and development of systems	To date, the majority of PEDs have been implemented with grant support from R&I projects.	The grants were essential in covering the additional costs of these measures (such as the battery and the energy management system).

Table 10. Barriers to energy system flexibility and observations on the impact of the PED approach in Amsterdam.

Barriers	Relevance to PED Approach	Has the PED Reduced This Barrier?
Technical Barriers and Grid Integration: High R&I costs for system development	The assumption is that the PED approach would improve the affordability and business case for flexibility-related products and services that are needed to allow more RESs.	The PED project has brought these systems to a higher TRL level. However, there are no clear signs that a significant market share with significant impact can be achieved soon.
Technical Barriers and Grid Integration: Affordability of flexibility systems for PEDs	Without affordable systems, flexibility options cannot be included in the PED.	To date, there is no evidence of products that would be affordable for potential users. Business models remain immature. Furthermore, development times are still too long. However, demand may increase with increasing congestion problems in Amsterdam.
Technical Barriers and Grid Integration: Lack of uptake of smart EMS by building owners and users	These systems are necessary to introduce energy flexibility in local energy systems. In PEDs, it is assumed that residents and other stakeholders take ownership of these systems and benefit from the advantages.	The main drivers of smart buildings are not specific to PEDs, the developing regulation on building management systems for utility buildings, or the increasing insights into costs and benefits.
Technical Barriers and Grid Integration: Lack of quantitative knowledge on the benefits impact of flexibility measures	Energy flexibility is assumed to be a key innovation in PEDs. In the EU, in the Netherlands and in Amsterdam, many R&I projects are ongoing, exploring and demonstrating the feasibility, impact and benefits of flexibility measures at building and district level.	The project has brought together researchers, product developers and other stakeholders. The investigation of grid balancing services was new. The project has provided insight into the business case for energy communities in PEDs to invest in flexibility measures. However, the final evaluation on technical and economic feasibility is still outstanding.
Technical Barriers and Grid Integration: Connection restriction due to congestion	The assumption is that the PED approach would improve the affordability and business case for flexibility-related products and services that are needed to allow more RESs.	The PED project has brought these systems to a higher TRL level. However, there are no clear signs that a significant market share with significant impact can be achieved soon.

Sustainability goals and urban transformation push cities to adapt to climate change targets, although historical cities often face challenges due to restrictive legislation. To overcome this, revisions in legislation are being considered to enable the installation of renewable energy systems compatible with cultural heritage. Financing from EU projects plays a crucial role in supporting the development of sustainable technologies, as many remain costly, and EU funds are essential in advancing their implementation. The limited options for increasing self-consumption highlight the need for more affordable and culturally compatible technologies to allow historical cities to participate fully in the energy transition.

High R&I costs for system development present a significant financial hurdle for the research into and development of advanced systems for PEDs. Similarly, the affordability of flexibility systems for PEDs remains a concern, as these systems are often costly to install and maintain, limiting widespread adoption. Additionally, the lack of uptake of smart EMS by building owners and users hinders the integration of smart technologies, largely due to unfamiliarity with their benefits. The lack of quantitative knowledge on the benefits and impact of flexibility measures further impedes progress, as stakeholders are often unsure about the tangible value these systems provide. Finally, connection restrictions due to grid congestion represent a physical limitation, as the outdated grid infrastructure struggles to accommodate the increased energy flow from renewable sources.

Impact area 3: Barriers and drivers related to citizen engagement in PED governance

The Amsterdam PED demonstration aims to establish this governance structure for the full local system and its assets, including not only renewable energy production, but also the use of energy storage and energy trading, both internally and externally. Alternative juridical models for this energy community have been investigated. The PED residents and other legal entities that occupy property in the PED, such as hotel, office space and food and beverage operators, which will become members of the energy communities, have been engaged and consulted. It is likely that residents would not have organized an energy community without the project developers' and R&I project's initiative and support.

The drivers contributing to the development of participatory governance in local energy systems include political and policy support at both national and local levels, including the municipality of Amsterdam (presented in Table 11). Additionally, many research and innovation projects are supporting the establishment of local energy communities. Strong motivation among a segment of the population to take initiatives is also a key factor. Furthermore, ongoing legal reform is aimed at facilitating the development of energy communities.

Table 11. Drivers of citizen engagement and observations on the impact of the PED approach in Amsterdam.

Drivers	Relevance to PED Approach	Has the PED Strengthened This Driver?
Sustainability Goals and Urban Transformation: Growing political and societal support for local governance of energy systems	In this initial phase, PED development in Amsterdam is strongly reliant on this support. The city of Amsterdam is project coordinator of the PED project, indicating their initial interest in PEDs.	There are no indications that this momentum is influenced by holistic approaches in PEDs in Amsterdam. PEDs are still very unknown to both politicians and civic society. In Amsterdam, the municipality has had a very limited role in the implementation of the PED project, which is indicative of the barriers to embedding the results of the R&I project into the city's policies and planning.
Local Renewable Energy and Efficiency: Foreseen economic benefits of self-production, consumption and energy sharing/trading	In Amsterdam, commercial parties, including SMEs, play a key role in initiating local energy projects, including PEDs, based on their expectations of positive future business cases. At the same time, a further hypothesis is that PED residents will profit financially from these new energy services.	The first indications, to be substantiated on the basis of final monitoring results, are that the business case for commercial parties offering products and services for PEDs are less favorable than initially expected. From the residents' perspective, the additional PED measures and investments in PV were not reflected in m^2 purchase price for their apartment. It is not yet clear how satisfied residents are with the energy bills, which were initially higher than expected.

On the other hand, several hurdles have been identified in the functioning of the energy community (presented in Table 12). These include low resident knowledge or interest, a lack of feelings of ownership and the different statuses of residents, such as renters, owners and commercial parties. Administrative challenges without a dedicated budget also pose a problem, as does the option for residents to opt out of the concept.

Table 12. Barriers to citizen engagement and observations on the impact of the PED approach in Amsterdam.

Barriers	Relevance for the PED Approach	Has the PED Reduced This Barrier?
Resident and Community Engagement Challenges: Lack of knowledge/motivation among residents	The energy community governing the PED strongly depends on voluntary contributions from the residents and other inhabitants of the PED.	The residents in the PED demonstration have been informed by project researchers. It remains to be seen whether they are motivated enough to engage in PED governance in the longer term. Not everyone is prepared to dive deep into energy systems. Research into their motivation is ongoing.
Resident and Community Engagement Challenges: Lack of opportunities for residents to become engaged and interact with the energy systems and markets	The influence of residents on their living environment is considered an important pre-requisite for a successful PED.	In República, the energy community members have been able to influence the operation of the energy community. However, many decisions on the design of the Amsterdam PED and the community were made before residents moved in and offered their views. This has restricted the engagement.
Regulatory and Contractual Barriers: Complex and restrictive regulations for (energy) contracting	Without an enabling regulatory environment, the flexibility of the PED energy system cannot be achieved.	Options for (legal) governance were explored in the project, but systemic barriers remain. This is subject of future energy law reform that is debated at EU and national level. It cannot be influenced by developments on a local level, such as PEDs.
Resident and Community Engagement Challenges: Strong dependence on the project support in involving residents and setting up the energy community organization	The majority of PEDs have been implemented in the Netherlands with grant support from R&I projects.	It is unlikely that it is possible to establish a new energy community for PEDs and have the necessary systems in place for them to interact with the energy systems and markets in a future PED with any additional project support. This seriously restricts replication in the short term.

Growing political and societal support for the local governance of energy systems empowers local authorities and communities to manage their energy resources, fostering initiatives such as self-production, consumption and energy sharing or trading. The foreseen economic benefits of self-production, consumption and energy sharing/trading motivate stakeholders to invest in decentralized energy systems, enhancing economic resilience and reducing energy costs.

The lack of knowledge and motivation among residents acts as a barrier, as many citizens are unaware of how they can contribute to or benefit from energy systems and markets. Additionally, there is a lack of opportunities for residents to become engaged and interact with the energy systems and markets, which limits their involvement in energy-saving initiatives. Complex and restrictive regulations for energy contracting further complicate participation, making it difficult for citizens and local businesses to engage in energy-sharing or other market-based systems. The strong dependence on project support in involving residents and setting up energy community organization emphasizes the challenge of building a self-sustaining community without continued external backing.

5. Espoo Case Study

The Espoonlahti district in Espoo, Finland, has recently undergone significant development, with the addition of a new shopping center and central services block. This dense mixed-use area is supported by a geothermal heating and cooling plant aimed at optimizing energy use. The block spans a heated floor area of approximately 150,000 m^2 and includes 550 residential apartments, a senior house with around 120 apartments and a shopping center with a leasable gross area of 44,000 m^2.

The development incorporates renewable energy solutions, including geothermal energy, solar PV panels and battery storage, in an effort to provide a more sustainable energy supply. Public transportation access, including a new metro line and bus terminal, has been expanded to encourage reduced reliance on personal vehicles. The buildings and infrastructure in the area have been designed to focus on energy efficiency, with an emphasis on minimizing energy consumption.

Community engagement has been part of the development process, with input from local residents considered in the planning stages. However, like with many large-scale urban projects, the success of these initiatives in meeting the community's long-term needs and sustainability goals will depend on continued monitoring and adjustments.

Impact area 1: High energy efficiency and local renewable energy production to achieve energy positivity

The RES production on the demonstration site in Espoo includes a 4.4 MW regenerative ground source heat pump plant, with approximately 50 kilometers of bore holes. The regenerative geo-energy system also supplies 100% of the cooling demand and most of the heating demand for the center with the stores and public services, as well as the residential buildings. During the monitoring period (2 years at the time of writing) in which the building has been in operation, practically no external heat has been needed from district heating, even during the cold winter days, although the system was designed to use heat from the district heating grid in cases of prolonged cold weather. An electric boiler also exists as a reserve system. Excess heat recovered during summer from air conditioning and from the refrigeration systems of grocery stores is injected into the geo-energy field, increasing the capacity and efficiency of the system. In essence, this means that the geo-energy system is used for seasonal thermal energy storage. The building block has PV panels on the roof and building façades, with peak power of 634 kWp. The panel area is 2400 m^2.

In thermal energy, a positive energy balance can technically be achieved in this demonstration site, but with the current pricing of the district heating, it is not economically viable in practice to sell heating energy outside of the demonstration block. The drivers of and barriers to energy efficiency and local RES are summarized in Tables 13 and 14.

Table 13. Drivers of energy efficiency and local renewable and observations on the impact of the PED approach in Espoo.

Drivers	Relevance to PED Approach	Has the PED Strengthened This Driver?
Sustainable Building Projects: Building owners targeted to build a new shopping center, with public services, to be as sustainable and efficient as possible	High. Building a new shopping center within a PED framework can provide numerous benefits, including increased energy efficiency, reduced environmental impact, enhanced community engagement and economic development. By aligning with PED principles, the shopping center can contribute to a more sustainable and resilient future.	No. Some integrated energy system advances were made, but on the other hand, the building would also have been at Gold LEED (Leadership in Energy and Environmental Design) level [39] without the PED approach.
Local Renewable Energy and Efficiency: Increase the use of local RES	High. Increasing the use of local RES is a crucial component of achieving the goals of PEDs. By harnessing renewable energy resources, PEDs can become more sustainable, resilient and economically viable communities.	Partly (for the dimensioning of the PV plant). Additional RES sources were studied, but the choices for the RES production were made without the PED driver.

Table 14. Barriers to energy efficiency and local renewable energy production and observations on the impact of the PED approach in Espoo.

Barriers	Relevance to PED Approach	Has the PED Reduced This Barrier?
Market and Financial Barriers: Business case not profitable for selling excess heat outside of PED, to the district heating grid	High. Selling excess heat to the grid can still be a valuable strategy for PEDs in certain circumstances.	No.
Regulatory and Contractual Barriers: Contractual difficulties: not possible to make an electricity-buying contract with a specific plant that the operator owns (outside of the PED demonstration area)	High. Without a direct contract, PEDs may be exposed to fluctuations in energy prices, which can impact their financial viability.	No. Practically, being able to have such a contract does not directly increase local RES production. Achieving energy positivity balance purely by using local RES sources does not seem feasible in the dense urban environment in Finland.

Building owners targeting sustainable and efficient developments, such as new shopping centers with public services, drive the integration of sustainability into urban planning. Efforts to increase the use of local RES reduce dependence on external energy supplies and support local economies.

Economic and contractual barriers play a role in delaying the energy transition. Business cases not being profitable for selling excess heat outside of PEDs to district heating grids limit financial incentives for developers. In many cases, the economic model does not support selling surplus energy to broader networks, reducing profitability. Additionally, contractual difficulties, such as not being able to make an electricity-buying contract with a specific plant owned by the operator outside of the PED demonstration area, create legal and logistical challenges, stalling efforts to scale up energy sharing across different regions.

Impact area 2: Increasing flexibility of the local energy system

Electric batteries, together with smart energy management systems, are used to optimize and control the energy systems of the building block in the Espoo PED case. In Nordic countries, newly built complex building entities typically already have very good energy efficiency, which often requires rather advanced energy management systems. With the electric battery, this demo site has large enough energy flexibility capacity to make it possible to participate in the national electricity grid's flexibility markets. Minimized electricity costs are drivers that provide financial incentives for adopting energy-saving measures, making sustainable investments more attractive (Table 15).

Table 15. Drivers of energy flexibility and observations on the impact of the PED approach in Espoo.

Drivers	Relevance to PED Approach	Has the PED Strengthened This Driver?
Local Renewable Energy and Efficiency: Minimized electricity costs	High. By implementing energy-efficient strategies and promoting renewable energy, PEDs can achieve significant savings for residents and businesses, while also contributing to a more sustainable and resilient future.	Yes. Based on the results, building owner decided to invest in battery storage and join Nordpool's reserve market. R&Is create the ground by showing the possibilities (as well as though sharing experiences from similar demonstrations).

Impact area 3: Citizen engagement in the governance of PEDs

The central block is an important mobility hub, connecting the metro station, bus terminal and facilities for cycling and e-mobility. The citizen engagement activities for engaging people in the PEDs were highly focused on sustainable mobility, including how to optimize people flow and user experience. Activities included studies on micro mobility and shared mobility, a 1.5-degree workshop, a mobile ethnographic user study in

Espoonlahti, test days for electric vehicles and engagement activities with local "buddy families", users of the shopping center and school children.

Citizens, particularly youths, have been engaged in the co-creation of the new shopping center. Additionally, a collaboration has taken place with a local elementary school class, with the students studying specific sustainability themes promoting sustainable lifestyles and energy solutions. The drivers and barriers are presented in Tables 16 and 17, respectively.

Table 16. Drivers of citizen engagement and observations on the impact of PED in Espoo.

Drivers	Relevance for PED Approach	Has the PED Strengthened This Driver?
Sustainability Goals and Urban Transformation: City wants to be a pioneer of sustainable lifestyle	Medium. Supporting sustainable lifestyle in the district by engaging people in the planning of their local mobility solutions.	Yes: by bringing resources and new tools for sustainable mobility development.

Table 17. Barriers to citizen engagement and observations on the impact of PED in Espoo.

Barriers	Relevance to PED Approach	Has the PED Reduced This Barrier?
Resident and Community Engagement Challenges: Behavioral change is difficult	Medium. By understanding the factors that make behavioral change difficult and implementing effective strategies, individuals can overcome obstacles and achieve their goals.	The project provided additional resources and channels for raising awareness of the behavioral changes, yet there is no information on whether this really changed the citizens' behavior.

The city's ambition to be a pioneer of a sustainable lifestyle drives a significant shift in urban planning and policymaking. Cities that aim to lead in sustainability often adopt policies that prioritize renewable energy, green transportation and resource-efficient infrastructure. Public awareness campaigns, green certifications and sustainability awards can further motivate citizens to adopt eco-friendly practices in their daily lives.

On the other hand, behavioral change is difficult and presents a significant barrier to the adoption of new energy-efficient practices. Even when the technology and infrastructure are in place, residents and businesses may resist changing long-standing habits, requiring ongoing education, incentives and engagement to achieve meaningful shifts in behavior toward more sustainable practices.

In the case of Espoo, the concept of additionality is demonstrated through the integration of renewable energy systems and sustainable urban development. The development incorporates advanced energy solutions, such as geothermal energy, solar PV panels and battery storage, optimizing energy use in this densely built environment. Additionally, expanded public transportation and ongoing community engagement are expected to drive behavioral shifts, encouraging reduced car usage and greater participation in local energy initiatives. Combined with energy-efficient building designs, these efforts are projected to yield long-term benefits, such as reduced energy costs and increased local energy resilience, clearly illustrating the future additionality of Espoo in promoting urban sustainability.

6. Discussion

This section delves into the concept of additionality within PEDs based on the case study findings. We note that the analysis is based on preliminary monitoring results, as these projects are still ongoing. The aim is to investigate whether the additionality perspective can potentially be of benefit in PED impact evaluations.

By analyzing these factors, we can identify the unique contributions of the PED approach and the extent to which it moves beyond "business as usual" practices. Additionally, we discuss broader drivers and barriers that can affect PED implementation and impact across different areas. Figure 5 summarizes the levels of additionality per impact area for both drivers and barriers.

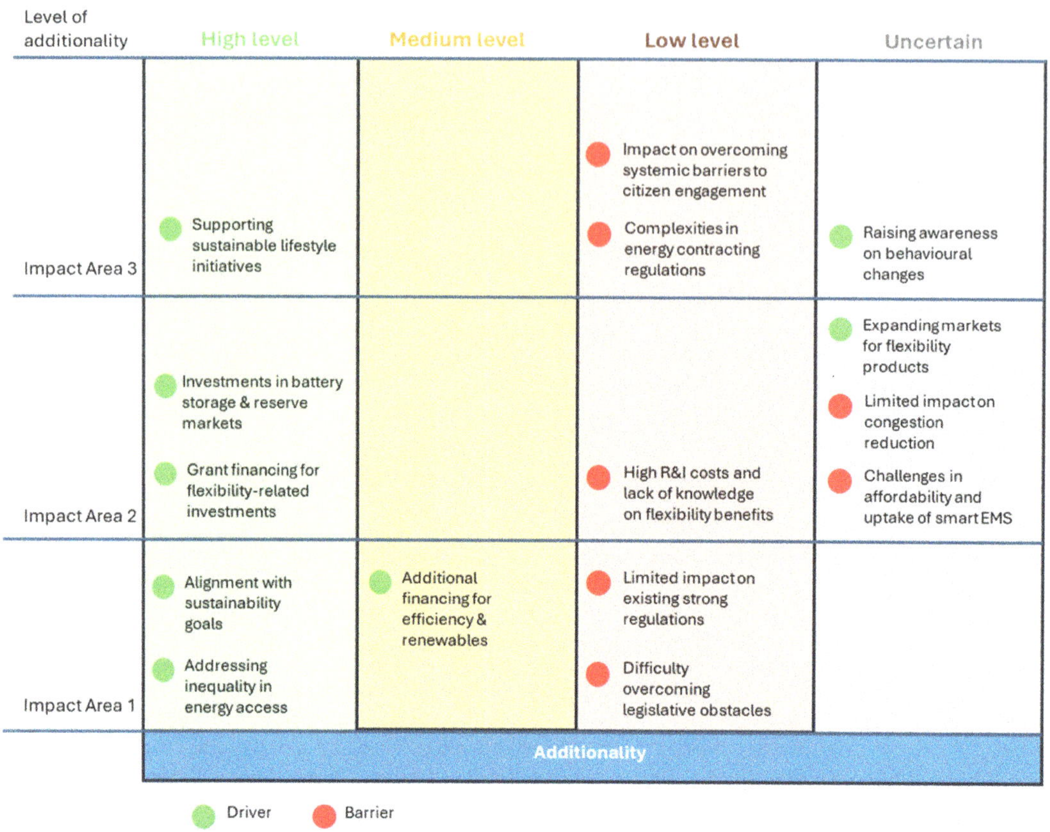

Figure 5. Levels of additionality in different impact areas.

6.1. Synthesis of Case Studies

Impact area 1: High energy efficiency and local renewable energy production

Assessing the drivers, it becomes evident that while PED initiatives align with sustainability targets and encourage increased use of local renewable energy sources, their direct impact on enhancing these aspects varies. PEDs demonstrate limited additionality in cases where existing regulations and quality standards are already in place, as their inclusion does not significantly enhance the regulatory framework or set higher standards.

However, PEDs exhibit moderate-to-high additionality when they facilitate additional financing, enabling the implementation of additional energy efficiency and renewable energy measures that might not have been feasible otherwise. Furthermore, PEDs role in addressing inequality issues within urban environments showcases their ability to offer additional benefits beyond what might have been achieved through conventional approaches.

On the other hand, when evaluating barriers, PEDs effectiveness in addressing these challenges and providing additional solutions is limited. While they aim to overcome obstacles such as contractual difficulties and injection-point unavailability, their impact remains constrained, as observed in cases where barriers persist despite PED implementation. Similarly, legislative obstacles and unprofitable business cases present significant challenges that PEDs are yet to effectively address, indicating limited additionality in resolving such barriers.

Impact area 2: Increasing flexibility of the local energy system

While PEDs have effectively strengthened certain drivers, such as minimizing electricity costs through investments in battery storage and participation in reserve markets, their impact on other drivers is less certain. For instance, while PEDs aim to address congestion bottlenecks in cities by increasing flexibility, their effectiveness in strengthening these drivers is uncertain without specific outcomes. Additionally, while PEDs contribute to expanding markets for energy flexibility products and services, their role in this regard is limited by the absence of a comprehensive PED approach.

However, PEDs have demonstrated clear additionality in providing grant financing for investments and system development. This funding has enabled crucial investments in equipment and human resources development.

When considering the barriers, PED initiatives have shown mixed levels of effectiveness.

While efforts have been made to address challenges such as high R&I costs and a lack of quantitative knowledge on the benefits of flexibility measures, the effectiveness of PEDs in fully overcoming these barriers is limited. For example, while PEDs aim to improve the affordability and uptake of smart energy management systems, evidence of their impact in this regard remains unclear. Nonetheless, PEDs have demonstrated clear additionality in addressing certain barriers. For instance, they have successfully overcome connection restrictions due to congestion by implementing flexible assets like EMS and batteries, which have facilitated grid connections.

Impact area 3: Citizen engagement in the governance of PEDs

PED initiatives have shown varying levels of impact, exemplified by their contribution to city's goal of pioneering sustainable lifestyles. By introducing new resources and tools for sustainable mobility development, PEDs have clearly supported initiatives within the district. However, the impact of engaging the community in the energy transition, although a medium-relevance driver of PEDs, remains uncertain. While PED initiatives may have increased awareness of behavioral changes, their translation into significant community engagement is not well documented.

Turning to barriers, PED initiatives have addressed certain challenges, but their effectiveness varies. For instance, while PED has bolstered residents' knowledge and motivation by providing additional support and resources, systemic barriers may persist. Additionally, while PEDs may have explored solutions for complex energy contracting regulations, systemic barriers endure. Furthermore, while citizen engagement serves as both a driver and a barrier, PED initiatives have not substantially strengthened this aspect, indicating ongoing challenges independent of the PED approach.

6.2. Cross-Cutting and Generic Drivers and Barriers in PEDs

Generic driver 1: Sustainability

The three cities involved have embraced climate mitigation goals and are intensifying their climate policies, driven by strong political and societal momentum for citizen participation in energy supply and demand. This driver underscores the importance of closely aligning projects with established climate goals and sustainability targets, emphasizing the need to reduce greenhouse gas emissions, mitigate climate change impacts and promote environmentally sustainable practices. The project aims to contribute to broader efforts in combating climate change and fostering sustainable development by prioritizing alignment with these goals.

Furthermore, this driver emphasizes the significance of raising awareness among stakeholders and the general public about projects aimed at transforming historical cities into smart cities. Educating stakeholders about the potential benefits, including improved quality of life, enhanced infrastructure and economic growth, is essential. By fostering understanding and appreciation for these initiatives, the projects aim to garner support and engagement from various stakeholders, facilitating smoother implementation and long-term success.

All three PED projects have developed long-term strategies towards sustainability, aiming to make their cities climate-neutral by 2050. Espoo and Amsterdam are EU Cities Mission cities, aiming to make their cities climate-neutral by 2030 [40].

The additionality of PEDs lies in their potential to create a positive energy balance, in which the energy produced exceeds the energy consumed, leading to a reduction in overall energy demand and greenhouse gas emissions.

Generic driver 2: Breaking silos between policy and R&I domains

Breaking silos between policy and R&I domains is crucial for the successful implementation of PED projects. This driver highlights the importance of fostering collaboration and communication between different stakeholders, including policymakers, researchers and industry experts, to ensure that PED projects are aligned with the latest research and technological advancements. It also emphasizes the need for continuous learning and adaptation to ensure that PED projects remain relevant and effective in achieving their goals.

PEDs have the potential to create a holistic approach to urban development, in which energy, infrastructure and environmental considerations are integrated into policy and research efforts.

Generic barrier 1: Planning and timing of the development of the PED and its components

The planning and timing of the development of a PED and its components are crucial for the success of the project. PED development is a cluster of different developments that are partly mutually related and subject to external factors. In Amsterdam, the main development was the construction of the buildings and related infrastructure, which encountered delays. The second main development was the design, development and installation of the smart grid infrastructure, including storage and related energy management, followed by the energy trading system. The third was the (legal) design of the energy community for governance, with its own timeline.

Developing a PED, the first of its kind in one city, is conceptually and organizationally challenging. Planning is of crucial importance, but developments rarely occur as planned, and many details are not foreseen. Amsterdam's example has shown that although innovative technology is available, to organize and implement it still poses a challenge.

The initial intention of having two different greenfield developments included in the PED was to increase the insights gained on the impact of building characteristics on the costs and benefits of PEDs, thus strengthening the empirical basis for the conclusions. Furthermore, it would be possible to investigate the use of storage within a virtual PED energy network. However, it has become clear that the implementations of these two developments have substantially diverged, which has made coordination more difficult.

PED development is complex and subject to external factors, which can lead to delays and challenges in coordinating various components, such as building construction, smart grid infrastructure and energy management systems. The additionality of PEDs lies in their potential to create a more efficient and integrated approach to urban development, where planning and coordination efforts are streamlined to minimize delays and maximize the benefits of PEDs. PEDs also allow greater flexibility in being able to deal with emerging challenges in energy transition, as with the case of grid congestion in Amsterdam.

Generic barrier 2: Limited uptake of PED concepts by municipality

The limited uptake of PED concepts by municipalities is a significant barrier to the successful implementation of these projects. The embedding of PEDs into cities' municipal energy and climate policies, and into the planning of energy infrastructure, is proceeding very slowly. After five years of project implementation, the question of how the knowledge gained on PEDs is effectively embedded into the municipal organization is still under evaluation. This is caused by (1) PEDs as holistic measures in energy and climate policies being too broad a concept to fit into current municipal policy and planning methods and processes, (2) a lack of capacity among policy makers and urban planners to actively engage in PED projects, as they are overburdened by their core business, and (3) a lack of political interest. This creates a contradiction between political ambition (see above) on the one hand and a lack of capacity to benefit from R&I projects on the other.

PEDs could create a more comprehensive approach to urban development, where PED concepts are integrated into municipal policies and planning efforts, leading to more sustainable and energy-efficient cities.

Generic Barrier 3: Complex administrative procedures, legal obstructions, and bureaucracy

This barrier refers to the bureaucratic hurdles and administrative complexities that often accompany large-scale urban development projects. It involves navigating multiple layers of government regulations, securing permits, and coordinating with various stakeholders, which can lead to delays and increased costs. This includes energy market regulation. Addressing this barrier requires streamlining administrative procedures, fostering inter-agency collaboration, and leveraging digital technologies to simplify processes and improve efficiency.

The additionality of PEDs lies in their potential to create a more streamlined and efficient approach to urban development, where administrative procedures are simplified and bureaucratic hurdles are identified and reduced, leading to the more cost-effective and timely implementation of PEDs.

Generic Barrier 4: Preservation Requirements in Heritage Listed Cities

This barrier highlights the challenges posed by stringent preservation requirements and regulations in cities with rich historical heritage. It involves navigating complex processes and ensuring that any smart city interventions are compatible with the architectural and cultural heritage of the city. Overcoming this barrier requires careful planning, stakeholder engagement, and innovative solutions that balance the preservation of heritage with the implementation of modern technologies and infrastructure.

PEDs have the potential to create innovative solutions that balance the preservation of heritage with the implementation of modern technologies and infrastructure, leading to a more sustainable and culturally sensitive approach to urban development.

7. Conclusions

This paper introduced the concept of additionality as a method to assess and understand the benefits of PEDs, examining the extent to which PEDs offer tangible additional benefits beyond what would have occurred without their implementation. To test whether this perspective could provide new insights into PED impacts, we applied this approach to three ongoing PED developments in Amsterdam, Évora, and Espoo. As final monitoring results are not yet available for these ongoing projects, the assessment was based on interim results in qualitative terms, which were mainly provided by the opinions of experts involved in these projects.

The main findings include the following: Understanding the additionality of PEDs is crucial for evaluating their effectiveness and justifying investments. Our findings suggest that PEDs have the potential to create a positive energy balance, in which the energy produced exceeds the energy consumed, leading to a reduction in overall energy demand and GHG emissions. This is achieved through a combination of high energy efficiency, local renewable energy production, and external renewable energy production integrated into the PED energy system. Furthermore, PEDs can increase the flexibility of local energy systems by maximizing self-consumption and minimizing peak power demand through the installation of smart (micro) grids, energy management and battery storage. This flexibility is crucial in locations where the existing grid infrastructure is becoming a constraint on new connections and the growth of renewable energy production.

In addition, PEDs aim to involve citizens and other local stakeholders in governance and provide opportunities for residents to influence and interact with energy systems, particularly through energy communities. This citizen engagement is essential for enhancing benefits for citizens and increasing the impact of PEDs societal goals.

However, our analysis also highlights several generic drivers and barriers that can affect PED implementation and impact different areas. These include alignment with climate goals, breaking silos between policy and R&I domains, the planning and timing of PED development, the limited uptake of PED concepts by municipalities, complex administrative procedures, and preservation requirements in heritage-listed cities. Our observations reveal a lack of differentiation between the drivers and barriers related to individual solutions, such as the energy efficiency of buildings, and those directly related to and influenced by the PED approach itself.

Furthermore, our analysis uncovers the absence of a detailed framework describing the essential components, desirable features, and the drivers of and barriers to PED development. A PED should not be merely an umbrella for policies and measures that could be implemented without a PED approach anyway, but it should bring added value and accelerate progress towards the societal goal of a sustainable energy transition. As PEDs evolve and address remaining challenges, their additionality in driving a more sustainable and resilient urban energy future is likely to increase.

Although our three case studies were based on preliminary and qualitative data and the results of the assessment are, therefore, indicative only, we can conclude that the additionality perspective could provide new insights and understanding on PED-impacts, beyond those resulting from the current practice of evaluation. We recommend that pro-jects include this approach in their final evaluation practices, based on the actual monitoring data.

Author Contributions: Conceptualization, A.N., P.P., M.v.W. and O.S.; methodology, A.N., M.v.W., M.H., J.K. and M.L.; writing—original draft preparation, A.N., P.P., M.v.W., D.Š., M.H. and M.L.; writing—review and editing, A.N., P.P., O.S., J.K. and M.L.; visualization, A.N.; supervision, A.N. and P.P. All authors have read and agreed to the published version of the manuscript.

Funding: This research was funded by the European Commission's Horizon 2020 Framework Programme for Research and Innovation for three projects ATELIER (grant agreement ID: 864374), POCITYF (grant agreement ID: 864400) and SPARCS (grant agreement ID: 864242).

Data Availability Statement: No new data were created in this study. All relevant information is discussed within the paper.

Conflicts of Interest: Author Mari Hukkalainen and Julia Kantorovitch were employed by VTT Technical Research Centre of Finland Ltd. The remaining authors declare that the research was conducted in the absence of any commercial or financial relationships that could be construed as a potential conflict of interest.

References

1. Salvia, M.; Reckien, D.; Pietrapertosa, F.; Eckersley, P.; Spyridaki, N.A.; Krook-Riekkola, A.; Olazabal, M.; Hurtado, S.D.G.; Simoes, S.G.; Geneletti, D.; et al. Will climate mitigation ambitions lead to carbon neutrality? An analysis of the local-level plans of 327 cities in the EU. *Renew. Sustain. Energy Rev.* **2021**, *135*, 110253. [CrossRef]
2. Bianchini, S.; Damioli, G.; Ghisetti, C. The environmental effects of the 'twin' green and digital transition in European regions. *Environ. Resour. Econ.* **2023**, *84*, 877–918. [CrossRef]
3. Lindholm, O.; Rehman, H.U.; Reda, F. Positioning Positive Energy Districts in European Cities. *Buildings* **2021**, *11*, 19. [CrossRef]
4. Energy Positive Neighborhoods and Smart Energy Districts: Methods, Tools, and Experiences from the Field. Available online: https://www.sciencedirect.com/book/9780128099513/energy-positive-neighborhoods-and-smart-energy-districts (accessed on 6 September 2024).
5. Bruck, A.; Ruano, S.D.; Auer, H. One piece of the puzzle towards 100 Positive Energy Districts (PEDs) across Europe by 2025: An open-source approach to unveil favourable locations of PV-based PEDs from a techno-economic perspective. *Energy* **2022**, *254*, 124152. [CrossRef]
6. Maestosi, P.C.; Salvia, M.; Pietrapertosa, F.; Romagnoli, F.; Pirro, M. Implementation of Positive Energy Districts in European Cities: A Systematic Literature Review to Identify the Effective Integration of the Concept into the Existing Energy Systems. *Energies* **2024**, *17*, 707. [CrossRef]
7. Horak, D.; Hainoun, A.; Neugebauer, G.; Stoeglehner, G. A review of spatio-temporal urban energy system modeling for urban decarbonization strategy formulation. *Renew. Sustain. Energy Rev.* **2022**, *162*, 112426. [CrossRef]
8. Derkenbaeva, E.; Vega, S.H.; Hofstede, G.J.; van Leeuwen, E. Positive energy districts: Mainstreaming energy transition in urban areas. *Renew. Sustain. Energy Rev.* **2022**, *153*, 111782. [CrossRef]

9. Krangsås, S.G.; Steemers, K.; Konstantinou, T.; Soutullo, S.; Liu, M.; Giancola, E.; Prebreza, B.; Ashrafian, T.; Murauskaitė, L.; Maas, N. Positive Energy Districts: Identifying Challenges and Interdependencies. *Sustainability* **2021**, *13*, 10551. [CrossRef]
10. Van Wees, M.; Revilla, B.P.; Fitzgerald, H.; Ahlers, D.; Romero, N.; Alpagut, B.; Kort, J.; Tjahja, C.; Kaiser, G.; Blessing, V.; et al. Energy Citizenship in Positive Energy Districts—Towards a Transdisciplinary Approach to Impact Assessment. *Buildings* **2022**, *12*, 186. [CrossRef]
11. Gouveia, J.P.; Seixas, J.; Palma, P.; Duarte, H.; Luz, H.; Cavadini, G.B. Positive Energy District: A Model for Historic Districts to Address Energy Poverty. *Front. Sustain. Cities* **2021**, *3*, 648473. [CrossRef]
12. Gregório, V.; Seixas, J. Energy savings potential in urban rehabilitation: A spatial-based methodology applied to historic centres. *Energy Build.* **2017**, *152*, 11–23. [CrossRef]
13. Mathew, D.; O'Hegarty, R.; Kinnane, O. Historic buildings to Positive Energy Buildings: A trilemma between energy efficiency, lifecycle carbon and architecture conservation. *J. Phys. Conf. Ser.* **2023**, *2654*, 012010. [CrossRef]
14. Pereira, L.D.; Tavares, V.; Soares, N. Up-To-Date Challenges for the Conservation, Rehabilitation and Energy Retrofitting of Higher Education Cultural Heritage Buildings. *Sustainability* **2021**, *13*, 2061. [CrossRef]
15. Homepage—POCITYF—POCITYF. Available online: https://pocityf.eu/ (accessed on 22 April 2024).
16. ATELIER—Positive Energy Districts. Available online: https://smartcity-atelier.eu/ (accessed on 22 April 2024).
17. Sustainable Energy Positive & Zero cARbon CommunitieS—Sparcs. Available online: https://sparcs.info/en/ (accessed on 22 April 2024).
18. Cartuyvels, P.; Ahlers, D.; Alpagut, B.; Cerna, V.; Cimini, V.; Haxhija, S.; Hukkalainen, M.; Kuzmic, M.; Livik, K.; Padilla, M.; et al. Action Clusters: Sustainable Districts and Built Environment; Business Models & Finance POSITIVE ENERGY DISTRICTS #EUSmartCities An Action Cluster of the Smart Cities Marketplace. Available online: www.jpi-urbaneurope.eu/ped (accessed on 11 September 2024).
19. Hedman, Å.; Rehman, H.U.; Gabaldón, A.; Bisello, A.; Albert-Seifried, V.; Zhang, X.; Guarino, F.; Grynning, S.; Eicker, U.; Neumann, H.M.; et al. IEA EBC Annex83 Positive Energy Districts. *Buildings* **2021**, *11*, 130. [CrossRef]
20. Marotta, I.; Guarino, F.; Longo, S.; Cellura, M. Environmental Sustainability Approaches and Positive Energy Districts: A Literature Review. *Sustainability* **2021**, *13*, 13063. [CrossRef]
21. Zhang, X.; Shen, J.; Saini, P.K.; Lovati, M.; Han, M.; Huang, P.; Huang, Z. Digital Twin for Accelerating Sustainability in Positive Energy District: A Review of Simulation Tools and Applications. *Front. Sustain. Cities* **2021**, *3*, 663269. [CrossRef]
22. Hajduk, M.; Jaššo, M.; Husár, M.; Ondrejička, V. Citizen engagement within the process of realisation of the city energy transition projects. *AIP Conf. Proc.* **2022**, *2574*, 160004. [CrossRef]
23. Kim, M.H.; Kim, D.; Heo, J.; Lee, D.W. Techno-economic analysis of hybrid renewable energy system with solar district heating for net zero energy community. *Energy* **2019**, *187*, 115916. [CrossRef]
24. Frieden, D.; Tuerk, A.; Neumann, C.; d'Herbemont, S.; Roberts, J. Collective Self-Consumption and Energy Communities: Trends and Challenges in the Transposition of the EU Framework. Available online: https://www.researchgate.net/publication/346975546 (accessed on 11 September 2024). [CrossRef]
25. Shnapp, S.; Paci, D.; Bertoldi, P. *Enabling Positive Energy Districts across Europe: Energy Efficiency Couples Renewable Energy*; Publication Office of the European Union: Luxembourg, 2020. [CrossRef]
26. Mavrigiannaki, A.; Pignatta, G.; Assimakopoulos, M.; Isaac, M.; Gupta, R.; Kolokotsa, D.; Laskari, M.; Saliari, M.; Meir, I.A.; Isaac, S. Examining the benefits and barriers for the implementation of net zero energy settlements. *Energy Build.* **2021**, *230*, 110564. [CrossRef]
27. Boll, J.R.; Dorizas, V.; Broer, R.; Toth, Z. WP5-Policy Mapping and Analysis of Plus-Energy Buildings and Neighbourhoods D5.1 Barriers and Opportunities of Plus-Energy Neighbourhoods in the National and Local Regulatory Framework. Available online: www.synikia.eu (accessed on 30 November 2023).
28. Tuerk, A.; Frieden, D.; Neumann, C.; Latanis, K.; Tsitsanis, A.; Kousouris, S.; Llorente, J.; Heimonen, I.; Reda, F.; Ala-Juusela, M.; et al. Integrating Plus Energy Buildings and Districts with the EU Energy Community Framework: Regulatory Opportunities, Barriers and Technological Solutions. *Buildings* **2021**, *11*, 468. [CrossRef]
29. Magrini, A.; Lentini, G.; Cuman, S.; Bodrato, A.; Marenco, L. From nearly zero energy buildings (NZEB) to positive energy buildings (PEB): The next challenge—The most recent European trends with some notes on the energy analysis of a forerunner PEB example. *Dev. Built Environ.* **2020**, *3*, 100019. [CrossRef]
30. D'Agostino, D.; Parker, D. A framework for the cost-optimal design of nearly zero energy buildings (NZEBs) in representative climates across Europe. *Energy* **2018**, *149*, 814–829. [CrossRef]
31. Uspenskaia, D.; Specht, K.; Kondziella, H.; Bruckner, T. Challenges and barriers for net-zero/positive energy buildings and districts—Empirical evidence from the smart city project SPARCS. *Buildings* **2021**, *11*, 78. [CrossRef]
32. Sareen, S.; Albert-Seifried, V.; Aelenei, L.; Reda, F.; Etminan, G.; Andreucci, M.B.; Kuzmic, M.; Maas, N.; Seco, O.; Civiero, P.; et al. Ten questions concerning positive energy districts. *Build. Environ.* **2022**, *216*, 109017. [CrossRef]
33. Gillenwater, M.W.; Gillenwater, M. What Is Additionality? Part 1: A Long Standing Problem. 2012. Available online: https://www.researchgate.net/publication/257945337 (accessed on 30 June 2024).
34. Smart Cities Market Place. Positive Energy Districts Fact Sheet. *European Commission*. Available online: https://smart-cities-marketplace.ec.europa.eu/insights/solutions (accessed on 22 April 2024).

35. EU Funding & Tenders Portal. Available online: https://ec.europa.eu/info/funding-tenders/opportunities/portal/screen/opportunities/topic-details/lc-sc3-scc-1-2018-2019-2020 (accessed on 24 April 2024).
36. Sustainability—Republica. Available online: https://www.republica.amsterdam/en/sustainability (accessed on 24 May 2024).
37. Diestelmeier, L.; Swens, J. Energy Communities in the Netherlands: Learning from Local Energy Initiatives. *Eur. Energy Law Rep. XIV* **2021**, *14*, 239–264. [CrossRef]
38. Williams, K.; Heller, R.; van Wees, M.; Vastenhout, T. Assessing the performance of Positive Energy Districts: The need for innovative methods. *IOP Conf. Ser. Earth Environ. Sci.* **2022**, *1085*, 012014. [CrossRef]
39. LEED Rating System | U.S. Green Building Council. Available online: https://www.usgbc.org/leed (accessed on 3 September 2024).
40. Climate-Neutral and Smart Cities—European Commission. Available online: https://research-and-innovation.ec.europa.eu/funding/funding-opportunities/funding-programmes-and-open-calls/horizon-europe/eu-missions-horizon-europe/climate-neutral-and-smart-cities_en (accessed on 24 April 2024).

designs

Article

Toward Positive Energy Districts by Urban–Industrial Energy Exchange

Erfan Shafiee Roudbari [1,*], Ramanunni Parakkal Menon [1], Ivan Kantor [2] and Ursula Eicker [1]

[1] Department of Building, Civil and Environmental Engineering, Gina Cody School of Engineering and Computer Science, Concordia University, Montreal, QC H3G 1M8, Canada
[2] Department of Chemical and Materials Engineering, Gina Cody School of Engineering and Computer Science, Concordia University, Montréal, QC H3G 1M8, Canada
* Correspondence: erfan.shafieeroudbari@mail.concordia.ca

Abstract: The concept of Positive Energy Districts (PEDs) has emerged as a promising approach to achieving sustainable urban development. PEDs aim to balance the energy demand and supply within a district while reducing the carbon footprint and promoting renewable energy sources. Urban–Industrial Symbiosis (UIS) is another approach that involves the exchange of energy and resources between industrial processes and nearby urban areas to increase efficiency and reduce waste. Combining the concepts of PED and UIS can create self-sufficient, sustainable, and resilient districts. As the analysis and implementation of such systems are barely studied in North America, this research study was structured to fill the gap by evaluating the financial and environmental advantages of this combination. This study proposes a methodology to design a heat transmission system; then, it is applied to the case of a paper-making factory and a multifunctional heritage building in Montreal, Canada. The results show that the building's new heating system can generate sufficient heat while emitting near-zero direct emissions. Overall, this paper argues that combining the concepts of PED and UIS can lead to a more sustainable and resilient urban area, and provides a roadmap for achieving this goal.

Keywords: urban industrial symbiosis; positive energy district; energy system design; energy exchange; heat recovery

1. Introduction

Industrialization and urbanization have risen over recent years, increasing greenhouse gas (GHG) emissions [1]. Aligned with rising population and demand, the industrial sector has increased the extraction of raw materials and solid waste generation [2]. Industrial development is essential for provision of a high quality of life for society, and has a high economic significance [3]. As a result, finding solutions that decrease negative consequences while maintaining sustainability with economic growth is critical. In general, economic growth is unsustainable if it is assumed that the only way to increase GDP is by increasing production quantity. However, an alternative approach to GDP growth is to improve production efficiency, using less input material/energy to produce more product [4]. In the same way, the building sector plays a significant role in emitting GHG emissions. Buildings are responsible for almost 40% of global carbon dioxide emissions [5]. For many years, collections of methods have been introduced and tested to decrease the negative impact of this sector on climate change. Reducing fuel consumption and implementing electrification strategies are promising methods of reaching positive energy districts [6].

In 2015, the Paris agreement formalized action against climate change while highlighting the importance of finding sustainable solutions. However, a single solution is not possible for the myriad contextual differences that exist globally. Energy exchange between industries is one of many potential approaches to addressing the adverse impacts of human activity; in the literature, this is called industrial symbiosis (IS) [7]. For many years, IS has been a promising approach to responding to climate change [8]. The concept of IS is derived from industrial ecology, which proposes that an industrial system can be interconnected in ways mimicking natural systems, and realizing benefits from integration. From the systems view, it is more efficient for industrial facilities to work jointly in concerted actions. Industrial ecology focuses on the facility, inter-firm, and regional or global levels. Typically, IS studies integration potential at the inter-firm level [9].

IS contributes to the sustainable development of societies by addressing its three pillars (economic, environmental, and social), making IS a novel vehicle to achieve the United Nations Sustainable Development Goals (SDGs) [10]. Energy exchange in IS contributes to achieving SDG 7 by recovering excess heat of factories and supplying the energy demand of nearby facilities. Implementing IS decreases the waste and greenhouse gas emissions related to energy use. Generally, IS is limited to eco-industrial parks, because short distances are vital to IS; however, some studies have considered symbiosis between distant enterprises [11]. Although geographic proximity enables symbiosis between industries, if industrial facilities are close enough to the urban area, another term could emerge: urban–industrial symbiosis (UIS) [12,13].

Setting up an energy symbiosis network in a UIS provides the nearby building with ready-to-use excess heat. Such a network accelerates the process of reaching a PED in urban areas which are located near factories. However, this connection requires some prerequisite evaluation. First, technical feasibility examines the methods and technologies necessary to handle recovered energy. The second requirement explores economic, environmental, and social sustainability. Eventually, resiliency evaluates the long-term operation of the network under uncertainty [7]. This study focuses on the first and second steps.

Process heat in industries uses the largest share of the energy input, and almost half of this heat is above 500 °C [14]. High energy costs led to considerable improvements in energy efficiency in these industries over the past century. However, many industrial sites still release substantial amounts of lower-temperature energy into the environment. The excess heat released into the environment from the industrial sector should be further decreased, as far as possible, by energy efficiency measures in the individual industrial sites. It could also be utilized as a heat source in low-temperature district heating networks [15]. After analysis of relevant papers in UIS, below are two essential features of the recently published articles, methodology, and types of UIS:

Mathematical programming, pinch-based analysis, and evaluation are the most frequently used methods. Studies with mathematical programming mostly use mixed-integer linear programming (MILP) to find optimal solutions of decision variables regarding each problem's characteristics. Among these studies, Kantor et al. [16] modeled and optimized material and energy symbiosis between factories, where urban areas could be considered as a sink for low-temperature heat. They also utilized the pinch analysis methodology to formulate heat cascade constraints while preserving the second law of thermodynamics. In 2022, Cunha et al. [17] evaluated the energy exchange of a waste-to-heat plant and a nearby district heating network (5 km distance). Their study claimed the proposed system achieved a 30% reduction in fuel consumption compared to the current condition. Simeoni et al. [18] developed a multi-objective optimization model to optimize the heat recovery decisions in the presence of factories and district heating networks. Ciotti et al. [19] developed a decision support system to facilitate decision makers in choosing the best heat recovery option. Nakama et al. [20] developed a dynamic optimization model to study the impact of thermal energy storage on system cost. Among others, some studies developed nonlinear models because of the hydraulic equations in optimization of the networks [21,22]. Finally, evaluation methods are broadly employed in the literature to measure the effectiveness of an existing or potential IS or UIS. For example, Kim et al. [23] measured the potential of using excess heat in an industrial park in Korea. They calculated the peak heat load of a regional urban area and assumed it as a sink site of excess heat. Ates and Ozcan [24] examined the excess heat in Turkey's industries, and then studied the potential of power conversion technologies to decrease that heat loss.

Water, energy, and material are generally exchanged in IS and UIS. Energy can be decomposed into three forms: heating, cooling, and electricity. Material is usually defined as urban waste in a UIS network [17]; however, materials also include industrial co-products when considering IS networks. Yong et al. [25] integrated heating, cooling, and electricity exchange in UIS. The proposed method aimed to increase energy efficiency by 34% while utilizing the excess heat of the power plant in fulfilling residential demand. In 2021, Misrol et al. [26] designed a wastewater treatment network to reuse the recovered water from the urban area in the processes of factories.

Although many research projects have studied the role and impact of buildings in reaching a PED, this study aims to answer the question: how do industries take part in achieving a PED? Thus, this study is structured to study technical know-how and evaluate UIS' capacity to realize a PED. This work is structured as follows: Section 2 explains the methodology and introduces the principles of all calculations. Section 3 introduces the case study and represents multiple heat recovery methods and their associated considerations. In addition, the current condition is compared to the proposed structures. Section 4 discusses the impact and limitations of this work. Finally, Section 5 concludes the study and provides potential directions for future research.

2. Materials and Methods

The proposed methodology evaluates the impact of recovering excess heat from industries and exchanging it with nearby buildings, which requires a heat transmission network and the installation of new units. The result is compared to the current state throughout the calculation of network variables. In this regard, Figure 1 shows the steps of the proposed methodology. The overall procedure consists of four steps:

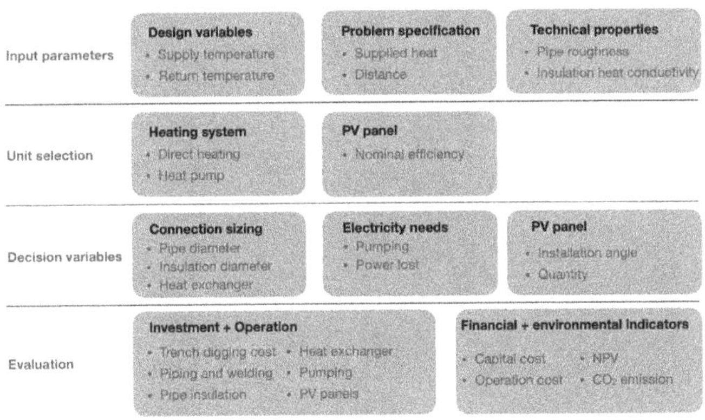

Figure 1. Overview of the methodology.

In the first step, the required input data are gathered. Then, new energy systems are designed to reuse heat from industries; the goal is to reach a positive energy building that does not emit onsite emissions (Scope 1) and has lower Scope 2 emissions compared to the current state. Scope 1 refers to direct onsite emissions. For example, emissions associated with fuel combustion in a boiler are Scope 1 emissions. Scope 2 accounts for emissions from upstream activities. For instance, emissions that originate from a power plant are considered as Scope 2 emissions [27]. Next, the system is sized, and decision variables are estimated or optimized. Finally, the financial and environmental aspects are evaluated in the last phase. Later, five sub-sections explain the calculation of five essential parts: heat exchanger, connection pipes, photovoltaic (PV) panels, financial analysis, and environmental analysis.

2.1. Heat Exchanger Variables

Designing a heat exchanger is accompanied by defining many variables, which are shown in Figure 2. In most cases, the objective is to transfer the maximum amount of heat while preserving the required temperature difference on both sides of the heat exchanger.

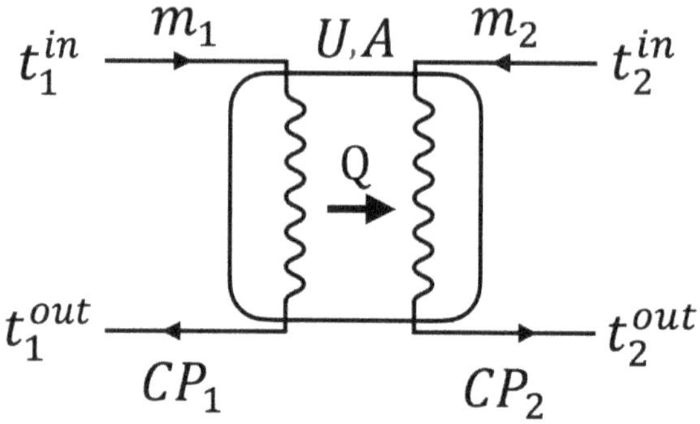

Figure 2. Schematic view of a heat exchanger.

As shown in Figure 2, a simple heat exchanger consists of three major sections. The hot stream (index 1 in Figure 2) is the first component, which must cool down from t_1^{in} to t_1^{out}. In addition, m_1 and CP_1 are the hot stream's mass flow rate and specific heat capacity, respectively. The second part is the cold stream (index 2 in Figure 2), which must be heated from t_2^{in} to t_2^{out}. Finally, the heat exchanger unit is the last component, which is accompanied by two critical parameters, heat transfer coefficient (U) and surface area (A).

As mentioned previously, a heat exchanger which transfers the maximum amount of heat (Q) is the main objective of the design process. As a result, the following mathematical model, nonlinear programming (NLP), is structured, which finds a feasible solution (t_1^{out}, t_2^{out}) while satisfying the linear and non-linear constraints.

$$Q = U \cdot A \cdot \frac{\left(t_1^{in} - t_2^{out}\right) - \left(t_1^{out} - t_2^{in}\right)}{\ln \frac{\left(t_1^{in} - t_2^{out}\right)}{\left(t_1^{out} - t_2^{in}\right)}} \tag{1}$$

$$Q = m_1 \cdot CP_1 \cdot \left(t_1^{in} - t_1^{out}\right) = m_2 \cdot CP_2 \cdot \left(t_2^{in} - t_2^{out}\right) \tag{2}$$

$$U \cdot A \cdot \frac{\left(t_1^{in} - t_2^{out}\right) - \left(t_1^{out} - t_2^{in}\right)}{\ln \frac{\left(t_1^{in} - t_2^{out}\right)}{\left(t_1^{out} - t_2^{in}\right)}} = m_2 \cdot CP_2 \cdot \left(t_2^{in} - t_2^{out}\right) \tag{3}$$

$$t_1^{in} > t_1^{out} > t_2^{in},\ t_1^{out} \geq t_2^{out} + \Delta \mathrm{t} \tag{4}$$

Equation (1) is the objective function and maximizes the amount of transferable heat in the heat exchanger. Equations (2) and (3) deal with the law of conservation of energy. Ultimately, Equation (4) limits the decision variables (t_1^{out}, t_2^{out}) into a reasonable interval, and Δt is the minimum approach temperature in the heat exchanger.

2.2. Connection Variables

Transferring the excess energy from one site (e.g., a factory) to another (e.g., a building) requires installing pipes between them. The calculation of pipe diameter, pressure drop, and insulation thickness are further explained in this section. Table 1 introduces the parameters and variables utilized in the rest of this paper.

Table 1. Input parameters and design variables of the model.

	Design Variable	Unit		Parameter	Unit
V_i	Insulation used	m^3	Q	Recovered heat	KW
f	Friction factor	-	L	Required pipe length	m
D_p	Pressure drop along pipes	Pa	T_o	Soil temperature	$^\circ C$
D_h	Pressure drop, heat exchanger	Pa	T_s	Supply temperature	$^\circ C$
Q_l	Heat losses	W_{th}	T_r	Return temperature	$^\circ C$
υ	Kinematic viscosity	$\frac{m^2}{s}$	s	Insulator thickness	m
A_p	Cross-sectional area of pipe	m^2	μ_t	Viscosity, function of Temperature	$Pa{\cdot}s$
V	Flow rate	$\frac{m^3}{s}$	ρ	Density	$\frac{kg}{m^3}$
Re	Reynolds number	-	C_p	Specific heat capacity of medium	$\frac{kJ}{kg{\cdot}K}$
n_p	Pumping efficiency	%	h_i	Insulator conductivity	$\frac{W}{m{\cdot}K}$
P_p	Pumping power	W_e	ε	Pipe roughness	mm
H	Pipe's hydraulic diameter	m			

The required flow rate is calculated in the following equation:

$$V = \frac{Q}{\rho{\cdot}C_P{\cdot}(T_s - T_r)} \tag{5}$$

Genic et al. [28] suggested the following equation for calculating the near-optimal pipe diameters (mm).

$$H = 0.34V^{0.45}{\cdot}\rho(T)^{0.133} \tag{6}$$

Equations (7)–(15) are used to calculate the pressure drop over the transmission line. The Kinematic viscosity and cross-sectional pipe area are calculated in Equations (3) and (4), which will be used later to calculate the Reynolds number.

$$\upsilon = \frac{\mu_t}{\rho} \tag{7}$$

$$A_P = \pi\left(\frac{H}{2}\right)^2 \tag{8}$$

Reynolds number is calculated in the following equation:

$$Re = V{\cdot}\frac{H}{\upsilon{\cdot}A_P} \tag{9}$$

A numerical approximation method called Serghides's solution [29] is used to estimate the friction factor. Equations (10)–(13) calculate this parameter.

$$X = -2log_{10}\left(\frac{\varepsilon}{3.7D} + \frac{12}{Re}\right) \tag{10}$$

$$Y = -2log_{10}\left(\frac{\varepsilon}{3.7D} + \frac{2.51X}{Re}\right) \tag{11}$$

$$Z = -2log_{10}\left(\frac{\varepsilon}{3.7D} + \frac{2.51Y}{Re}\right) \tag{12}$$

$$f = \left(X - \frac{(Y-X)^2}{Z - 2Y + X} \right)^{-2} \tag{13}$$

The pressure drop is determined using the Darcy-Weisbach equation, considering the friction in the pipe, velocity of the fluid, and the length of the pipe:

$$D_p = f\left(\frac{2 \cdot L}{H}\right) \cdot \left(\frac{\rho \cdot \left(\frac{V}{A}\right)^2}{2}\right) \tag{14}$$

Finally, the required pump power is calculated based on the pressure drop, flowrate, and pump efficiency:

$$Pp = (D_p + D_h) \cdot \frac{V}{np} \tag{15}$$

Equations (16) and (17) calculate the transmission line's heat loss and the insulation volume required to preserve the heat.

$$Q_l = 2\pi \cdot \frac{h_i}{\ln(1 + 2 \cdot \frac{s}{H})} \cdot (T_s - T_o) \cdot L \tag{16}$$

$$V_i = \frac{\pi}{4} \cdot \left((H+s)^2 - s^2 \right) \cdot L \tag{17}$$

Equation (16) calculates heat losses over distance based on the conductivity of the insulator and the temperature difference between the pipes and the soil. Finally, Equation (17) calculates the volume of insulation needed to be installed. Insulation thickness is selected based on the work of Bahadori and Vuthaluru [30], which proposes a simple estimation model.

2.3. PV Panel

As a source of electricity generation, PV panels are among the most reliable and feasible solutions [31]. However, the installation location and available space strongly affect the output energy of this option. In this study, INSEL [32], an open-access simulation software, is applied to calculate the electricity generation of each PV panel. Weather data are an important input, providing various weather information specifically for each location; solar radiation is used, among the provided data. In the real world, solar radiation and temperature change quickly impact the power of PV panels. In this case, the maximum power point of a photovoltaic generator, MPP block, determines the maximum product of the naturally unimodal function $P = V \cdot I$. Figure 3 demonstrates other impactful parameters in energy generation. Tilt angle (α) and length (X) are calculated based on the installation latitude. Equation (18) calculates the distance (X) while β is the solar radiation angle, or sun elevation, at noon of the shortest day of the year. (Figure 3).

$$X = L \cdot \cos\alpha + \frac{L \cdot sin\alpha}{tan\beta} \tag{18}$$

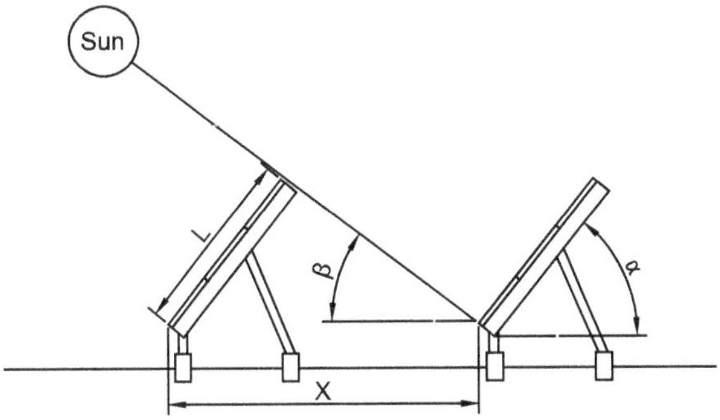

Figure 3. Schematic view of PV panel installation properties.

2.4. Financial Analysis

The financial analysis is based on cash flow analysis. Two main types of cost in cash flow analysis are capital expenditure (CapEx) and operating expenditure (OpEx). The principal difference between these two is the time of payment. Typically, the former is paid once at the beginning of the project; the latter is repeated for each operating period. For example, equipment purchase costs are classified as CapEx, while the electricity consumption of the equipment is classified as OpEx.

As described previously, PV panels are the only component in this problem that generates onsite electricity (Table 2). Matching the electricity generation with onsite electricity demand is out of the scope of this work; therefore, it is assumed that the annual generated electricity is firstly consumed in the building, and the rest is sold to the electricity grid as revenue. Surplus PV power is assumed to be dispatched at the same rate as the supply rate from Hydro Quebec. This study calculates operation costs based on 0.097 CAD/kWh for electricity [33] and 0.41 CAD/m^3 for natural gas consumption (including base, distribution, and service costs) [34].

Table 2. Components of cash flow analysis.

CapEx	OpEx	Revenue
Pump	Electricity (Pumping)	Electricity (PV panel)
PV panel	Electricity (Heat pump)	
Heat Pump	Electricity (Electric boiler)	
Electric Boiler	Natural gas (Steam boiler)	
Heat Exchanger	Annual service cost	
Heat transmission line		

Two metrics are evaluated through the financial analysis section. The first one is Net Present Value (NPV), which gives the current value of a future stream of payments. NPV is calculated by summing up the annual cash flow for the system's lifetime and discounting with the interest rate. The second metric is Discounted Payback Period (DPP), which gives the years to recover the upfront investment cost, accounting for decreasing currency value. Table 3 describes the nomenclature of financial analysis.

$$NPV_T = \sum_{u=1}^{U}\sum_{t=1}^{T}\left[E_{PV}^{t}\cdot C_{el}^{t}\cdot(1+i)^{-t}\right] - \sum_{u=1}^{U} CC_u - \sum_{u=1}^{U}\sum_{t=1}^{T}\left[\left(OC_u^t + SC_u^t\right)\cdot(1+i)^{-t}\right] \quad (19)$$

Table 3. Sets, parameters, and variables of financial analysis.

Description		Unit
	Sets and indices:	
$u : 1, \ldots, U$	Installed unit (e.g., PV panels)	-
$t : 1, \ldots, T$	Counter of Year	-
	Parameters:	
i	Interest rate	%
C_{el}^{t}	Electricity cost at period t	$\frac{CAD}{kWh}$
	Variables:	
CC_u	Capital cost of unit u	$CAD \times 10^3$
OC_u^t	Operation cost of unit u at period t	$CAD \times 10^3$
SC_u^t	Service cost of unit u at period t	$CAD \times 10^3$
E_{PV}^{t}	Excess electricity generaion of PV panels at period t	kWh

Equation (19) calculates NPV_T of positive and negative cash flows by discounting all inflows and outflows to the present (the base year).

DPP finds the minimum years required to gain a positive NPV. It shows how fast the investment will be recovered throughout the years. The following shows the steps to find the DPP:

1. Initial investment cost: this is the money required to invest.
2. Cash inflows: this is the amount of cash the investment is expected to generate over its useful life.
3. Discounted cash inflows: this involves discounting each cash inflow from the investment by the appropriate discount rate.
4. Calculate the cumulative discounted cash inflows: this involves adding the "number 3" until they equal the initial investment cost.
5. Determine the discounted payback period: this is the time it takes for the cumulative discounted cash inflows to equal the initial investment cost.

2.5. Environmental Analysis

This study evaluates Scope 1 and Scope 2 GHG emissions. The total GHG emissions are the sum of the two. This is because Scope 1 emissions are direct emissions from sources owned or controlled by the reporting organization. In contrast, Scope 2 emissions are indirect emissions from the generation of purchased electricity, heat, or steam consumed by the customer. This study converts all emissions to the same scale, CO_2 equivalents. The following equation calculates this parameter:

$$CO_2 = \sum_{k=1}^{K} \sum_{t=1}^{T} F_k^t \cdot GWP_k + \sum_{j=1}^{J} \sum_{t=1}^{T} E_j^t \cdot EER_j \tag{20}$$

In Equation (20), $k = 1, \ldots, K$ is the set of fuels (e.g., natural gas), and GWP_k is the global warming potential associated with fuel consumption k. In addition, $j = 1, \ldots, J$ is the set of electricity sources (e.g., hydro or solar), and EER_j is the electricity emission rate related to source j. Note that total emission is summed over T, which is determined based on the research goal.

3. Case Study

The proposed method is applied to a multifunctional building in Montreal. This building is pursuing zero emission by electrifying the heating system. Meanwhile, a paper-making factory (Kruger, Place Turcot Mill) is located near that building (800 m), which is an appropriate case for energy exchange. The building's heating load is provided by the owner based on a historical analysis, which had been derived from a simulation model. Figure 4 shows the heating load throughout the year. In this section, net present value is calculated for the period of 20 years, and discount rate is derived based on current and historical data in the Bank of Canada [35].

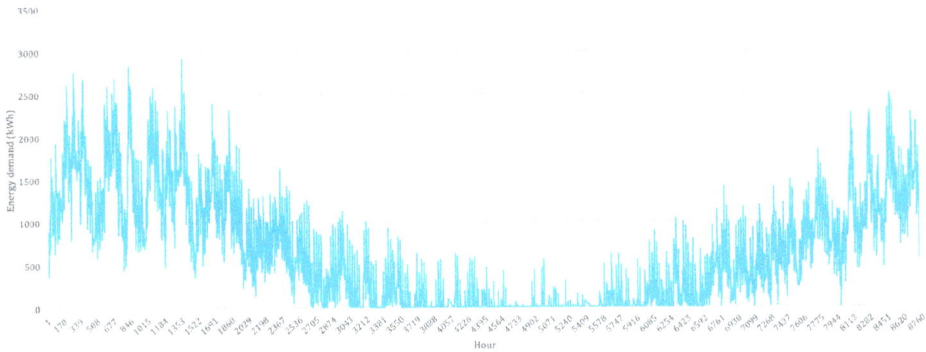

Figure 4. Hourly heating load.

3.1. *Unit Selection and Sizing*

Selecting the correct heating system for a building is crucial for achieving comfortable indoor temperatures while minimizing energy consumption and cost. While the local climate, unit availability, and serviceability are vital factors, the owner's strategy has more priority. Since the building aims to become carbon neutral, several units are considered appropriate candidates. Direct heating from the nearby factory, heat pump, and electric boiler are among the options which do not include Scope 1 emissions. In this study, different scenarios are generated based on the combination of units; then, the result is compared to assist key stakeholders in reaching a final decision. Note that the heat distribution network is excluded from the analysis. As such, the evaluation phase does not consider the modification of the piping network inside the building.

3.2. *Heating System Sizing*

3.2.1. Scenario 0, Present Condition

Two steam boilers and a backup boiler cover the building's heating demand. Natural gas is the fuel source; as a result, direct emissions are implied in this scenario. Based on the Canada energy conversion tables [36], and by assuming 88% efficiency of the steam boilers, 608, 000 m^3 natural gas is consumed in this scenario. Figure 5 represents the current condition, where the warm wastewater in the paper making factory has not been integrated.

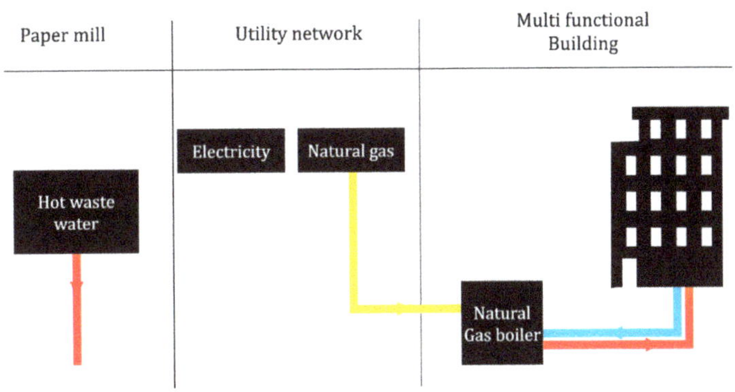

Figure 5. Configuration of the actual case.

3.2.2. Scenario 1, PV Installation

Figure 6 indicates the scope and detail of the network in the first scenario. It shows how PV panel installation will impact the energy consumption in the building. This scenario provides the principal calculation of PV panels and their capacity to generate electricity. Subsequently, the result phase shows the financial analysis of the PV panel installation.

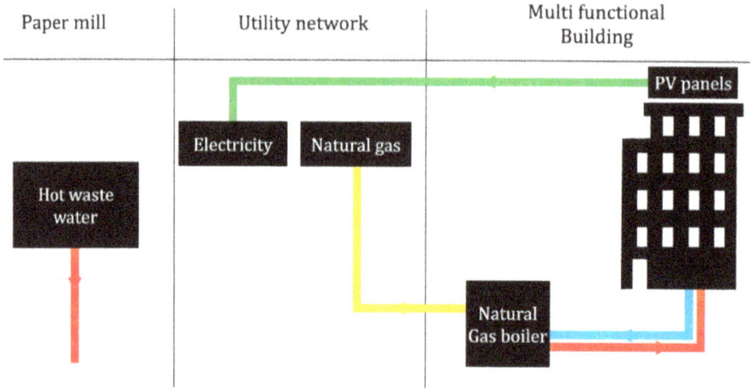

Figure 6. Configuration of Scenario 1.

Based on the Canadian PV panel market availability, model CS5P-240 manufactured by Canadian Solar Inc. is selected for installation [37]. The dimensions of the selected PV panel are 1.6 m × 1.0 m with nominal power of 240 W and nominal efficiency of 14.12%. Rowlands et al. [38] investigated the optimum PV tilt in Ottawa and Toronto. They found 38° as the optimal tilt in Ottawa. This study uses this number, as Montreal and Ottawa have almost the same latitude. The distance (X, in Figure 3) is calculated using Equation (18) to be 3.75 m; this prevents panels from shading each other in the winter.

The roof and installation layout, with 100% coverage, is shown in Figure 7. In this figure, each rectangle represents three bundled solar panels (width: 3 m, height: 1.6 m). In fact, the total roof area is not available for installation purposes. However, based on the site visit, almost 90% of the roof area is suitable. As a result, 2485 panels may be installed.

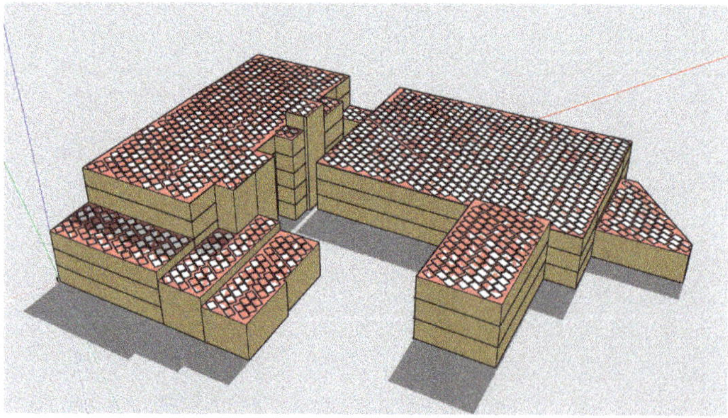

Figure 7. Panels installation layout based on 100% coverage.

After implementing the model in INSEL, the hourly energy of PV panels is calculated and presented in Figure 8. As an output, all panels generate 931 MWh annually.

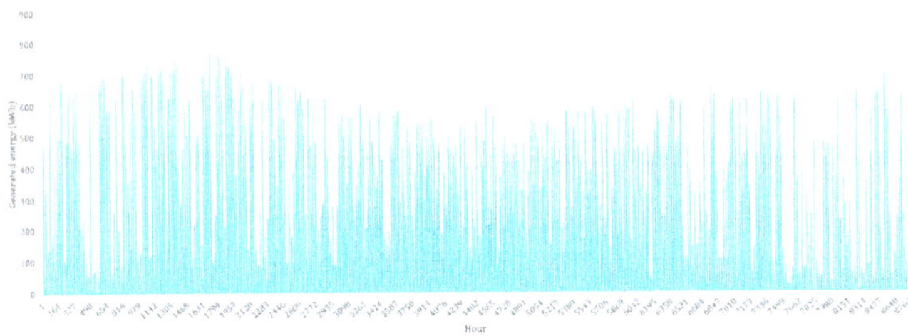

Figure 8. Electricity generation of PV panels over one year.

3.2.3. Scenario 2, Direct Heating

In this scenario, the heat exchanger in the paper-making factory transfers the excess heat to the medium in the pipeline, which will be directly supplied to the building. Figure 9 shows the new scope and configuration. IPOPT [39], an open-source solver, was selected to optimize the NLP model. Table 4 summarizes parameters and decision variables derived from the optimization model and evaluation. In this scenario, 700 kW of power is delivered to the building through four-inch pipes (supply and return).

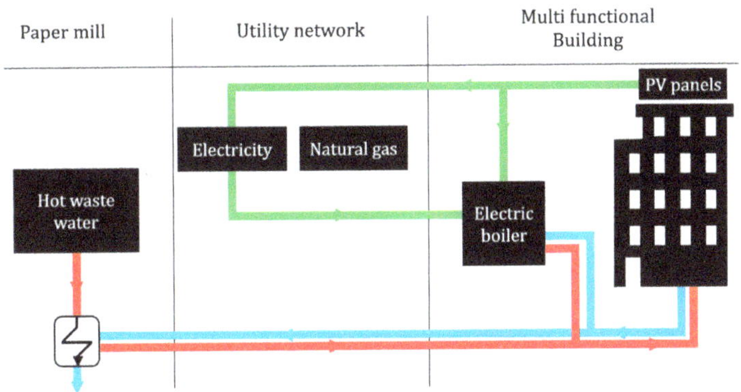

Figure 9. Configuration in Scenario 2.

Table 4. Design parameters of Scenario 2.

Parameter	Quantity	Unit	Reference	Parameter	Quantity	Unit	Reference
t_1^{in}	75	$°C$	[40,41]	t_2^{in}	40	$°C$	[42]
t_1^{out}	43	$°C$	-	t_2^{out}	58	$°C$	[42]
m_1	5.3	Kg/s	[40,41]	m_2	9	Kg/s	-
U	1700	$\frac{W}{m^2 \cdot K}$	[43]	A	50	m^2	-
H	101 (4)	$mm\ (in)$	-	Pp	2.5	kW	-
s	63	mm	[30]	V_i	10.7	m^3	-

In this scenario, direct heating serves as the primary source of the heating system. It is sufficient to provide enough heat for 60% of the heating demand. The remaining 40% is provided by integrating the electric boiler into the system. Figure 10 shows the status of two designed units. Direct heating requires energy for pumping the medium along the network, 15 MWh, and the electric boiler consumes 2116 MWh. After deduction, the electricity generated by PV panels, 1200 MWh, is imported from the utility grid.

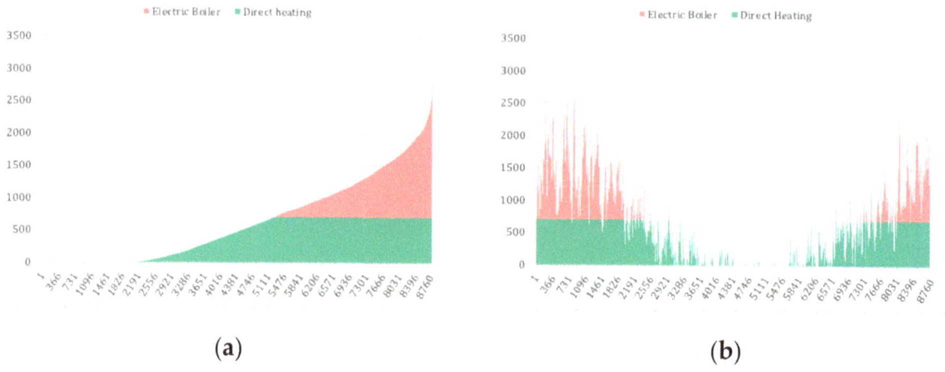

Figure 10. Heating demand supplied from industrial excess heat (60%) and electric boiler (40%): (**a**) sorted based on the energy; (**b**) sorted hourly (1 January–31 December).

3.2.4. Scenario 3, Heat Pump

Figure 11 shows the configuration for Scenario 3, which leverages the use of a heat pump. The heat exchanger in the paper-making factory transfers the excess heat to the water in the pipeline, which will be utilized in the heat pump. Table 5 summarizes parameters and decision variables derived from the optimization model and evaluation. In this scenario, the input temperature to the heat pump is far lower than in the previous scenario; as a result, more power is recovered. The connection network delivers 1506 kW to the heat pump evaporator.

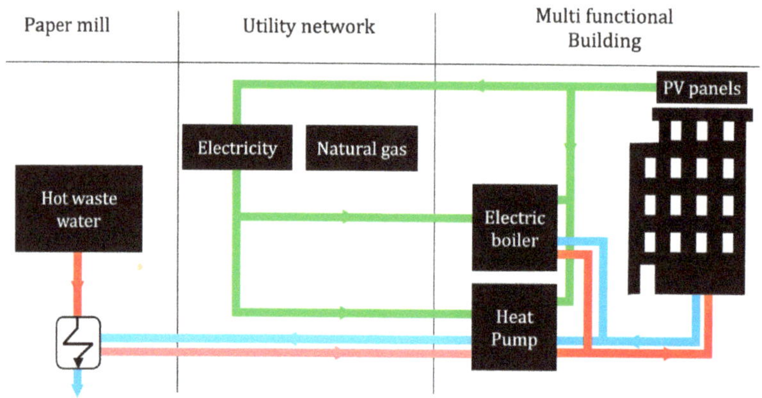

Figure 11. Configuration in Scenario 3.

Table 5. Design parameters of Scenario 3.

Parameter	Quantity	Unit	Reference	Parameter	Quantity	Unit	Reference
t_1^{in}	75	°C	[40,41]	t_2^{in}	4	°C	[44]
t_1^{out}	7	°C	-	t_2^{out}	22	°C	[44]
m_1	5.3	Kg/s	[40,41]	m_2	19	Kg/s	-
U	1700	$\frac{W}{m^2 \cdot K}$	[43]	A	50	m^2	-
H	148 (6)	$mm\ (in)$	-	Pp	4.8	kW	-
s	56	mm	[30]	V_i	12.4	m^3	-

In this scenario, the heat pump plays a central role in heating the building. A heat pump with 1800 kW capacity provides 94% of the energy demand in this case. In the same way as the first scenario, the rest, 6%, is provided by integrating the electric boiler into the system. Figure 12 shows the coverage of each proposed unit in a year. The advantage of this scenario is the presence of a heat pump with a high coefficient of performance (COP). Based on this observation, Johnson Controls company [44] manufactures industrial heat pumps that match the excess heat recovery network. Among various models, "NS Heat Pump 193 HP" is selected, which offers COP 3.9 in the designed condition. Scenario 3 includes more components than Scenario 2, but has lower electricity consumption. Here, the heat pump consumes 1383 *MWh*, the pumps consume 29 *MWh*, and the electric boiler consumes 150 *MWh*. After deducting the generated electricity, 631 *MWh* of energy is required from the utility grid.

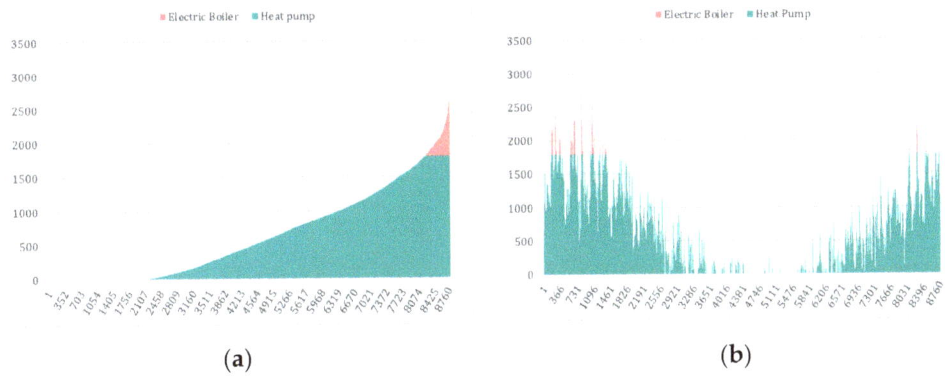

Figure 12. Heating demand supplied from heat pump (94%) and electric boiler (6%): (**a**) sorted based on the energy; (**b**) sorted based on date (1 January–31 December).

3.2.5. Scenario 4, Heat Pump without PV Panels

Scenario 4 resembles Scenario 3; however, the PV panels are excluded from the retrofit plan. This scenario examines the advantage of utilizing excess heat combined with the heat pump unit. In this system, Scope 1 emissions are zero regarding the electrification of the heating system. Moreover, due to the presence of Hydro Quebec electricity, Scope 2 emissions are far lower than the electricity generated by power plants. In contrast, the building loses the ability to generate electricity and does not dispatch the surplus electricity to the grid. In other words, the building cannot send energy into the district.

3.3. Evaluation

Each scenario was evaluated for financial performance based on the defined economic metrics. The first scenario does not require a circulation pump, and therefore does not consume electricity as part of the heating system. Based on this fact, Table 5 summarizes all scenarios' CapEx, OpEx, and Revenue. The most critical parameter of the operation phase is the utility rate, introduced in the methodology section, which is then cross-checked with the actual bills. In addition, the "Heat transmission line" is made of these items: pipe cost, insulation cost, and installation cost. Table 6 demonstrates the PV panel's considerable capital cost, which is necessary to achieve a zero emission building.

Table 6. Financial analysis of all scenarios.

Item	Reference	Scenario 0	Scenario 1	Scenario 2	Scenario 3	Scenario 4
Investment (CapEx)						
Pump	[45]	- *	-	11,680	17,520	17,520
PV panel	[46]	-	1,789,200	1,789,200	1,789,200	-
Heat Pump	[44]	-	-	-	243,000	243,000
Electric Boiler	[47]	-	-	100,000	50,000	50,000
Heat Exchanger	[48]	-	-	15,000	15,000	15,000
Heat transmission line		-	-	359,418	487,579	487,579
Operation (OpEx)						
Electricity (Pumping)	[33]	-	-	1453	2778	2778
Electricity (Heat pump)	[33]	-	-	-	134,142	134,142
Electricity (Electric boiler)	[33]	-	-	205,227	14,544	14,544
Natural gas (Steam boiler)	[34]	249,266	249,266	-	-	-
Annual service cost	[49]	9970	39,922	32,382	34,017	4065
Revenue						
Electricity (PV panel)	[33]	-	90,320	90,320	90,320	-

* All prices are based on Canadian dollar (CAD).

In order to calculate the financial metrics, NPV and DPP, the definition of the revenue is important. In this case, operation costs in Scenario 0 are considered the base of calculation, and economization by implementing other scenarios is considered the revenue, aside from the revenue of selling generated electricity. Table 7 concludes the result of the evaluation phase and reports all metrics introduced in the methodology section.

Table 7. Financial and environmental metrics.

Item	Unit	Scenario 0	Scenario 1	Scenario 2	Scenario 3	Scenario 4
NPV *	*CAD*	-	−431,708	−7234	631,676	1,063,384
DPP	*Year*	-	>25	20	16	8
CO_2 **	*Tonnes/Year*	1168	1136	41	21.7	54

* Net present value is calculated over a period of 20 years. ** GWP and EER parameters for the emission calculation are from [50,51].

4. Discussion

The final evaluation showed that reaching nearly zero emission is possible, not only through the electrification of the heating system, but also by recovering excess heat from factories and transferring it to nearby buildings or districts. Scenario 1 represents that PV panel installation alone does not seem appealing for owners, with a payback period of more than 25 years. However, when the self-generation strategy is aligned with the zero-emission building strategy, Scenario 3 becomes more interesting. In this case, the building's CO_2 footprint for heating approaches zero. Moreover, surplus electricity from PV panels is injected into the grid, providing electricity to the district. Scenario 2 may be selected with a short-term planning horizon, as the heating system has a lower capital cost (32%) compared to Scenario 3. However, for long-term planning, it performs worse than Scenario 3, based on the high OpEx. Finally, Scenario 4 demonstrates better financial metrics compared to the others. Electrification of the heating system emits less than 5% of the current emissions because of the low-emission electricity source (hydro). The capital and operation costs are recovered in only 8 years, which motivates the building's owner to invest in this scenario.

There are two limitations associated with this study. First, the wastewater temperature at the factory is uncertain. There is a possibility that Scenario 2 cannot be implemented because of low-temperature wastewater. For example, when the temperature drops to 40 °C, the potential for direct heating declines. In that case, Scenario 3 still works with a lower heat capacity, and electricity consumption decreases because of the lower COP. However, the heat pump specification and its capacity to utilize a low temperature medium impact decision-making. Physical realization of the systems, such as piping placement, is another general limitation. The shortest distance between supplier and demand should be used, though site-specific restrictions (public or private land) may cause the piping length to increase. The distance of 800 m considered in the case study includes foreseeable conditions, but further pursuit of the project may yield additional complications. Finally, examination of the heat distribution system in the building was excluded from this study, which needs further investigation on the supply temperature inside the building.

5. Conclusions

This study aimed to investigate the feasibility of exchanging industrial excess heat with nearby buildings for space heating. The study proposed a technical evaluation method to analyze the feasibility of such a network. The method was applied to a case study of a paper-making factory near a large building with a high heating demand in a cold climate. One motivation to examine this case study was the zero-emission strategy of the building.

Through a series of optimization problems, along with technical and economic analyses, this study demonstrated that exchanging industrial waste heat with a nearby building could result in significant energy savings for the building. The calculation showed that the excess heat could be used to provide up to 60% of the heating demand in the direct heating scenario, and up to 94% with the installation of a heat pump.

Economic analyses also indicated that implementing such a system (a heat pump without PV panels) would be financially feasible for the building, with payback periods of 8 years. Moreover, the possibility of a nearly zero-emission building was shown, which increased the investment payback period to 16 years. However, the study also highlighted some technical and operational challenges that must be addressed to implement such a system successfully. These include ensuring a match between the temperature and quality of the excess heat and the building's heating demand, as well as establishing a reliable and efficient heat transfer system.

In conclusion, the study demonstrated that exchanging industrial waste heat with a nearby building can be a viable and cost-effective way to meet energy demands while reducing greenhouse gas emissions. Future research should address the technical and operational challenges associated with the energy storage concept for both electricity and heat, which ultimately impacts the system's operation. In this study, different scenarios were defined and evaluated; however, in a complex network with multiple factories and buildings, the number of scenarios increases drastically, so formulating an optimization model to find a set of optimal solutions under different contextual settings would provide a more generalizable approach. The optimization model can include one or multiple objectives: economic, environmental, and social. The solutions may be analyzed with different methods, for example, the Pareto frontier, which provides decision-makers an overview of the optimum solutions.

Author Contributions: Conceptualization, E.S.R.; methodology, E.S.R.; validation, E.S.R.; writing—original draft preparation, E.S.R.; writing—review and editing, E.S.R., U.E., I.K. and R.P.M.; supervision, U.E. and I.K.; project administration, U.E.; funding acquisition, U.E. All authors have read and agreed to the published version of the manuscript.

Funding: This research was funded by Canada Excellence Research Chair in Smart, Sustainable and Resilient Communities and Cities and funded by Tri-Agency Institutional Program Secretariat.

Data Availability Statement: Data will be available upon request.

Acknowledgments: The authors would like to express their appreciation to the Canada Excellence Research Chairs Program. In addition, the authors would like to extend their sincere gratitude to Ahmad Tavajohi Birjandi, whose feedback and suggestions were immensely helpful in refining the content.

Conflicts of Interest: The authors declare no conflict of interest. The funders had no role in the design of the study; in the collection, analyses, or interpretation of data; in the writing of the manuscript; or in the decision to publish the results.

References

1. Dong, F.; Wang, Y.; Su, B.; Hua, Y.; Zhang, Y. The Process of Peak CO2 Emissions in Developed Economies: A Perspective of Industrialization and Urbanization. *Resour. Conserv. Recycl.* **2019**, *141*, 61–75. [CrossRef]
2. Guan, Y.; Huang, G.; Liu, L.; Zhai, M.; Zheng, B. Dynamic Analysis of Industrial Solid Waste Metabolism at Aggregated and Disaggregated Levels. *J. Clean. Prod.* **2019**, *221*, 817–827. [CrossRef]
3. Haraguchi, N.; Martorano, B.; Sanfilippo, M. What Factors Drive Successful Industrialization? Evidence and Implications for Developing Countries. *Struct. Chang. Econ. Dyn.* **2019**, *49*, 266–276. [CrossRef]
4. Feldstein, M. Underestimating the Real Growth of GDP, Personal Income, and Productivity. *J. Econ. Perspect.* **2017**, *31*, 145–164. [CrossRef]
5. Directorate-General for Energy (European Commission). *Clean Energy for All Europeans*; Publications Office of the European Union: Luxembourg, 2019; ISBN 978-92-79-99835-5.
6. Hedman, Å.; Rehman, H.U.; Gabaldón, A.; Bisello, A.; Albert-Seifried, V.; Zhang, X.; Guarino, F.; Grynning, S.; Eicker, U.; Neumann, H.-M.; et al. IEA EBC Annex83 Positive Energy Districts. *Buildings* **2021**, *11*, 130. [CrossRef]
7. Afshari, H.; Tosarkani, B.M.; Jaber, M.Y.; Searcy, C. The Effect of Environmental and Social Value Objectives on Optimal Design in Industrial Energy Symbiosis: A Multi-Objective Approach. *Resour. Conserv. Recycl.* **2020**, *158*, 104825. [CrossRef]
8. Fan, Y.; Qiao, Q.; Fang, L.; Yao, Y. Emergy Analysis on Industrial Symbiosis of an Industrial Park—A Case Study of Hefei Economic and Technological Development Area. *J. Clean. Prod.* **2017**, *141*, 791–798. [CrossRef]
9. Chertow, M.R. Industrial Symbiosis: Literature and Taxonomy. *Annu. Rev. Energy. Environ.* **2000**, *25*, 313–337. [CrossRef]
10. Envision2030: 17 Goals to Transform the World for Persons with Disabilities | United Nations Enable. Available online: https://www.un.org/development/desa/disabilities/%20envision2030.html (accessed on 5 March 2023).
11. Yu, C.; de Jong, M.; Dijkema, G.P.J. Process Analysis of Eco-Industrial Park Development—The Case of Tianjin, China. *J. Clean. Prod.* **2014**, *64*, 464–477. [CrossRef]
12. Chen, X.; Fujita, T.; Ohnishi, S.; Fujii, M.; Geng, Y. The Impact of Scale, Recycling Boundary, and Type of Waste on Symbiosis and Recycling. *J. Ind. Ecol.* **2012**, *16*, 129–141. [CrossRef]
13. Berkel, R.V.; Fujita, T.; Hashimoto, S.; Fujii, M. Quantitative Assessment of Urban and Industrial Symbiosis in Kawasaki, Japan. *Environ. Sci. Technol.* **2009**, *43*, 1271–1281. [CrossRef]
14. Rehfeldt, M.; Fleiter, T.; Toro, F. A Bottom-up Estimation of the Heating and Cooling Demand in European Industry. *Energy Effic.* **2018**, *11*, 1057–1082. [CrossRef]
15. Xu, Z.Y.; Wang, R.Z.; Yang, C. Perspectives for Low-Temperature Waste Heat Recovery. *Energy* **2019**, *176*, 1037–1043. [CrossRef]
16. Kantor, I.; Robineau, J.-L.; Bütün, H.; Maréchal, F. A Mixed-Integer Linear Programming Formulation for Optimizing Multi-Scale Material and Energy Integration. *Front. Energy Res.* **2020**, *8*, 49. [CrossRef]
17. Cunha, J.M.; Faria, A.S.; Soares, T.; Mourão, Z.; Nereu, J. Decarbonization Potential of Integrating Industrial Excess Heat in a District Heating Network: The Portuguese Case. *Clean. Energy Syst.* **2022**, *1*, 100005. [CrossRef]
18. Simeoni, P.; Ciotti, G.; Cottes, M.; Meneghetti, A. Integrating Industrial Waste Heat Recovery into Sustainable Smart Energy Systems. *Energy* **2019**, *175*, 941–951. [CrossRef]
19. Ciotti, G.; Cottes, M.; Mazzolini, M.; Sappa, A.; Simeoni, P. A decision Support System for Industrial Waste Heat Recovery: The CE-HEAT Project. In Proceedings of the Summer School F. Turco—Industrial Systems Engineering, Brescia, Italy, 11–13 September 2019.
20. Nakama, C.S.M.; Knudsen, B.R.; Tysland, A.C.; Jäschke, J. A Simple Dynamic Optimization-Based Approach for Sizing Thermal Energy Storage Using Process Data. *Energy* **2023**, *268*, 126671. [CrossRef]
21. Pan, M.; Sikorski, J.; Akroyd, J.; Mosbach, S.; Lau, R.; Kraft, M. Design Technologies for Eco-Industrial Parks: From Unit Operations to Processes, Plants and Industrial Networks. *Appl. Energy* **2016**, *175*, 305–323. [CrossRef]
22. Pang, K.Y.; Liew, P.Y.; Woon, K.S.; Ho, W.S.; Wan Alwi, S.R.; Klemeš, J.J. Multi-Period Multi-Objective Optimisation Model for Multi-Energy Urban-Industrial Symbiosis with Heat, Cooling, Power and Hydrogen Demands. *Energy* **2023**, *262*, 125201. [CrossRef]
23. Kim, H.-W.; Dong, L.; Choi, A.E.S.; Fujii, M.; Fujita, T.; Park, H.-S. Co-Benefit Potential of Industrial and Urban Symbiosis Using Waste Heat from Industrial Park in Ulsan, Korea. *Resour. Conserv. Recycl.* **2018**, *135*, 225–234. [CrossRef]
24. Ates, F.; Ozcan, H. Turkey's Industrial Waste Heat Recovery Potential with Power and Hydrogen Conversion Technologies: A Techno-Economic Analysis. *Int. J. Hydrogen Energy* **2022**, *47*, 3224–3236. [CrossRef]
25. Yong, W.N.; Liew, P.Y.; Woon, K.S.; Wan Alwi, S.R.; Klemeš, J.J. A Pinch-Based Multi-Energy Targeting Framework for Combined Chilling Heating Power Microgrid of Urban-Industrial Symbiosis. *Renew. Sustain. Energy Rev.* **2021**, *150*, 111482. [CrossRef]
26. Misrol, M.A.; Wan Alwi, S.R.; Lim, J.S.; Manan, Z.A. An Optimal Resource Recovery of Biogas, Water Regeneration, and Reuse Network Integrating Domestic and Industrial Sources. *J. Clean. Prod.* **2021**, *286*, 125372. [CrossRef]
27. US EPA. Scope 1 and Scope 2 Inventory Guidance. Available online: https://www.epa.gov/climateleadership/scope-1-and-scope-2-inventory-guidance (accessed on 8 March 2023).

28. Genić, S.B.; Jaćimović, B.M.; Genić, V.B. Economic Optimization of Pipe Diameter for Complete Turbulence. *Energy Build.* **2012**, *45*, 335–338. [CrossRef]
29. Serghides, T.K. Estimate Friction Factor Accurately. *Chem. Eng.* **1984**, *91*, 63–64.
30. Bahadori, A.; Vuthaluru, H.B. A Simple Correlation for Estimation of Economic Thickness of Thermal Insulation for Process Piping and Equipment. *Appl. Therm. Eng.* **2010**, *30*, 254–259. [CrossRef]
31. Panwar, N.L.; Kaushik, S.C.; Kothari, S. Role of Renewable Energy Sources in Environmental Protection: A Review. *Renew. Sustain. Energy Rev.* **2011**, *15*, 1513–1524. [CrossRef]
32. INSEL—Homepage—INSEL En. Available online: https://www.insel.eu/en/home_en.html (accessed on 5 March 2023).
33. Rate D. Available online: https://www.hydroquebec.com/residential/customer-space/rates/rate-d.html (accessed on 9 March 2023).
34. Price and Evolution | Énergir. Available online: https://www.energir.com/en/major-industries/natural-gas-price/price-and-evolution/ (accessed on 3 April 2023).
35. Policy Interest Rate. Available online: https://www.bankofcanada.ca/core-functions/monetary-policy/key-interest-rate/ (accessed on 3 April 2023).
36. Energy Conversion Tables—Canada.Ca. Available online: https://apps.cer-rec.gc.ca/Conversion/conversion-tables.aspx#s1ss2 (accessed on 8 March 2023).
37. CSI Solar–Global. Available online: https://www.csisolar.com/ (accessed on 19 April 2023).
38. Rowlands, I.H.; Kemery, B.P.; Beausoleil-Morrison, I. Optimal Solar-PV Tilt Angle and Azimuth: An Ontario (Canada) Case-Study. *Energy Policy* **2011**, *39*, 1397–1409. [CrossRef]
39. Ipopt: Documentation. Available online: https://coin-or.github.io/Ipopt/ (accessed on 8 March 2023).
40. Bengtsson, C.; Nordman, R.; Berntsson, T. Utilization of Excess Heat in the Pulp and Paper Industry—A Case Study of Technical and Economic Opportunities. *Appl. Therm. Eng.* **2002**, *22*, 1069–1081. [CrossRef]
41. Manz, P.; Kermeli, K.; Persson, U.; Neuwirth, M.; Fleiter, T.; Crijns-Graus, W. Decarbonizing District Heating in EU-27 + UK: How Much Excess Heat Is Available from Industrial Sites? *Sustainability* **2021**, *13*, 1439. [CrossRef]
42. Ovchinnikov, P.; Borodiņecs, A.; Strelets, K. Utilization Potential of Low Temperature Hydronic Space Heating Systems: A Comparative Review. *Build. Environ.* **2017**, *112*, 88–98. [CrossRef]
43. Bergman, T.L. (Ed.) *Introduction to Heat Transfer*, 6th ed.; Wiley: Hoboken, NJ, USA, 2011; ISBN 978-0-470-50196-2.
44. Johnson Controls. Available online: https://www.johnsoncontrols.com:443/ (accessed on 9 March 2023).
45. Horizontal Close Coupled Centrifugal Water Pump and Standardized F40/125A 2.2 kW. Available online: https://www.tomeiwatersolutions.com/en/home/water-pumps/pedrollo/centrifugal/standard-centrifugal-according-to-en733/horizontal-close-coupled-centrifugal-water-pump-and-standardized-f40-125a-2-2kw.2.5.231.gp.38427.uw (accessed on 3 April 2023).
46. Cost of Solar Power In Canada. 2021. Available online: https://www.energyhub.org/cost-solar-power-canada/ (accessed on 3 April 2023).
47. Complete Boiler Room Solutions | Cleaver-Brooks. Available online: https://cleaverbrooks.com/ (accessed on 3 April 2023).
48. More Positions in Canada. Available online: https://www.alfalaval.ca/ (accessed on 3 April 2023).
49. Jossen, A.; Garche, J.; Sauer, D.U. Operation Conditions of Batteries in PV Applications. *Solar Energy* **2004**, *76*, 759–769. [CrossRef]
50. Environment and Climate Change Canada (ECCC). Emission Factors and Reference Values. Available online: https://www.canada.ca/en/environment-climate-change/services/climate-change/pricing-pollution-how-it-will-work/output-based-pricing-system/federal-greenhouse-gas-offset-system/emission-factors-reference-values.html (accessed on 9 March 2023).
51. GHG Emissions and Electricity | Hydro-Québec. Available online: https://www.hydroquebec.com/sustainable-development/specialized-documentation/ghg-emissions.html (accessed on 9 March 2023).

designs

MDPI

Article

The Modeling and Simulation of Non-Isolated DC–DC Converters for Optimizing Photovoltaic Systems Applied in Positive Energy Districts

Tohid Hashemi and Hamed Jafari Kaleybar *

Department of Energy, Politecnico di Milano, 20156 Milan, Italy; tohid.hashemi@mail.polimi.it
* Correspondence: hamed.jafari@polimi.it

Abstract: DC–DC converters are critical for energy management in positive energy districts (PEDs) because they allow for efficient conversion between different voltage levels, enabling the integration of various renewable energy sources, energy storage systems, and loads. The demand for high-voltage gain DC–DC converters in photovoltaic power systems has surged in recent times. Despite the numerous converter topologies reported, there is a focused effort to streamline components, particularly switching devices, passive elements, and overall converter losses. This paper introduces the single switching impedance network (SSIN)-based converter as a unique DC–DC converter topology, designed in both one-stage and double-stage configurations for photovoltaic applications. One of the main characteristics of the SSIN converter is that it needs just one switch and three capacitors for the n-stage. A comparative analysis with conventional boost converter topology demonstrates the SSIN-based converter's capability to achieve a desirable output voltage that closely approximates an ideal sine waveform. Furthermore, the application of advanced control strategies to the proposed converter highlights its superior performance and robustness in maintaining output voltage stability under varying conditions. These characteristics make the SSIN-based converter particularly well-suited for PED applications, where efficiency, reliability, and the seamless integration of renewable energy sources are crucial.

Keywords: non-isolated DC–DC converter; high DC voltage gain; photovoltaic (PV); maximum power point tracking (MPPT); positive energy districts (PEDs)

1. Introduction

Due to population development and industrialization, there is a huge increase in the need for electrical energy, and fossil fuels continue to be the most popular energy source despite their role in global warming [1]. To address the challenges posed by the increasing energy demand and the environmental impact of fossil fuels, the concept of positive energy districts (PEDs) has emerged as a transformative solution [2]. PEDs are urban areas designed to produce more energy than they consume by leveraging local renewable energy sources, smart grids, and energy storage systems [3]. By prioritizing sustainability, PEDs aim to contribute to the global energy transition by reducing carbon emissions, enhancing energy efficiency, and promoting decentralized energy management [4]. One of the critical technologies enabling energy optimization in PEDs is the DC–DC converter. DC–DC converters play a fundamental role in managing the energy generated from various renewable sources, such as solar photovoltaic (PV) panels, and coordinating energy storage systems, like batteries. By efficiently converting DC voltages to match the needs of different energy components within the district, these converters help minimize energy losses and enhance overall system efficiency. Nevertheless, the utilization of fossil fuels poses environmental concerns, making alternative sources like solar energy increasingly attractive due to its inherent noise-free and environmentally friendly characteristics [5]; this is while the efficiency of photovoltaic (PV) panels is significantly influenced by factors such as sun exposure, temperature variations, and shadow fluctuations [6].

This study's objective is fully consistent with the positive energy district (PED) concept, which seeks to achieve a positive energy balance. A conventional solar energy harvesting system comprises key components, such as a photovoltaic (PV) array, a maximum power point tracking (MPPT) controller, as well as a DC–DC converter, a battery, a load (AC/DC), and an inverter, as illustrated in Figure 1 [7]. To cater to the voltage requirements of the load, intermediate circuits, such as DC–DC converters, are crucial for transforming the low direct current (DC) output voltage from a solar photovoltaic (PV) energy-producing system into a voltage that is appropriate for use [8]. Some of the rectifiers' problems, such as harmonics, are passed on to the DC–DC converters. Implementing active power filters solves this issue [9]. The power converters also include a controller that is set up to manage the switching of the many switching units [10]. Peripheral pulse-width modulated (PWM) boost converters have facilitated the development of switched-mode DC–DC converters aimed at producing output DC voltages greater than the input DC voltage. These setups use magnetic field storage components, like inductors, or electric field storage components, like capacitors, with active or passive switching elements to temporarily store and distribute energy from the input [11,12].

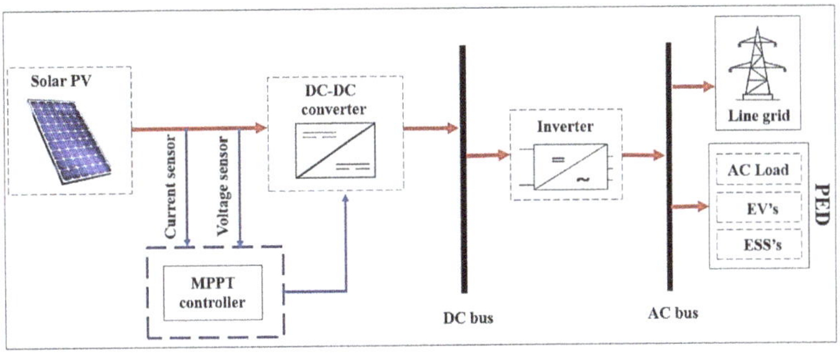

Figure 1. Conventional solar energy harvesting system.

In response to the drawbacks of conventional PWM boost converters, considerable efforts have been devoted to alleviating these limitations through non-isolated high step-up DC–DC converters (NHSDCs) [13]. These converters can be categorized into isolated and non-isolated types. Isolated converters, featuring electrical separation between input and output using transformers, offer protection against high-power flows [14]. However, they come with drawbacks, such as transformer core saturation, increased component count, and lower efficiency, particularly within applications requiring low and medium power [15,16]. Conversely, non-isolated DC–DC converters, lacking a transformer, boast fewer components, reduced energy loss, and a smaller overall size [17]. Utilizing strategies like interleave boost (IB), coupled inductor (CI), switched capacitor (SC), cascaded boost (CB), switched inductor (SI), and voltage multiplier (VM), a number of non-isolated topologies have surfaced in recent years to achieve high step-up voltage gains [18]. These topologies change the duty ratio in an effort to get high-voltage gains [19].

Earlier, the switch voltage stress in DC–DC boost converters was equivalent to the output voltage, necessitating the use of high-rating switches. This requirement significantly increased conduction losses and restricted the achievable voltage gain. To overcome these limitations, researchers introduced advanced non-isolated DC–DC converters with higher voltage gains, expressed by the relation $(1 + \delta)/(1 - \delta)$, where δ represents the duty ratio of the switch [20]. These converters allow for improved performance by reducing switch stress and enhancing energy conversion efficiency. However, despite these advancements, there remains a substantial scope for further improvements, particularly in developing high step-up DC–DC converters with better efficiency, higher voltage gain, and reduced component count.

In response to these challenges, Kumaravel Sundaramoorthy et al. [21] proposed a novel single switching impedance network (SSIN)-based converter. This topology offers the advantage of structural scalability, allowing for an extension with multiple stages (n-stage), which increases the number of diodes and inductors by 3(1 + n) and 1 + n, respectively. The key innovative aspects of this paper include:

1. New Topology: The SSIN converter topology is unique, offering a design that achieves higher or more efficient voltage conversion with a single switch, potentially simplifying circuitry and reducing cost.
2. Dual-Mode Operation Analysis: The study covers both continuous and discontinuous conduction modes, providing a comprehensive understanding of the converter's operation across different load conditions.
3. Impact of Parasitic Elements: By analyzing how parasitic elements affect the converter's DC voltage gain and efficiency, the paper addresses real-world performance considerations, enhancing the converter's practical applicability.

However, while this work introduces a valuable contribution to converter design, it does not explicitly address key challenges, such as stability in grid-connected systems or the integration of MPPT controllers under fault conditions. Our study builds upon this foundation by simulating a grid-connected PV system that incorporates MPPT and a three-level inverter for enhanced energy management. Furthermore, we introduce and analyze the system's response to short-circuit faults to assess the converter's stability and suitability for real-world renewable energy systems. The ability to integrate MPPT and maintain system stability during fault conditions is critical for optimizing the performance of grid-connected PV systems. This becomes especially important in the context of PEDs, where energy efficiency, reliability, and seamless integration with renewable energy sources are paramount.

By simulating a grid-connected PV system that incorporates MPPT and a 3-level inverter for improved energy management, and introducing and analyzing the system's response to two short-circuit faults, this study aims to evaluate the converter's stability and its suitability for real-world renewable energy systems. The ability to integrate MPPT and maintain system stability in fault conditions is crucial for optimizing the performance of grid-connected PV systems, particularly in the context of PEDs, where energy efficiency and reliability are paramount.

The application of advanced control strategies to the proposed converter highlights its superior performance and robustness in maintaining output voltage stability under varying conditions. In the context of the converter proposed for optimizing photovoltaic systems, control strategies refer to the methodologies and algorithms deployed to regulate and optimize the converter's operation. These strategies are designed to achieve critical objectives, such as maximizing energy conversion efficiency, maintaining stability in output voltage or current despite fluctuations in photovoltaic input (e.g., shading or varying solar irradiance), and dynamically tracking the MPPT to ensure optimal energy extraction under all conditions. Additionally, advanced control approaches ensure robust performance by adapting to environmental variations like temperature changes, minimizing output ripple to enhance reliability, and safeguarding connected loads. Techniques such as feedback loops, PID controllers, model predictive control (MPC), and artificial intelligence (e.g., neural networks for MPPT optimization) are employed to implement these strategies effectively. These advanced control methods enable the converter to efficiently manage energy flow in photovoltaic systems, making them particularly suitable for integration into PEDs where reliability, efficiency, and seamless integration with renewable energy sources are crucial.

The format of this document is as follows: After a brief overview of PV systems and converter topology in Section 1, Section 2 describes how to operate the recommended SSIN converter. The SSIN converter's design element is presented in Section 3. Section 4 presents the comparative simulation results, which highlight the superiority of the proposed converter and Section 5 describes the discussion and comparison. The article is concluded in Section 6.

2. Functioning of the Converter

The operation of the proposed one-stage converter, depicted in Figure 2, involves the interaction of several components: a single semiconductor switch, six diodes, two inductors (L_1 and L_2), and three capacitors (C_1, C_2, and an additional one). When the switch S is turned on, inductors L_1 and L_2 are charged by the input source. The energy stored in the inductors is transferred to the cross-connected capacitors C_1 and C_2 when the switch is turned off. These capacitors act as a voltage doubler, thereby enhancing the voltage boost achieved by the converter. This design facilitates efficient energy conversion and voltage amplification.

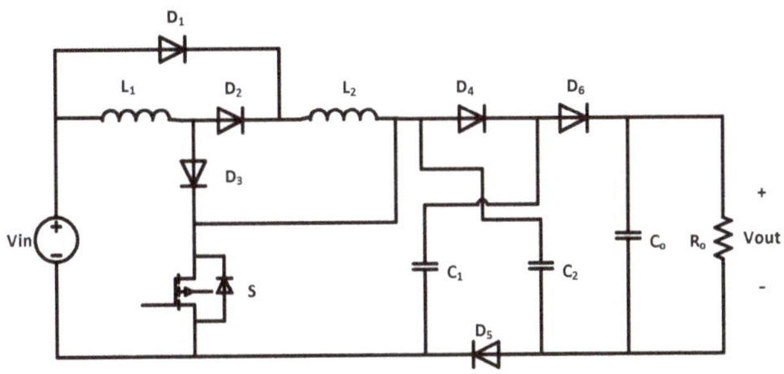

Figure 2. Topology of the converter with one-stage.

To understand the fundamental working concept of one-stage high step-up DC–DC converter, the following procedures are given [21]:

As shown in Figure 3a, in Mode 1, this operating mode appears when:

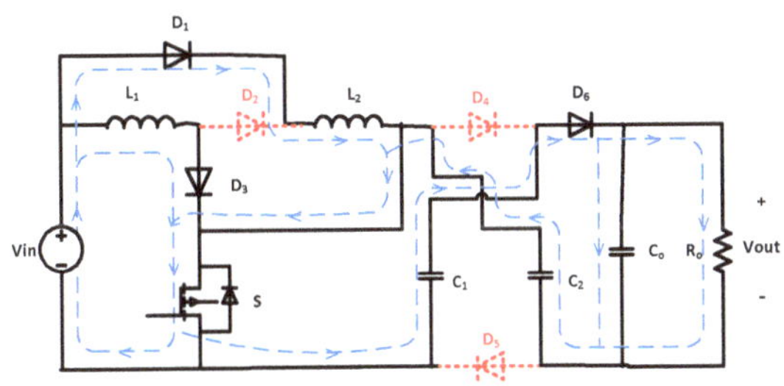

(a)

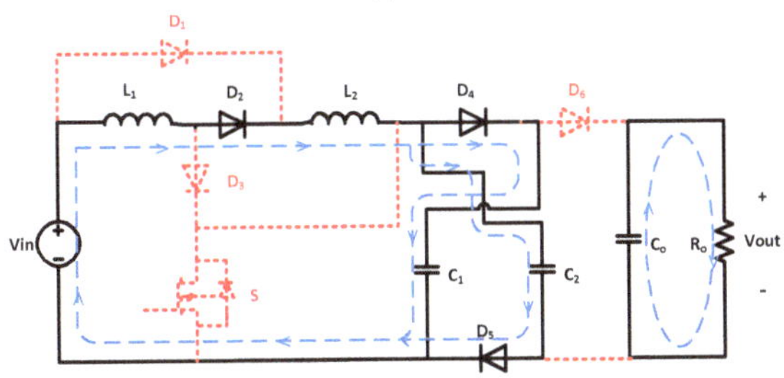

(b)

Figure 3. Equivalent circuit of proposed converter: (**a**) Mode 1; (**b**) Mode 2.

Switch S On: For the interval $0 < t \leq \delta Ts$, the switch S is closed.

Diodes Forward Biased: D_1, D_3, and D_6.

Diodes Reverse Biased: D_2, D_4, and D_5.

Inductors Charging: Input voltage V_{in} charges L_1 and L_2 in parallel.

Energy Supply: Capacitors C_1 and C_2 supply energy to the output capacitor C_o and the load.

Voltage Across Inductors: V_{in}.

The output voltage of the load is the summation of the voltages provided by the voltage between the cross-connected capacitors as $V_0 = V_{C1} + V_{C2}$.

As shown in Figure 3b, in Mode 2, this operating mode appears when:

Switch S Off: For $\delta Ts \leq t < (1 - \delta)$ Ts, the switch S is open.

Diodes Forward Biased: D_2, D_4, and D_5.

Diodes Reverse Biased: D_1, D_3, and D_6.

Inductors Discharging: L_1 and L_2 discharge energy into capacitors C1 and C2, which are cross connected, in series. The inductors L1 and L2 release their stored energy into the cross-connected capacitors in a series configuration.

Voltage Across Capacitors: At steady-state, $V_{C1} = V_{C2} = V_{out}/2$.

As illustrated in Figure 3b, the voltage between inductors L_1 and L_2 is given by (1) when the output voltage is equal to the capacitor voltage:

$$V_{L1} = V_{L2} = \frac{2V_{in} - V_{out}}{4} \tag{1}$$

The average voltage of the inductor is zero according to the volt-second balance of the inductor L1, as stated:

$$V_{in}\delta + \frac{2V_{in} - V_{out}}{4}(1 - \delta) = 0 \tag{2}$$

By solving Equation (3), the DC voltage gain for the converter operating in continuous conduction mode (CCM) is derived as follows:

$$\frac{V_{out}}{V_{in}} = 2\frac{1 + \delta}{1 - \delta} \tag{3}$$

3. Design Components of Converter

The suggested converter design is made up of various essential parts, each playing a crucial role in its operation. Here is a detailed breakdown of these components that is given by [21]:

3.1. Inductors

The current ripple (ΔiL) of the inductors L_1 and L_2 during the switch S's on-state is calculated as,

$$\Delta i_{L1} = \frac{V_{in}}{L_1}\delta T_s; \Delta i_{L2} = \frac{V_{in}}{L_2}\delta T_s \tag{4}$$

From the aforementioned equations of the current ripple, the deduced inductance values for the inductors L1 and L2 are provided as,

$$L_1 \geq \frac{V_{in}\delta}{f_s \Delta i_{L1}}; L_2 \geq \frac{V_{in}\delta}{f_s \Delta i_{L2}} \tag{5}$$

As stated in (6), the average currents flowing through the inductors L1 and L2 are.

$$I_{L1} = \frac{2I_0}{(1 - \delta)}; I_{L2} = \frac{2I_0}{(1 - \delta)} \tag{6}$$

3.2. Capacitors

When the switch S is in the off-state, or mode 2, the capacitors C_1 and C_2 are charged. When the switch S is in the on position, the capacitor C_0 is charged. The expression provided in Equation (4) represents the energy stored in the capacitors.

$$\begin{cases} Q_{C1} = C_1 \Delta V_{C1} = \frac{I_{L1}}{2} \delta T_s \\ Q_{C2} = C_2 \Delta V_{C2} = \frac{I_{L1}}{2} \delta T_s \\ Q_{C0} = C_0 \Delta V_0 = I_0 (1-\delta) \delta T_s \end{cases} \tag{7}$$

where I_{in} and I_{out} show, respectively, the average input and output currents. Equation (8) provides the expressions for these average currents.

$$I_{in} = 2\left(\frac{1+\delta}{1-\delta}\right) I_0; I_0 = \frac{V_{out}}{R_0} \tag{8}$$

Equations (6) and (7) yield the final equations for the capacitance values of the capacitors C_1, C_2, and C_0, which are as follows:

$$\begin{cases} C_1 \geq \frac{\delta V_{out}}{(1-\delta) R_0 f_s \Delta V_{C1}} \\ C_2 \geq \frac{\delta V_{out}}{(1-\delta) R_0 f_s \Delta V_{C2}} \\ C_0 \geq \frac{V_{out}(1-\delta)}{R_0 f_s \Delta V_{C0}} \end{cases} \tag{9}$$

where ΔV_C is the voltage ripple over the capacitor.

3.3. Semiconductor Devices

The switch's current and voltage stresses are indicated by the variables V_S and I_S, and are listed below.

$$\frac{V_S}{V_{in}} = \frac{1+\delta}{1-\delta} \tag{10}$$

$$I_{S,RMS} = \frac{4\sqrt{\delta} I_0}{1-\delta} \tag{11}$$

The following are the diodes' RMS current and voltage stress expressions.

$$\frac{V_{D1}}{V_{in}} = \frac{V_{D3}}{V_{in}} = \frac{\delta}{(1-\delta)} \tag{12}$$

$$I_{D1,RMS} = I_{D3,RMS} = \frac{2\sqrt{\delta} I_0}{(1-\delta)} \tag{13}$$

$$\frac{V_{D4}}{V_{in}} = \frac{V_{D5}}{V_{in}} = \frac{V_{D6}}{V_{in}} = \frac{1+\delta}{1-\delta} \tag{14}$$

$$\frac{V_{D2}}{V_{in}} = 1 \tag{15}$$

$$I_{D2,RMS} = \frac{2\sqrt{1-\delta} I_0}{(1-\delta)} \tag{16}$$

where V_{D1}, V_{D2}, V_{D3}, V_{D4}, V_{D5}, and V_{D6} represent the voltage stress on diodes D_1, D_2, D_3, D_4, D_5, and D_6, respectively, and I_{D1}, I_{D2}, I_{D3}, I_{D4}, I_{D5}, and I_{D6} denote the current stress on diodes D_1, D2, D_3, D_4, D_5, and D_6, respectively.

4. Simulation Result

4.1. Comparison of Converter Performance

In this section, the simulation results for boost converters as a various type of non-isolated converters for solar PV panels integrated through VDC micro grids are presented and compared.

The simulated circuit of the high step-up DC–DC converter in one stage, based on the converter specification given in Table 1, is shown in Figure 4.

Table 1. Component specification.

Parameter	Specifications
Input voltage	24 V
Inductors	200 μH
Capacitors	100 μF
Switching frequency	5 KHZ
Switch	IGBT

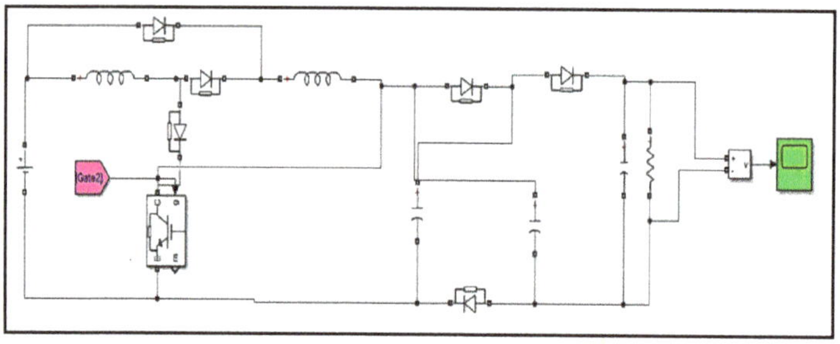

Figure 4. One-stage high step-up DC–DC converter.

For Vin = 24 and duty cycle (δ = 0.5), it was calculated that the gain of this converter will result in an output voltage of 144. The output voltage plot of this converter is depicted in Figure 5. This output voltage typically reflects how well the converter regulates the voltage level after converting it from the input source, and it also indicates the converter's efficiency in maintaining a stable output.

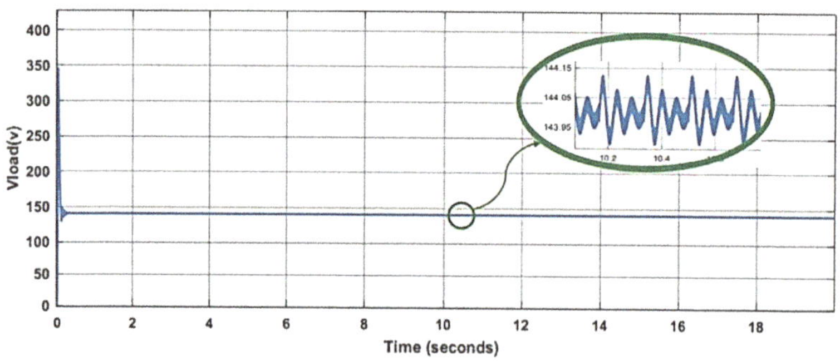

Figure 5. Output voltage of one-stage converter.

The simulated circuit of the high volDC–DC converter is shown in Figure 6.

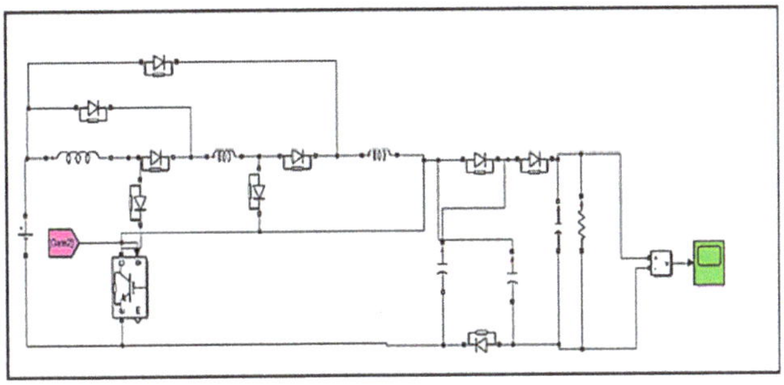

Figure 6. High volDC–DC converter.

The output voltage of the double-stage high step-up DC–DC converter is shown in Figure 7, which also indicates the converter's efficiency in maintaining a stable output.

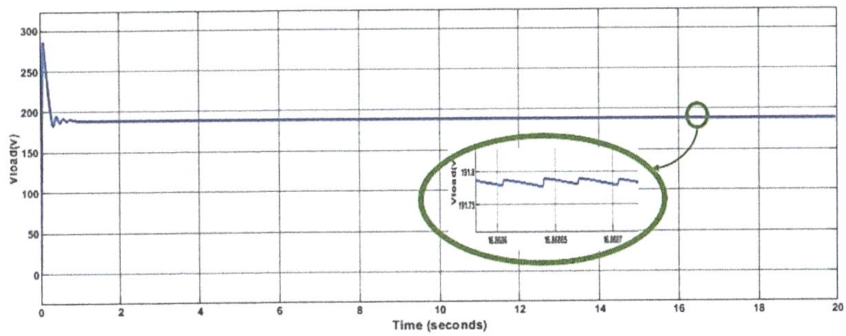

Figure 7. Output voltage of double-stage converter.

The comparative DC voltage gain for different configurations from previous works [22–30] are shown in Figure 8. In comparison to other documented converters, the suggested SSIN converter in its double-stage configuration offers a significantly higher voltage gain, as illustrated in the figure. This improvement is particularly evident at higher duty cycles, where the voltage gain increases substantially due to the optimized design of the converter. The double-stage architecture effectively combines the benefits of enhanced energy transfer and reduced losses, allowing the converter to achieve superior performance under demanding operating conditions.

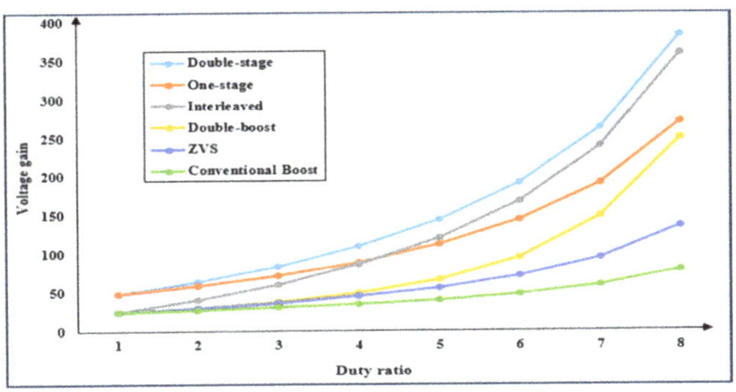

Figure 8. Comparative DC voltage gain for double-stage (suggested), One-stage [21], Interleaved [22–25], Double-boost [26,27], ZVS [15,28–30] and Conventional Boost.

At higher duty cycles, the system leverages the extended range of energy conversion, enabling greater output voltage levels without compromising efficiency. This feature makes the SSIN converter particularly suitable for applications requiring high-voltage gains, such as photovoltaic systems and other renewable energy integrations, where achieving maximum power extraction and system scalability are critical.

In Table 2, five different boost topologies were compared. As can be seen, the n-stage (n = 1.2) has a reasonable complexity and a high-voltage gain, making it a good candidate to use with low voltage DC sources, such as ultra-capacitors, solar PV, fuel cells, and batteries.

Table 2. Boost converters comparison.

Converters	No. of switch	No. of Diode	No. of Capacitor	No. of Inductor	Topology Complexity	Control Complexity	Voltage Gain
Conventional boost	1	1	1	1	low	low	low
Interleaved [22]	3	3	1	3	medium	high	high
Double boost [15]	1	3	3	2	low	low	low
Zero switching [29]	2	1	2	2	medium	high	medium
SSIN one-stage	1	6	3	2	low	low	medium
SSIN double-stage	1	9	3	3	medium	low	high

4.2. Grid-Connected PV System for Feeding Electrical Energy into the PED

In this paper, a PED-connected PV system was considered. The PV solar array in a grid-connected PV system, sometimes referred to as a "grid-tied" solar system, is "tied" to the local mains' electricity grid and supplies electrical energy to the PED. The configuration of the solar PV system under study fed to either the PED or the grid, as shown in Figure 9.

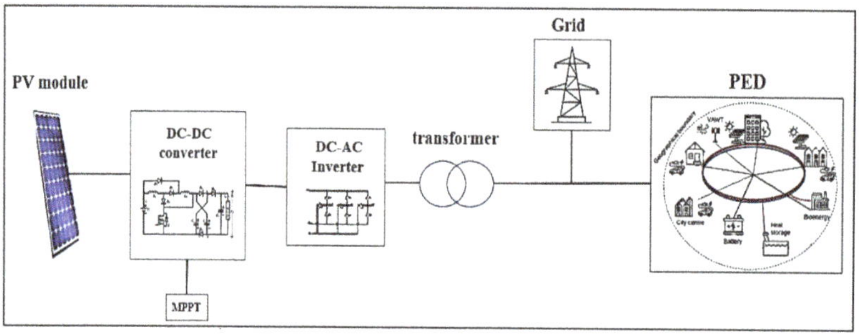

Figure 9. Grid-connected PV system which feeds electrical energy into the PED.

As shown in Figure 10, the photovoltaic (PV) system connected to the one-stage high step-up DC–DC converter with MPPT, was integrated into a grid-connected system using a 3-level inverter. MPPT is implemented in the DC–DC boost converter by means of a state-flow implementation using the incremental conductance technique as previously demonstrated in [31,32].

In the PED-connected PV system under study, 66 strings of five series modules connected in parallel are present, and a high DC–DC step-up boost converter increases the DC voltage from the PV maximum natural voltage, ranging from 273 V to 500 V.

The incremental conductance approach was used to implement MPPT in this DC–DC converter. The MPPT controller changes the switching duty cycle to produce the voltage required for extracting the greatest power. Figure 11 shows the results of a solar radiation experiment conducted to examine the DC side of the system. In this figure, the radiation changed, and in t = 0.4 s, the MPPT is active. As can be seen, the voltage increased to the

PV maximum natural voltage of 273 V, and the maximum power according to radiation was extracted.

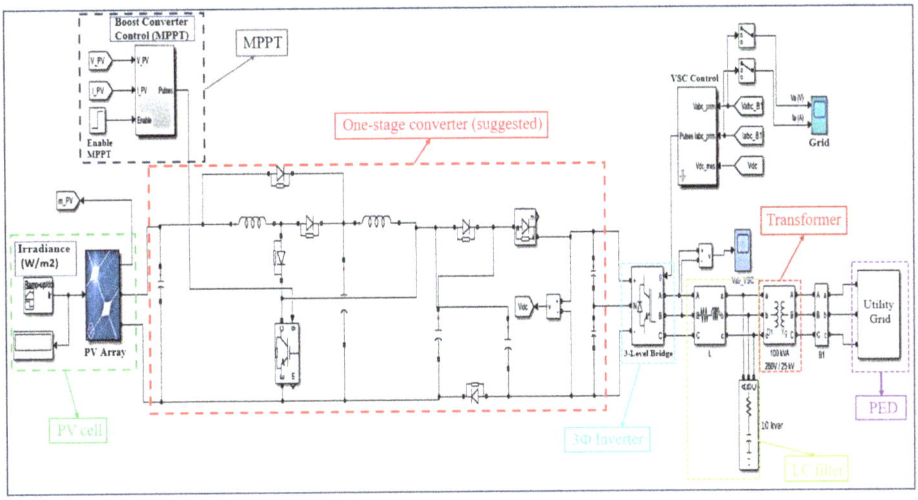

Figure 10. PED-connected PV system with a one-stage DC–DC converter.

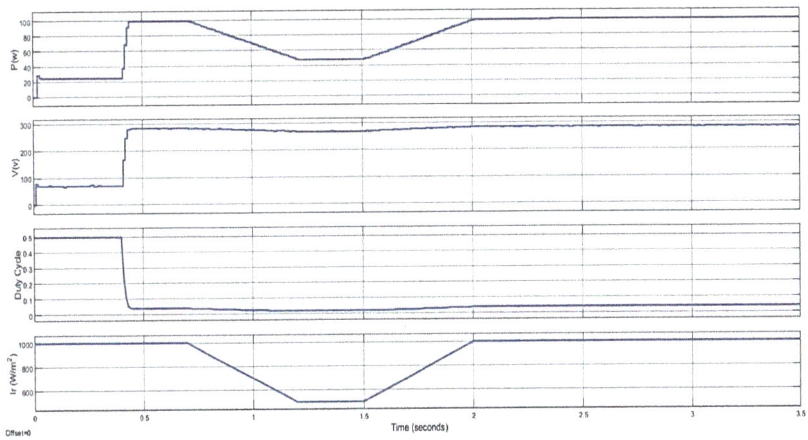

Figure 11. PV output power, voltage, duty cycle voltage, and radiation.

Figure 12 shows the current of the first inductor of the high DC–DC step-up boost converter, and it represents the current flowing through the inductor connected to the converter's input stage. This current is crucial because it affects the converter's performance, efficiency, and stability.

As stated before, the output of the high step-up boost converter was connected to voltage source inverter. The input voltage of the inverter was considered to be 500 v. Consequently, 500 v should be the output voltage of the high DC–DC step-up boost converter. The output voltage of the high DC–DC step-up boost converter is shown in Figure 13. This figure illustrates the voltage level achieved by the converter after it had increased (or "boosted") the input voltage.

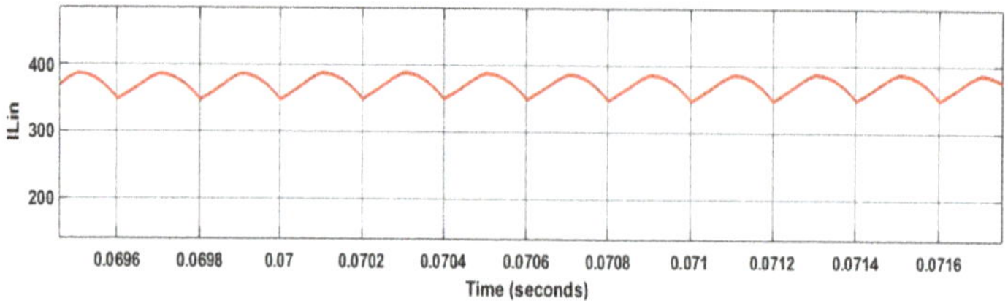

Figure 12. Input inductor current in proposed converter.

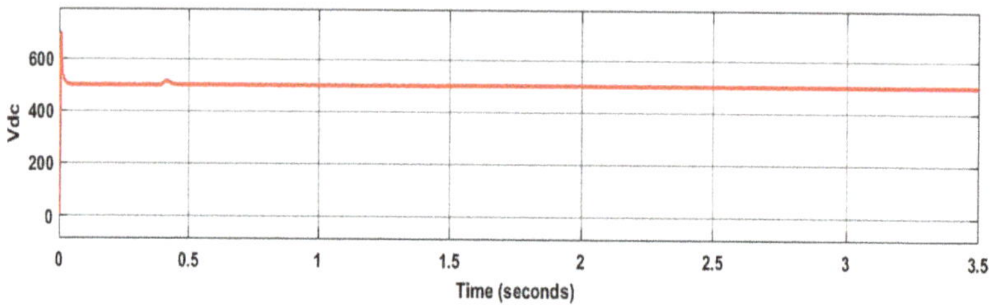

Figure 13. Output voltage of DC–DC boost converter.

The voltage source inverter generates alternative current (AC) voltage at 50 Hz. In order to inject this voltage into the grid, a three-phase transformer increases the voltage to a three-phase, line-to-line 25 kV distribution network voltage level. Figure 14 shows the AC output phase voltage and the current of the transformer, and it refers to the electrical parameters produced at the transformer's secondary winding after it has transferred energy from the primary winding. Understanding these parameters is crucial for evaluating the transformer's performance and its suitability for various applications.

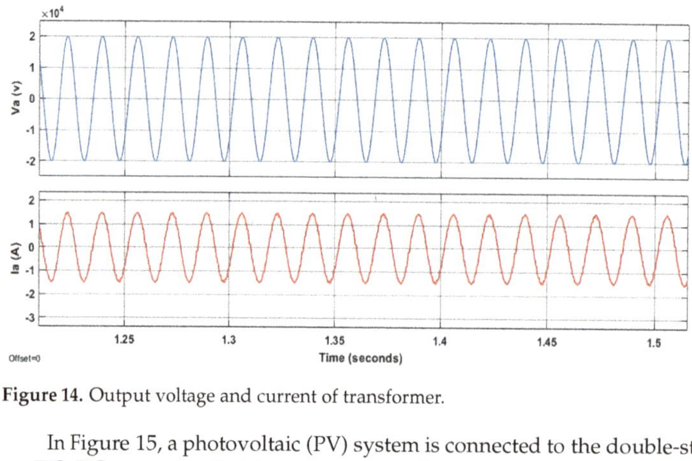

Figure 14. Output voltage and current of transformer.

In Figure 15, a photovoltaic (PV) system is connected to the double-stage high step-up DC–DC converter that incorporates maximum power point tracking (MPPT). This integrated system is designed to be PED-connected and utilizes a 3-level inverter.

In Figure 16, the graph displays the current flowing through the first inductor of the high DC–DC step-up boost converter. The results confirm the low input inductor current of converter.

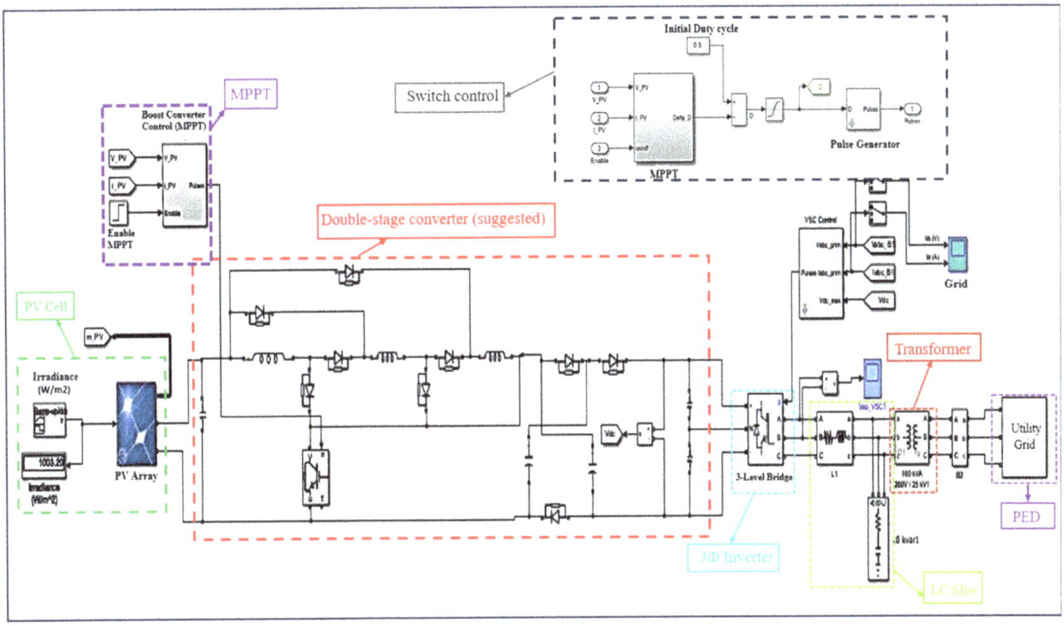

Figure 15. PED-connected PV system with double-stage DC–DC converter.

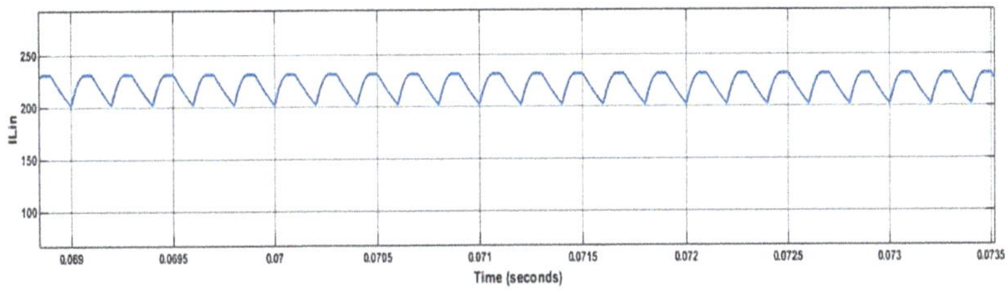

Figure 16. Input inductor current.

Figure 17 shows the output voltage of the double-stage boost converter and it's increased voltage level, produced at the converter's output, which is higher than the input voltage. This output voltage is a key indicator of the converter's ability to step up voltage to meet specific requirements.

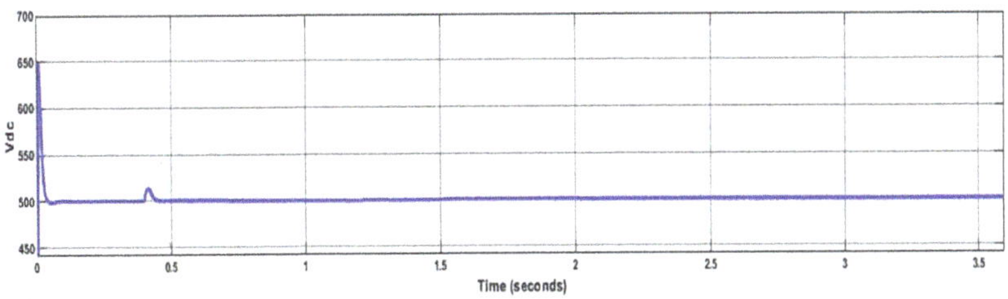

Figure 17. Output voltage of DC–DC boost converter.

Figure 18 illustrates the AC output phase voltage and the current of the transformer using the double-stage configuration, highlighting their sinusoidal nature. This demonstrates that the proposed approach maintains a low total harmonic distortion (THD) at the point of common coupling (PCC), ensuring compliance with power quality standards. The sinusoidal form of the voltage and current, achieved through advanced modulation techniques and optimized control strategies, minimizes harmonic content, thereby preventing any significant increase in THD at the PCC. This highlights the effectiveness of the double-stage topology in preserving power quality in grid-connected systems.

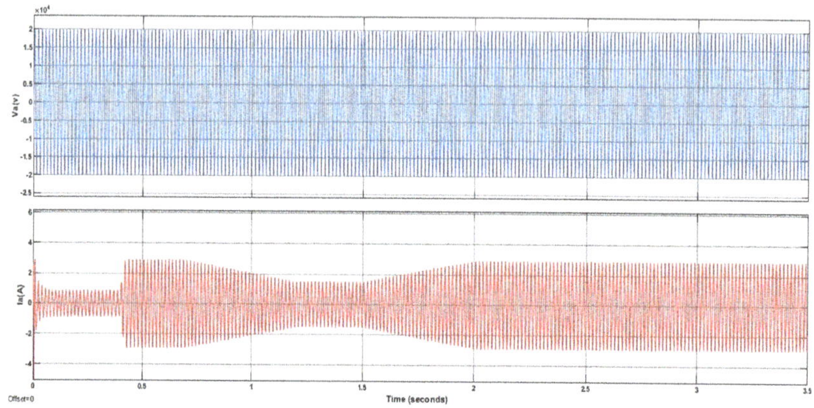

Figure 18. Output voltage and current of transformer.

To analyze the impact of grid disturbances on the PV system, two short-circuit fault scenarios were applied to the one-stage converter.

Scenario 1: A single line-to-ground fault was applied at t = 2.5 s and cleared after 100 ms. During this fault, the system experienced a temporary voltage dip, but the MPPT algorithm quickly adjusted the duty cycle to restore the system's operating point. This demonstrated the system's capability to maintain voltage stability and output power recovery despite transient disturbances.

Scenario 2: A three-phase short-circuit fault at t = 3 s, a more severe three-phase short-circuit fault, was introduced and cleared after 100 ms. Despite the significant disruption caused by this fault, the MPPT algorithm effectively re-adjusted the duty cycle, enabling the PV system to return to stable operation. The system maintained output power and voltage stability, underscoring its resilience under more critical fault conditions.

In order to consider the power grid on the PV system, two short circuit faults were applied to the one-stage converter. The first fault type was a single line-to-ground, which was applied at t = 2.5 s and cleared after 100 ms; the second one was at t = 3 s, and a three-phase short circuit fault was introduced and cleared after 100 ms. Figure 19 illustrates the output characteristics of the PV system under these fault scenarios. The results validate that the proposed configuration ensures operational reliability by maintaining stable performance during faults, highlighting its robustness and fault-tolerant design for real-world applications. This resilience is critical for ensuring the integration of PV systems into power grids with minimal disruption to performance.

Figure 20 illustrates the DC link voltage under fault conditions, demonstrating the system's stability even after severe faults. The results show that the DC link voltage remains unaffected during the single line-to-ground fault, maintaining its nominal value without any noticeable deviation. However, during the three-phase short-circuit fault, the DC link voltage temporarily increases due to the abrupt energy imbalance in the system.

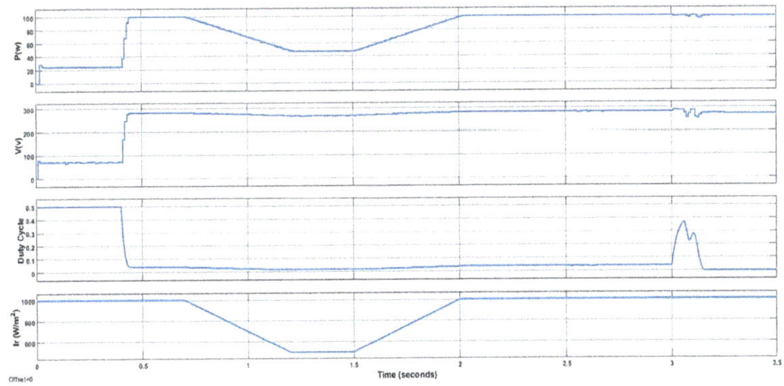

Figure 19. PV output power, voltage, duty cycle voltage, and radiation after short circuit faults.

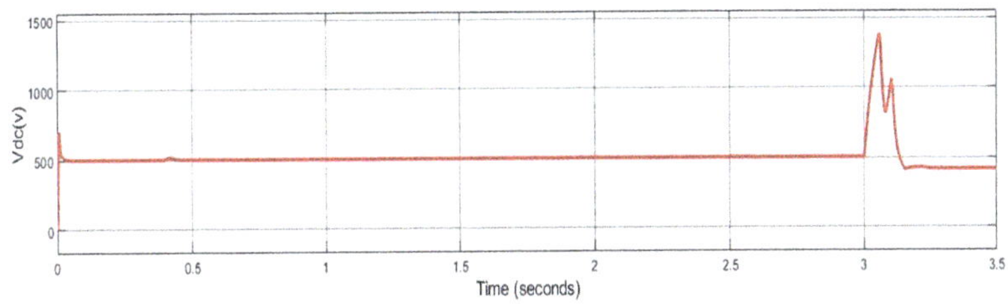

Figure 20. Output voltage of converter after short circuit faults.

This transient rise in voltage highlights the importance of designing the DC link capacitor to withstand higher voltage stresses under contingency scenarios, particularly during severe faults like three-phase short circuits. Ensuring that the capacitor can handle these elevated voltage levels is critical for maintaining the overall reliability and durability of the system under fault conditions.

Figure 21 depicts the output voltage and current of the transformer following short-circuit faults. The results show that the output current experiences a sharp spike immediately after the fault, as expected due to the abrupt disturbance. However, the system's robust control mechanisms quickly mitigate this spike, bringing the current back to its nominal levels within a short duration, thereby preserving system stability.

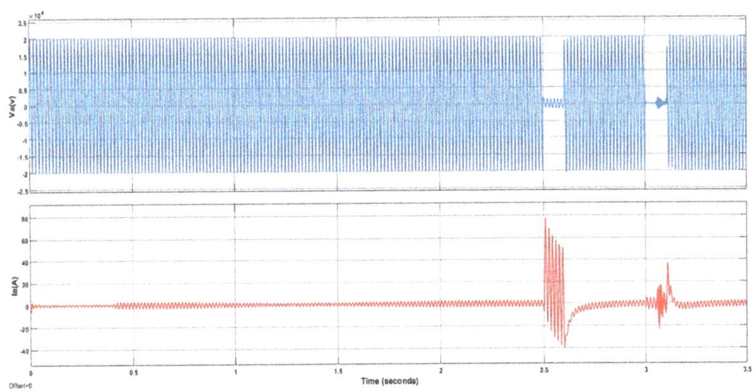

Figure 21. Output voltage and current of transformer after short circuit faults.

Similarly, the output voltage exhibits transient fluctuations during the fault, particularly following the short circuit. These fluctuations are effectively damped as the fault is cleared, with the voltage gradually stabilizing to its nominal value. This response highlights the robustness and resilience of the system's design, demonstrating its capability to handle sudden and severe disruptions while ensuring reliable and stable operations under adverse conditions.

5. Discussion and Comparison

The simulation results presented in this study demonstrate the performance and effectiveness of various non-isolated high step-up DC–DC converters, with a particular focus on their integration into a grid-connected PV system for PEDs. The primary goal was to compare different converter topologies and evaluate their suitability for low-voltage DC sources like solar PV panels in LVDC microgrids. The results offer key insights into the behavior of one-stage and double-stage high step-up converters, as well as their interaction with grid-connected systems under fault conditions. The voltage gain of the one-stage and double-stage converters was calculated as $2(1 + \delta)/(1 - \delta)$ and $2(1 + 2\delta)/(1 - \delta)$, respectively. For a duty cycle of $\delta = 0.5$, the one-stage converter produced an output voltage of 144 V, with an input of 24 V, and the double-stage converter exhibited a higher gain, delivering even greater output voltages. The performance of these converters under different load conditions suggests that the double-stage configuration provides a significantly higher voltage gain while maintaining a reasonable level of complexity.

The comparative DC voltage gain for different converter topologies, highlights the trade-offs between complexity and performance. While conventional boost converters offer simplicity in design and control, their voltage gain is limited, making them less suitable for high-gain applications in PEDs. On the other hand, the SSIN-based converters, particularly the double-stage configuration, achieve superior voltage gains while keeping the component count and control complexity at a manageable level. This makes the double-stage SSIN converter a strong candidate for applications requiring high-voltage step-up with minimal energy losses, such as in PV systems and battery storage in PEDs.

According to the study, it is evident that while interleaved and zero-switching converters offer high-voltage gains, they come with higher control complexity and an increased number of components. The SSIN-based converters, particularly the one-stage and double-stage configurations, stand out for their balance between topology complexity and voltage gain. The single-switch design in both stages ensures that switching losses are minimized, thereby enhancing overall efficiency. This is especially beneficial in systems like PEDs, where energy efficiency and stability are paramount. The results from the grid-connected PV system simulations with MPPT, further emphasize the robustness and flexibility of the proposed SSIN-based converters. The ability of the one-stage and double-stage converters to maintain voltage stability and system efficiency under varying solar radiation and fault conditions showcases their reliability.

When subjected to line-to-ground and three-phase short-circuit faults, the system remained stable, with minimal impact on PV output power and voltage. The sharp rise in output current after the faults, followed by the rapid stabilization of both voltage and current, underscores the effectiveness of the system's control mechanisms in managing sudden disruptions. This resilience is crucial for grid-connected systems in PEDs, where an uninterrupted power supply is essential for maintaining the energy balance of the district.

This behavior underscores the efficiency of the SSIN converter's control mechanisms in maintaining stability during fault conditions, which is essential for systems integrated into Positive Energy Districts (PEDs), where uninterrupted power supply and energy balance are critical. Unlike previous works such as Kumaravel Sundaramoorthy et al. [21], which focused on the converter's design and performance under normal conditions, our analysis emphasizes the converter's ability to handle fault conditions and maintain system stability, a critical factor for real-world grid-connected systems. It should be noted that stability analysis and energy efficiency under fault conditions were not specifically addressed in [21], whereas our study has provided an in-depth evaluation of these aspects. Our findings indicate that the SSIN converter performs reliably under fault disturbances, making it an excellent candidate for high-efficiency, fault-tolerant applications. This aligns with the need for robust energy systems in PEDs, as highlighted in the literature on fault-tolerant converter designs and their importance for grid stability, such as the studies by Gouveia et al. [24] and Zhang et al. [25].

6. Conclusions

This paper addresses the growing need for efficient power conversion in positive energy districts, which rely on the integration of renewable energy sources like photovoltaics, energy storage systems, and smart grids. PEDs require advanced DC–DC converters to manage energy flows efficiently, as renewable sources typically generate energy in DC form. High-performance DC–DC converters ensure minimal energy losses during conversion, optimizing the district's overall energy production and consumption. In this context, the development of novel converter topologies is essential to maximize energy efficiency and system reliability.

Novel one-stage and double-stage SSIN converters are proposed, with the DC voltage gain of the suggested converters derived analytically. The gain was determined to be $2(1 + \delta)/(1 - \delta)$ for the one-stage converter and $2(1 + 2\delta)/(1 - \delta)$ for the double-stage converter. An innovative approach was adopted by modifying the one-stage and double-stage methodologies to simulate a PV power system. This system integrates maximum power point tracking within a grid-connected system using a 3-level inverter. The incremental conductance method was employed in the state flow implementation of MPPT in the DC–DC boost converter. To evaluate the PV system's interaction with the power grid, two short-circuit faults were applied. The results indicate that, following both line-to-ground and three-phase faults, the current increases; however, the system remains stable even under severe fault conditions. Furthermore, the use of a single semiconductor switch in the SSIN topology, regardless of the number of stages, was found to enhance the converter's overall efficiency. This makes it an excellent candidate for integrating low-voltage DC sources, such as batteries and fuel cells, particularly in applications like PEDs. The converter can also be adapted for use with other inverter types, such as the S3 inverter.

Author Contributions: In the process of preparing this manuscript, T.H. was responsible for executing the methodology, managing data curation, conducting software-based simulations, and drafting the original written content. On the other hand, H.J.K. undertook the tasks of editing, validation, visualization, handling resources, and funding acquisition, as well as contributing to editing and supervision. All authors have read and agreed to the published version of the manuscript.

Funding: This research received no external funding.

Data Availability Statement: Some data are contained within the article.

Conflicts of Interest: The authors declare no conflicts of interest.

References

1. Meshael, H.; Elkhateb, A.; Best, R. High Gain Isolated Switched-Boost DC-DC Converter for PV Application. In Proceedings of the 2023 IEEE International Conference on Environment and Electrical Engineering and 2023 IEEE Industrial and Commercial Power Systems Europe (EEEIC/I&CPS Europe), Madrid, Spain, 6–9 June 2023; pp. 1–6.
2. Castillo-Calzadilla, T.; Oroya-Villalta, J.; Borges, C.E. Energy Management System for a Residential Positive Energy District Based on Fuzzy Logic Approach (RESTORATIVE). *Smart Cities* **2024**, *7*, 1802–1835. [CrossRef]
3. Kitsopoulou, A.; Pallantzas, D.; Bellos, E.; Tzivanidis, C. Mapping the Potential of Zero-Energy Building in Greece Using Roof Photovoltaics. *Designs* **2024**, *8*, 68. [CrossRef]
4. Ali, M.F.; Sarker, N.K.; Hossain, M.A.; Alam, M.S.; Sanvi, A.H.; Syam Sifat, S.I. Techno-economic feasibility study of a 1.5 MW grid-connected solar power plant in Bangladesh. *Designs* **2023**, *7*, 140. [CrossRef]
5. Jafari Kaleybar, H.; Hafezi, H.; Brenna, M.; Faranda, R.S. Smart AC-DC Coupled Hybrid Railway Microgrids Integrated with Renewable Energy Sources: Current and Next Generation Architectures. *Energies* **2024**, *17*, 1179. [CrossRef]

6. Espinoza-Trejo, D.R.; Castro, L.M.; Bárcenas, E.; Sánchez, J.P. Data-Driven Switch Fault Diagnosis for DC/DC Boost Converters in Photovoltaic Applications. *IEEE Trans. Ind. Electron.* **2024**, *71*, 1631–1640. [CrossRef]
7. Sutikno, T.; Samosir, A.S.; Aprilianto, R.A.; Purnama, H.S.; Arsadiando, W.; Padmanaban, S. Advanced DC–DC converter topologies for solar energy harvesting applications: A review. *Clean Energy* **2023**, *7*, 555–570. [CrossRef]
8. Ahmadi, M.; Kaleybar, H.J.; Brenna, M.; Castelli-Dezza, F.; Carmeli, M.S. DC Railway Micro Grid Adopting Renewable Energy and EV Fast Charging Station. In Proceedings of the 2021 IEEE International Conference on Environment and Electrical Engineering and 2021 IEEE Industrial and Commercial Power Systems Europe (EEEIC/I&CPS Europe), Bari, Italy, 7–10 September 2021; pp. 1–6. [CrossRef]
9. Castillo-Calzadilla, T.; Cuesta, M.A.; Quesada, C.; Olivares-Rodriguez, C.; Macarulla, A.M.; Legarda, J.; Borges, C.E. Is a massive deployment of renewable-based low voltage direct current microgrids feasible? Converters, protections, controllers, and social approach. *Energy Rep.* **2022**, *8*, 12302–12326. [CrossRef]
10. Zhou, Z.; Liu, J.; Motwani, J.K.; Dong, D.; Burgos, R. Power Converter. U.S. Patent 20240275286, 29 August 2024. Available online: https://www.freepatentsonline.com/y2024/0275286.html (accessed on 29 August 2024).
11. Hashemi, T.; Mahboobi Esfanjani, R.; Jafari Kaleybar, H. Composite Switched Lyapunov Function-Based Control of DC–DC Converters for Renewable Energy Applications. *Electronics* **2023**, *13*, 84. [CrossRef]
12. Sutikno, T.; Purnama, H.S.; Widodo, N.S.; Padmanaban, S.; Sahid, M.R. A review on non-isolated low-power DC–DC converter topologies with high output gain for solar photovoltaic system applications. *Clean Energy* **2022**, *6*, 557–572. [CrossRef]
13. Tarzamni, H.; Gohari, H.S.; Sabahi, M.; Kyyrä, J. Non-isolated high step-up dc-dc converters: Comparative review and metrics applicability. *IEEE Trans. Power Electron.* **2024**, *39*, 582–625. [CrossRef]
14. Raghavendra, K.V.G.; Zeb, K.; Muthusamy, A.; Krishna, T.N.V.; Kumar, S.V.P.; Kim, D.H.; Kim, M.S.; Cho, H.G.; Kim, H.J. A comprehensive review of DC–DC converter topologies and modulation strategies with recent advances in solar photovoltaic systems. *Electronics* **2019**, *9*, 31. [CrossRef]
15. Kumar, K.H.; Rao, G.V.S.K. A review of various dc-dc converter topologies for photovoltaic applications. In Proceedings of the 2021 6th International Conference on Communication and Electronics Systems (ICCES), Coimbatre, India, 8–10 July 2021.
16. Suryadevara, R.; Parsa, L. Full-bridge ZCS-converter-based high-gain modular DC-DC converter for PV integration with medium-voltage DC grids. *IEEE Trans. Energy Convers.* **2018**, *34*, 302–312. [CrossRef]
17. Jotham Jeremy, L.; Ooi, C.A.; Teh, J. Non-isolated conventional DC-DC converter comparison for a photovoltaic system: A review. *J. Renew. Sustain. Energy* **2020**, *12*, 013502. [CrossRef]
18. Schmitz, L.; Martins, D.C.; Coelho, R.F. Comprehensive conception of high step-up DC–DC converters with coupled inductor and voltage multipliers techniques. *IEEE Trans. Circuits Syst. I Regul. Pap.* **2020**, *67*, 2140–2151. [CrossRef]
19. Meshael, H.; Elkhateb, A.; Best, R. Topologies and Design Characteristics of Isolated High Step-Up DC–DC Converters for Photovoltaic Systems. *Electronics* **2023**, *12*, 3913. [CrossRef]
20. Li, Q.; Huangfu, Y.; Xu, L.; Wei, J.; Ma, R.; Zhao, D.; Gao, F. An improved floating interleaved boost converter with the zero-ripple input current for fuel cell applications. *IEEE Trans. Energy Convers.* **2019**, *34*, 2168–2179. [CrossRef]
21. Shanthi, T.; Prabha, S.U.; Sundaramoorthy, K. Non-isolated n-stage high step-up DC-DC converter for low voltage DC source integration. *IEEE Trans. Energy Convers.* **2021**, *36*, 1625–1634. [CrossRef]
22. Nahar, S.; Uddin, M.B. Analysis the performance of interleaved boost converter. In Proceedings of the 4th International Conference on Electrical Engineering and Information & Communication Technology (iCEEiCT), Dhaka, Bangladesh, 13–15 September 2018.
23. e Silva Aquino, R.N.A.L.; Tofoli, F.L.; Praca, P.P.; Oliveira, D.D.S., Jr.; Barreto, L.H.S.C. Soft switching high-voltage gain dc–dc interleaved boost converter. *IET Power Electron.* **2015**, *8*, 120–129. [CrossRef]
24. Ahmed, N.A.; Alajmi, B.N.; Abdelsalam, I.; Marei, M.I. Soft switching multiphase interleaved boost converter with high voltage gain for EV applications. *IEEE Access* **2022**, *10*, 27698–27716. [CrossRef]
25. Rahimi, T.; Hosseini, S.H.; Sabahi, M.; Abapour, M.; Gharehpetian, G.B. Three-phase soft-switching-based interleaved boost converter with high reliability. *IET Power Electron.* **2017**, *10*, 377–386. [CrossRef]
26. Li, K.; Hu, Y.; Ioinovici, A. Generation of the large DC gain step-up nonisolated converters in conjunction with renewable energy sources starting from a proposed geometric structure. *IEEE Trans. Power Electron.* **2016**, *32*, 5323–5340. [CrossRef]
27. Wang, F. A novel quadratic boost converter with low current and voltage stress on power switch for fuel-cell system applications. *Renew. Energy* **2018**, *115*, 836–845. [CrossRef]
28. Biswas, P.; Halder, D.; Halder, T. A Zero Voltage Switching (ZVS) Boost Converter Suitable for Power Factor Correction. In Proceedings of the 14th IEEE India Council International Conference (INDICON), Roorkee, India, 15–17 December 2017.
29. Moschopoulos, G.; Jain, P.; Joos, G. A novel zero-voltage switched PWM boost converter. In Proceedings of the PESC'95-Power Electronics Specialist Conference, Atlanta, GA, USA, 18–22 June 1995.
30. Kaleybar, H.J.; Brenna, M.; Llaneza, I.M.G.; Zanineli, D. Interleaved Four Quadrant Converter-Based Active Traction Substation for Recuperating Regenerative Braking Energy in DC Railway Line. In Proceedings of the 2024 International Symposium on Power Electronics, Electrical Drives, Automation and Motion (SPEEDAM), Napoli, Italy, 19–21 June 2024; pp. 1099–1104. [CrossRef]

31. Banu, I.V.; Istrate, M. Modelling of maximum power point tracking algorithm for PV systems. In Proceedings of the 2012 International Conference and Exposition on Electrical and Power Engineering (EPE), Iasi, Romania, 25–27 October 2012.
32. Darvishpour, Y.; Mousavi Gazafrudi, S.M.; Jafari Kaleybar, H.; Brenna, M. Integration of Rooftop Solar PV on Trains: Comparative Analysis of MPPT Methods for Auxiliary Power Supply of Locomotives in Milan. *Electronics* **2024**, *13*, 3537. [CrossRef]

designs

MDPI

Article

Declaration-Ready Climate-Neutral PEDs: Budget-Based, Hourly LCA Including Mobility and Flexibility

Simon Schneider [1,*], Thomas Zelger [1], Raphael Drexel [1], Manfred Schindler [1], Paul Krainer [1] and José Baptista [2]

[1] Department of Industrial Engineering, University of Applied Sciences Technikum Wien, 1200 Vienna, Austria
[2] Department of Engineering, School of Science and Technology, University of Trás-os-Montes and Alto Douro, INESC-TEC, UTAD's Pole, 5000-801 Vila Real, Portugal
* Correspondence: simon.schneider@technikum-wien.at

Abstract

In recent years, Positive Energy Districts (PEDs) have been interpreted in many—and often conflicting—ways. We recast PEDs as a vehicle for verifiable climate neutrality and present a declaration-ready assessment that integrates (i) a cumulative, science-based GHG budget per m^2 gross floor area (GFA), (ii) full life-cycle accounting, and (iii) time-resolved conversion factors that include everyday motorized individual mobility and quantify flexibility. Two KPIs anchor the framework: the cumulative GHG LCA balance (2025–2075) against a maximum compliant budget of 320 $kg_{CO_2e} \cdot m^{-2}{}_{GFA}$ and the annual primary energy balance used to declare PED status with or without mobility. We follow EN 15978 and apply time-resolved emission factors that decline to zero by 2050. Its applicability is demonstrated on six Austrian districts spanning new builds and renovations, diverse energy systems, densities, and mobility contexts. The baseline scenarios show heterogeneous outcomes—only two out of six meet both the cumulative GHG budget and the positive primary energy balance—but design iterations indicate that all six districts can reach the targets with realistic, ambitious packages (e.g., high energy efficiency and flexibility, local renewables, ecological building materials, BESS/V2G, and mobility electrification). Hourly emission factors and flexibility signals can lower import-weighted emission intensity versus monthly or annual factors by up to 15% and reveal seasonal import–export asymmetries. Built on transparent, auditable rules and open tooling, this framework both diagnoses performance gaps and maps credible pathways to compliance—steering PED design away from project-specific targets toward verifiable climate neutrality. It now serves as the basis for the national labeling/declaration scheme klimaaktiv "Climate-Neutral Positive Energy Districts".

Keywords: Positive Energy Districts (PEDs); climate neutrality assessment; greenhouse gas (GHG) emissions; life-cycle assessment (LCA); embodied and operational emissions; building energy systems; mobility emissions

1. Introduction

Achieving climate neutrality by mid-century, as stipulated under the Paris Agreement [1], requires rapid and deep decarbonization across all sectors. The IPCC Sixth Assessment Report confirms that limiting warming to 1.5 °C necessitates stringent carbon budgets, with global net-zero CO_2 emissions reached around 2050 [2–5]. The building sector is pivotal in this transition, contributing significantly to both operational greenhouse gas (GHG) emissions and embodied emissions from construction and refurbishment [6].

Over the past decade, ref. [7] have identified more than 25 different concept names for climate-friendly or energy-positive districts or neighborhoods. Most of these approaches emphasize improved efficiency and renewable integration at the energy-system level but vary widely in scope and lack consistent definitions—a critique common to many studies. As a result, many remain difficult to apply in practice, are hard to replicate, and have seen limited regulatory uptake. Among this broad landscape, four framings are of particular interest for this study: NZEB definitions, their extensions into renewable energy communities, the Zero-Emission Neighborhood (ZEN) framework, and Positive Energy Districts (PEDs). Each of these has shaped the discourse on carbon-neutral or energy-positive districts, but all remain closely tied to annual balances or ambition levels rather than finite, Paris-aligned carbon budgets:

- **NZEB and Net-Zero Definitions.** Classic NZEB frameworks focus on annual site/source primary energy balances, with variants on cost and emissions neutrality [8]. Their common core is a 12-month energy bookkeeping, sometimes coupled to EPB standards (EN/ISO 52000-1) for primary energy weighting [9].
- **NZEBs in Renewable Energy Communities (RECs) and energy sharing.** Recent work extends NZEB logic to building-to-building exchange, where prosumer surpluses buffer consumer deficits inside a REC [10]. While this can improve local self-consumption and grid friendliness, it remains annual-balance driven and typically does not enforce finite, Paris-compatible CO_2 budgets.
- **Zero-Emission Neighborhoods (ZEN).** The Norwegian ZEN program defines neighborhood-scale goals "towards zero life-cycle GHG emissions," explicitly emphasizing life-cycle accounting, high energy efficiency, high shares of local renewables, and smart/flexible operation. ZEN provides ambition levels and guidance but leaves budget allocation and declaration rules to localization [11]
- **Positive Energy Districts (PEDs).** IEA EBC Annex 83 positions PEDs as districts with net annual energy surplus and operational flexibility. Many PED implementations, however, still report annual net balances without translating them into cumulative, per-m^2 CO_2 budgets aligned with Paris [12].

Positive Energy Districts (PEDs) are increasingly promoted as a transformative model to support this transition, integrating high energy efficiency, local renewable generation, and energy flexibility to achieve an annual net-positive energy balance [13–16].

The PED Database compiled well over 50 projects on a self-reporting basis and showed that PED projects across Europe differ widely in energy performance, approach, and ambition and considered system boundaries [17]. This highlights the need for further clarification of the definition of PEDs, especially taking into account the specific context of the district (see, e.g., [18]). The increasing integration of volatile renewable energy (VRE) could significantly benefit from widespread demand-side response [19], but many actors in the field consider the assessment of energy flexibility in the context of a PED an important, yet unresolved question [20].

By contrast, climate neutrality requires aligning emissions with a finite, science-based GHG budget that limits global warming to 1.5 °C (or well below 2 °C), necessitating a lifecycle-based approach that captures embodied emissions, biogenic carbon flows, and time-resolved operational emissions [6]. Additionally, ref. [21] point out that low or zero carbon buildings require clear decarbonization targets as early as in the design phase.

Recent work has also examined the broader challenges of implementing energy-efficient and climate-neutral districts. For instance, ref. [22] identify governance structures, financing models, and integration of building- and district-scale technologies as critical enablers of high-performance districts. While their focus is on efficiency strategies, the findings underscore that achieving Paris-compatible Positive Energy Districts requires not only carbon-budget compliance, as proposed here, but also systemic strategies to overcome institutional and financial barriers.

At the PED scale, three main challenges emerge:

1. Defining system boundaries that distinguish emissions under the direct control of developers from those managed at municipal or national scales [14,23–25].
2. Allocating GHG budgets consistently with global and national targets [4,26,27].
3. Accounting for temporal aspects, including the disproportionate climate impact of early emissions [6,28].

Life cycle assessment (LCA) standards, particularly EN 15978:2024 [29], provide a robust structure for defining system boundaries and modules for embodied and operational emissions in buildings and districts. Previous work [30–32] highlights the need to adapt these methods to the district scale while ensuring compatibility with sectoral carbon budgets. Dynamic emission factors—which reflect the projected decline in grid and district heating carbon intensities—are essential for accurately representing operational impacts over time [33]. International studies [34,35] demonstrate the influence of such temporal factors on lifecycle results, and highlight differences in the treatment of embodied emissions of energy infrastructure.

Novelty and Contribution of This Study

This study extends the established Positive Energy District (PED) assessment framework [24,36] by operationalizing climate neutrality through life-cycle carbon accounting and science-based carbon budgets. It (i) applies a life cycle GHG accounting method consistent with national decarbonization pathways, (ii) incorporates time-differentiated operational emission factors for multiple energy carriers, and (iii) integrates everyday motorized individual mobility as part of project-level design.

Building on EN 15978 system boundaries, it integrates dynamic, energy-carrier-specific emission factors and operationally includes everyday motorized individual mobility (EMIM) and embodied emissions from construction and maintenance from now until 2050. Within the broader assessment framework, this paper addresses the climate neutrality and life-cycle carbon accounting dimension, complementing the framework's energy system modeling and flexibility evaluation capabilities presented in earlier work [24,36]. The shared objective is to provide a methodologically rigorous yet practice-ready approach that captures both (i) the operational responsiveness of PEDs to system needs and (ii) their compliance with carbon budgets consistent with Austria's 2040 climate neutrality target.

The framework presented here is designed to be robust for research purposes—allowing sensitivity analyses with hourly emission factors—while remaining compatible with the monthly conversion factors of the Austrian building code for practical declarations of climate-neutral PEDs as part of the Austrian sustainability label family klimaaktiv [37]. Its core contribution is the derivation of a quantifiable carbon budget per square meter of gross floor area (GFA), integrating both operational and embodied emissions in a manner that is directly linkable to the performance outcomes of the broader PED assessment framework.

The novelty of this work consists of these three parts:

1. The integration of operational, embodied, and mobility energy and emissions into a combined assessment system with quantitative target budgets
2. The assembly of these factors into a nationally applicable declaration system.
3. The demonstration of achievability of this declaration for a diverse range of ambitious districts.

2. Materials and Methods

The methods build on previous work on primary energy-based PED definitions and extend them with a life-cycle, budget-oriented approach. To structure the framework, the development process and the subsequent application are separated. The development side describes the design choices, boundaries, and assumptions that define the assessment logic, while the application side illustrates how the framework is implemented using project data and simulation tools. This dual view ensures the transparency of our methodological decisions and shows how the framework translates into declaration-ready results. Figure 1 provides a schematic overview of the framework development and application.

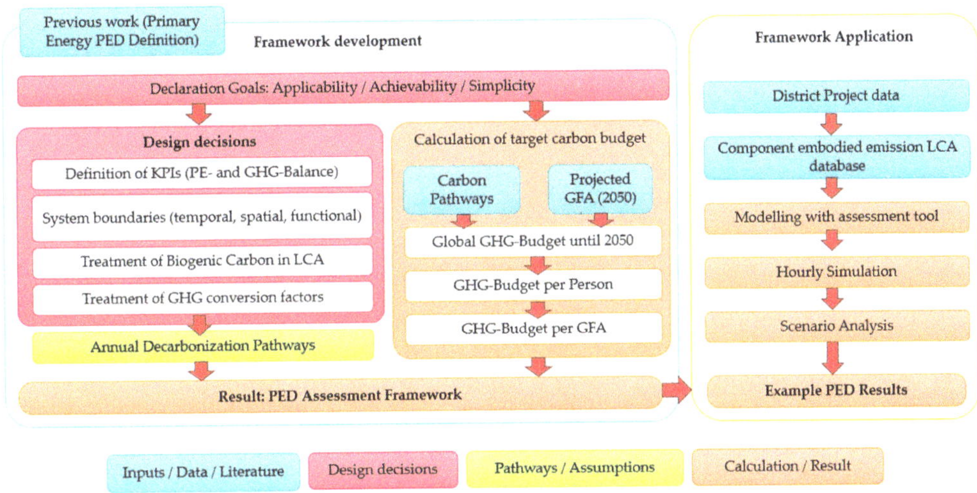

Figure 1. Overview of the climate neutrality assessment framework. The left-hand side illustrates the framework development process, including declaration goals, design decisions, carbon budget derivation, and integration of decarbonization pathways. The right-hand side shows the application to district projects, linking data inputs, modeling, and simulation to example results. Colors indicate categories: inputs/literature, design decisions, assumptions, and results.

2.1. Methodological Principles and Declaration Goals

The proposed climate neutrality assessment framework is grounded in a set of guiding principles that combine practical applicability with scientific rigor, ensuring relevance for both practitioners and researchers.

- Practical applicability and stakeholder relevance—The method is designed to be appliable for real estate developers, municipalities, and planning professionals, enabling its integration into certification schemes and project assessments [38,39]. Its scope is limited to emissions that can be directly influenced through planning and construction decisions—namely, operational energy use, EMIM, and embodied emissions—ensuring decision-making relevance.
- Integrated and adaptable system boundaries—Assessment boundaries extend beyond individual buildings to the full Positive Energy District (PED) system, covering on-site generation, storage, flexibility, and interconnections with surrounding grids. The framework retains adaptability in data sources, emission factors, and system configurations, allowing regional customization while ensuring methodological comparability.
- Temporal differentiation—Emissions are assessed over a 50-year timeframe (2025–2075), with the assumption that climate neutrality is achieved for all services by 2050. As a result, all conversion factors for both operational and embodied energy are assumed

to be zero by 2050, reflecting the expected decarbonization of energy carriers and building systems. This approach ensures alignment with long-term climate goals while considering the gradual reduction in emissions over the assessment period.

- Science-based, budget-oriented compliance thresholds—The method is anchored in downscaled CO_2 budgets consistent with Austria's 2040 climate neutrality target and the 1.5 °C global carbon budget [1]. It establishes a hierarchy of compliance thresholds: an ambitious target value representing a precautionary level for high-confidence Paris alignment, and a maximum compliance limit that must not be exceeded for a project to qualify as "climate-neutral PED."

These principles acknowledge that PED definitions inherently allow for multiple interpretations, particularly regarding system boundaries, included emission sources, and treatment of temporal dynamics. As noted in prior work [24,40], these degrees of freedom can lead to inconsistent assessments if not addressed explicitly. The framework therefore applies a design thinking approach to close such gaps: key methodological choices—such as which life cycle stages to include, how to allocate renewable generation, and how to integrate mobility—are made transparent, justified against policy and scientific references, and consistently applied. Figure 2 summarizes the methodological approach:

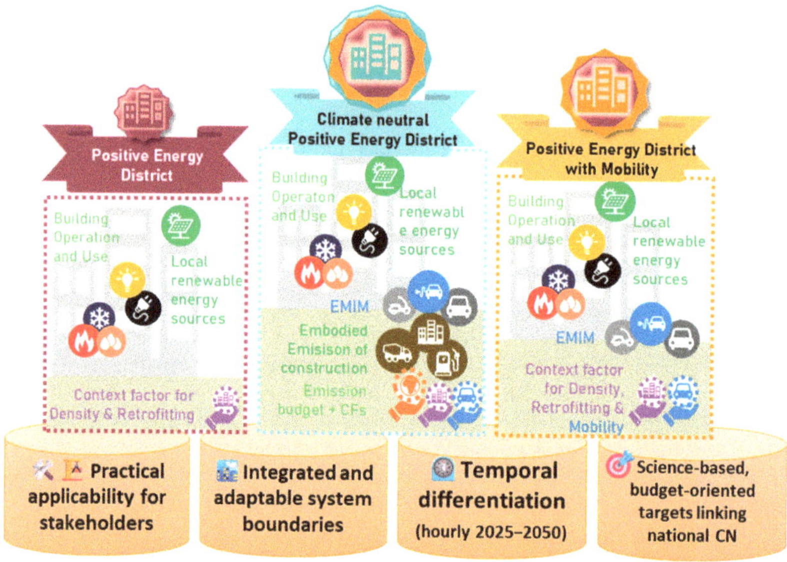

Figure 2. Methodological approach to design of climate neutrality definition for PEDs is based on four pillars, on which three practical PED definitions are based (boxes show conceptualized functional system boundaries).

The following section details how these methodological principles are operationalized within the proposed framework, from KPI definition and the derivation of target carbon budgets to the application of dynamic emission factors and the inclusion of mobility and embodied emissions.

2.2. Key Performance Indicators (KPIs) and PED Levels

We base our assessment framework on two main indicators as defined in [24]:

1. **Cumulative GHG balance** (2025–2075, $kg_{CO_2e} \cdot m^{-2}_{GFA}$)—as defined in the framework above: cradle-to-grave embodied emissions (A1–A3; buildings, TES, PV, batteries, vehicles, C1–C4 before 2050), time-resolved operational emissions (B6; grid electricity

and thermal carriers), operational mobility (B8 EMIM; ICE/EV), and avoided grid emissions (D2; PV exports and EMIM substitution). Negative components reduce the balance. Compliance bands are L3 = 320, L2 = 196, L1 = 72 $kg_{CO_2e} \cdot m^{-2}{}_{GFA}$.

2. **Primary energy balance** ($kWh_{pe} \cdot m^{-2}{}_{NFA} \cdot a^{-1}$)—computed per [24,36] as the difference between annual primary energy grid equivalence from local renewable supply and flexible grid use (PV self-use and feed-in, flexible TES and EMIM operation) and the district's primary energy demand (grid equivalence, HVAC and electricity; EMIM included or excluded depending on the level). A positive value denotes a "positive balance". Context factors for density, mobility and renovation may apply as virtual demands or supplies.

These two KPIs are reported consistently across case studies; figures annotate overshoot/undershoot against L-levels, and tables list the corresponding PE balances. Operational energy KPIs are expressed per conditioned Net Floor Area (NFA) because heating, cooling, and end-use electricity demands are specified for conditioned area in design practice and codes, so NFA best matches how these targets are set and tracked. Lifecycle GHG indicators are expressed per Gross Floor Area (GFA) to maintain comparability across typologies and density/fit-out choices; NFA↔GFA conversions are straightforward via use-type factors that are routinely applied (and refined from early planning estimates). Based on these indicators, we distinguish three PED levels as shown in Table 1:

Table 1. Types of PED as defined in [24].

Type of PED	Acronym	Requirements
Positive Energy District	PED	PE Balance > 0 excl. EMIM
Positive Energy District with Mobility	PED + M	PE Balance > 0 incl. EMIM
Climate-Neutral Positive Energy District (reported in this study)	CN PED	PE Balance > 0 incl. EMIM GHG LCA ≤ L3 over 2025–2075

Results are compiled with the scenario analysis tools provided by the framework (publicly available here [41]).

2.3. Carbon Budget for Climate-Neutral Districts

District-level carbon budgets are derived by downscaling a finite, Paris-compatible global CO_2 budget to Austria and further allocating a sectoral share to buildings, including construction, operation, and everyday motorized mobility (EMIM). This ensures consistency between global climate targets and district-scale declarations.

The calculation follows three steps. First, we adopt the +1.5 °C scenario with 66% probability of success, both with and without temporary temperature overshoot, and subtract materialized national emissions (2022–2024) from the 2022–2075 budget to obtain the available 2025–2075 budget. Second, we allocate a fixed share of the national budget (50%) to buildings and EMIM, expressed on a per capita basis. Finally, per capita values are translated into gross floor area (GFA)-specific thresholds using projected Austrian floor space statistics for 2050 (906.7 million m^2).

Natural carbon sequestration is included as a sensitivity, with additional sinks of 0.5 and 1.0 t_{CO_2e} cap^{-1} a^{-1}. These variants yield three levels of compliance:

- **L1** (Paris-compatible target): stringent limit without sequestration,
- **L2** (precautionary limit): moderate sequestration (0.5 t_{CO_2e} cap^{-1} a^{-1}),
- **L3** (absolute compliance limit): generous sequestration (1.0 t_{CO_2e} cap^{-1} a^{-1}).

Intermediate values and assumptions underlying this derivation are documented in Appendix A Table A1 for transparency. The resulting compliance thresholds, which serve as methodological reference points for district assessment, are summarized in Table 2.

Table 2. Target and compliance carbon budget limits for climate-neutral Positive Energy Districts (PEDs) in Austria for 2025–2075, expressed per gross floor area (GFA). These values represent the main methodological result of the budget downscaling process, derived from global Paris-compatible CO_2 pathways and Austrian building stock data.

Limit	Value	Unit	Description and Compliance Note
Paris-compatible target limit (GHG L1)	72	$kg_{CO_2e} \cdot m^{-2}{}_{GFA}$	Stringent, science-based carbon budget based on [2,27] ensuring a safe pathway to 1.5 °C. Assumes no contribution from biogenic carbon sequestration. Achieving this limit provides the highest certainty of Paris alignment but is challenging in practice.
Precautionary compliance limit (GHG L2)	196	$kg_{CO_2e} \cdot m^{-2}{}_{GFA}$	Upper bound for climate-neutral declaration, assuming 0.5 $t_{CO_2e} \cdot a^{-1} \cdot cap^{-1}$ sequestration. Offers flexibility while retaining a safety buffer; exceeding this value increases risk of overshooting Paris-compatible budgets.
Absolute compliance limit (GHG L3)	320	$kg_{CO_2e} \cdot m^{-2}{}_{GFA}$	Maximum permissible budget, assuming 1 $t_{CO_2e} \cdot a^{-1} \cdot cap^{-1}$ global sequestration. Exceeding this value disqualifies a PED from climate-neutral status under this framework.

2.4. LCA System Boundary and Scope of District

This section defines the temporal and functional boundaries for life-cycle assessment (LCA) within this proposed climate-neutrality assessment framework for PEDs. Unless stated otherwise, the assessment horizon is 2025–2075. All operational conversion factors decline to zero by 2050 for carriers on a national decarbonization pathway (e.g., electricity, district heating,); end-of-life and post-2050 processes are assumed carbon-neutral. As illustrated in Figure 3, the system boundaries include both operational and construction emissions, with a focus on the decarbonization of energy carriers and the treatment of biogenic carbon storage.

The assessment adapts life-cycle boundaries based on EN 15978 [29], with notable alterations: Unlike EN 15978, which references the EN 15804 [42] product standard that books stored biogenic CO_2 as an emission, this framework employs a time-dynamic LCA method for detailed declarations (Section 2.5.1). For practicability, a second, simplified method (Section 2.5.2) can also be used, which credits biogenic carbon in building materials in part as carbon storage (negative emissions, 55–100% depending on its origin and turnover period) over the horizon.

For mobility emissions, the framework includes parts of B8 that correspond to EMIM travels to the district and embodied emissions for vehicle construction (Section 2.4.2). Table 3 gives an overview of the concrete system boundary inclusions and exclusions.

To ensure consistent coverage of embodied and operational impacts, the framework uses building balance boundary BB6 (see Appendix A Table A2 for an overview of boundaries as defined by [43]) for A1–A3 (product stage), B2–B5 (use stage), and End-of-life prior to 2050 (C1–C4). BB6 covers the complete building: thermal envelope, interior partitions, basement structures, unheated buffer zones, technical systems, and exterior facilities (carports, bicycle shelters, auxiliary buildings).

Table 3. System boundaries of Life Cycle Stage Inclusion in the assessment framework.

Life Cycle Stage	Sub-Stage	Module	Considered	Notes/Comments
A—Pre-construction	Pre-construction	A0	No	Outside EN 15978 scope here
A—Product stage	Product stage	A1	Yes	Raw material supply (embodied)
		A2	Yes	Transport to manufacturing
		A3	Yes	Manufacturing emissions

Table 3. *Cont.*

Life Cycle Stage	Sub-Stage	Module	Considered	Notes/Comments
A—Construction process stage	Construction	A4	Yes	Assessed if practical/data available
		A5	Yes	Assessed if practical/data available
B—Use stage	Use (non-energy)	B1	No	Not considered
	Maintenance	B2	Yes	Included; post-2050 assumed carbon-neutral
	Repair	B3	Yes	Same as B2
	Replacement	B4	Yes	Same as B2
	Refurbishment	B5	Yes	Same as B2
	Operational energy use	B6	Yes	Dynamic/Monthly factors (to 0 by 2050)
	Operational water use	B7	No	Water-related GHG excluded
	Other building-related activities	B8	Partly	EMIM B8.1 and B8.2 included (Section 2.4.2) Operational mileage + one-time vehicle embodied (≤2050)
C—End-of-life stage	End-of-life	C1	Yes	Included; post-2050 carbon-neutral
		C2	Yes	Same as C1
		C3	Yes	Same as C1
		C4	Yes	Same as C1
D—Beyond system boundary	Reuse/recycling recovery	D1	Yes	
	Exported utilities	D2	Yes	PV grid feed in credited as substitution with negative hourly/monthly factors
Contextual	Off-plot public infrastructure	—	No	Optional reporting only (not for compliance)

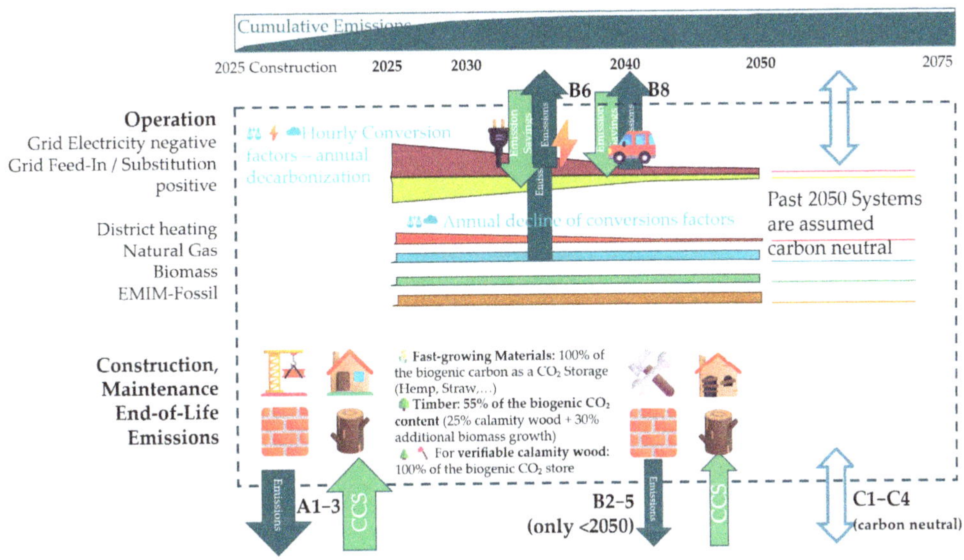

Figure 3. System boundaries and scope of the life-cycle assessment (LCA), covering 2025–2075 and including both operational and construction emissions. Operational emissions comprise grid electricity use and substitution as well as other carriers (natural gas, biomass, fossil, district heating and cooling), with annual decarbonization of conversion factors to zero by 2050 where applicable. Construction emissions (A1–A3, B2–B5) account for certain biogenic carbon storage in bio-based material.

2.4.1. Operational Energy Flows, Local PV, and Exported Electricity (Module B6/D2)

Local PV is handled operationally only: on-site generation replaces grid electricity using the framework's conversion factors (hourly for analysis and optimization; monthly building-code factors for declarations). Surplus feed-in is credited as grid substitution potential (Module D2) using the same operational factors. This means that for declaration GHG and PE accounting purposes, feed in and self-consumption are equivalent. For research and future iterations of declarations, this can be detailed by setting the feed-in grid substitution factor to an appropriate/time-dependent fraction of the grid import factor.

Embodied impacts of PV systems are not counted at project level, as they are allocated to the national energy sector that underlies both the budget derivation and the operational factors used here (which exclude power-plant embodied impacts). This avoids double counting and maintains consistency with sectoral budgets. Crucially, it defuses the paradoxical tension of PEDs being penalized in the GHG balance for their role as energy providers: in our framework, EE from RES is already allocated and budgeted for in the energy sector, giving a clear energy and GHG incentive to include as much RES as is viable.

Waste heat imports are accounted for with zero direct emission intensity, since the associated emissions are allocated to the primary process generating the waste heat. District cooling supplied from a central city network is treated analogously to other imported carriers by applying the emission intensities provided by the utility or national statistics. Where district cooling is produced within the assessed district itself, the corresponding emissions are calculated based on the energy carriers used for its generation, most commonly electricity. See Figure 4 for a schematic overview of the energy flows, including the treatment of waste heat imports and district cooling emissions.

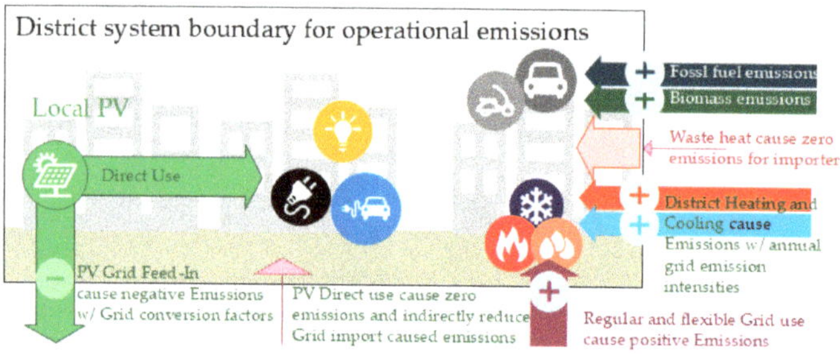

Figure 4. Schematic overview of emissions caused by operational energy flows in the climate-neutral PED framework, illustrating the treatment of local PV direct use and grid feed in, waste heat imports, district cooling emissions, and other energy carriers.

2.4.2. Mobility (Module B8)

Operational mobility emissions (Modules B8.1 for ICE and B8.2 for EV) are accounted for only for EMIM trips whose destination lies within the district. Public transport and active modes (walk/bike/other) are not directly modeled—either immaterial at district scale or outside district control and therefore allocated to the transport sector for GHG budgeting (as per [24]). However, district mobility measures increasing active or public modes of transport can decrease EMIM mileage, and thus, shifts in modal split are also considered under this framework. Trips whose destination lies outside the district are modeled, but considered the responsibility of the target location, omitting it in the district balance to avoid double counting. For EVs we cannot perfectly map charging emissions to discharge purpose; therefore we (i) compute emissions from all EV charging metered inside and outside the district using time-resolved grid factors, (ii) deduct a grid-equivalent credit for trips whose destination is outside the district (treated as an avoided-emission transfer to the destination system), and (iii) treat vehicle-to-building (V2B) as battery operation to the district with its charging emissions accounted for. This yields a net, destination-consistent EV operational footprint for the district, ensures fair effort towards sharing EMIM emissions between all buildings and districts, while retaining physical model accuracy and research trajectories regarding charging infrastructure, where actual expected loads are critical. For ICE vehicles, B8.1 is computed from destination-based trip mileage and carrier-specific $g_{CO_2e} \cdot km^{-1}$ factors. Table 4 shows the resulting inclusion matrix.

Table 4. B8 Mobility emission inclusion matrix.

B8 Emission Cause/Use	Accounted	Modeled	Emission Balance Impact	Treatment/Rationale
Public Transport	✗	✗	None	Allocated to transport sector; outside district control.
Pedestrian, Bike, Other	✗	✗	None	Negligible at district scale. (Included if not)
EMIM **ICE** Trips to District	✓	✓	+	Included.
EMIM **ICE** Trips elsewhere	✗	✗	None	Not included here; attributed to destination system.
EMIM EV On-site Charging	(✓)	✓	+	Included using hourly source emission factors (grid, local PV = 0)
EMIM EV Off-site Charging	(✓)	✓	+	Included using hourly grid emission factors.
→ Trips to District	✓	✓	(+)	Implicitly included
→ Trips elsewhere	✗	✓	−	Deduct grid-equivalent credit to remove out-of-district travel from EV charging OE
→ V2B discharge	✓	✓	(+)	Like BESS: Charging included, but lowering building OE at discharge

Notes: "✓": Accounted/Modelled, "✗ ": Not accounted/modelled, "Emission balance impact": "+" adds emissions; "−" is a credit.

Embodied vehicle impacts are included once at purchase using static average factors (ICE $\approx$ 5 t_{CO_2e} per car; EV $\approx$ 9 t_{CO_2e} per car) without replacements or manufacturing decarbonization; this is a conservative snapshot and should be revisited in future work. Sensitivities can vary lifetimes and embodied factors, which should be further discussed; however, the mobility case study (Section 3.4) indicates that EMIM EE only plays a secondary role in district compliance.

2.4.3. Maintenance and End-of-Life: Post-2050 Assumptions (Modules B2–B5; C1–C4; Vehicles)

Building Maintenance (B2–B5) and End-of-life processes (C1–C4) are included. Since all processes taking place after 2050 are assumed to be carbon-neutral, in practice only those with a lifetime of <25 years need to be considered (or preferably avoided). The same convention applies to vehicles.

2.4.4. Public Infrastructure and Contextual Assets

For on-plot infrastructure within BB6 (e.g., internal access roads, on-plot sewers, bike shelters), impacts are included. Off-plot public infrastructure (municipal roads, trunk sewers, transit systems) is outside project control and not included in the core balance; where data are available, such items may be reported contextually but are not part of compliance checks. This keeps project-level accounting aligned with municipal/national responsibilities.

2.5. Treatment of Biogenic Carbon and Carbon Storage in Building Materials

The framework offers two admissible EE accounting options for declarations:

1. An optional but preferred dynamic LCA method, for time-explicit treatment where results materially depend on biogenic storage and timing effects (e.g., rotation periods, storage duration, end-of-life timing);
2. A default simplified LCA method provided for practicality and comparability, where biogenic carbon contained in bio-based building materials is partly credited as negative emissions (carbon storage) over the assessment horizon, based on carbon origin and use but independent of emission time.

The framework tooling accommodates both simplified and dynamic biogenic carbon methods, with the latter recommended where data availability allows.

2.5.1. Dynamic LCA Method

The dynamic method evaluates the climate impact of emissions and removals as a function of when they occur over the assessment horizon (here: 2025–2075). In contrast to static accounting, time-dependent characterization captures that (i) early emissions induce higher radiative forcing (RF) and contribute more to peak warming, whereas (ii) delayed emissions exert a smaller effect. Carbon storage in bio-based building materials is therefore credited according to its duration and timing [44].

Methodologically, dynamic LCA proceeds as follows. First, for each GHG, the atmospheric decay of a pulse emission is represented via an impulse response function (IRF): for CO_2 with the Bern carbon cycle model; for CH_4 and N_2O with first-order exponential decay. Second, instantaneous dynamic characterization factors (DCF_{inst}) are obtained by integrating the gas-specific radiative efficiency over the decay curve from the time of emission/removal to any time t. Third, cumulative warming impact is computed by summing the DCF_{inst} contributions over all time steps and flows, and results are expressed as dynamic CO_2 equivalents by normalizing to the absolute GWP of a 1 kg CO_2 pulse over the same horizon. This yields time-differentiated CO_2e values for both positive emissions and negative fluxes (e.g., biogenic uptake/storage) [45,46].

For biogenic carbon, two regrowth timing conventions must be declared transparently because they affect results: (a) "growth before harvest" (uptake occurs prior to material use) or (b) "regrowth after harvest" (uptake occurs during/after the building life). The latter typically yields smaller near-term credits due to delayed removals. Fast-rotation materials (e.g., straw, hemp) provide stronger near-term mitigation than slow-rotation timber; dynamic LCA makes these differences explicit. Compared with static "0/0" or "−1/+1" treatments, dynamic accounting reduces bias from ignoring timing and avoids net-negative artifacts when only early-stage credits are counted. The literature consistently finds the dynamic approach more robust and transparent for bio-based construction [44,47].

2.5.2. Simplified LCA Method

The simplified method accounts for biogenic carbon storage based on the origin of the material and its expected service life without differentiating the timing of emissions. All emissions are considered with equal weighting, regardless of whether they occur in the near or distant future. The following rules apply as described in [48], with mass-weighted averages for composite materials:

- **Fast-growing plant-based materials** (e.g., straw, hemp, flax; rotation period ~1 year): → 100% of the biogenic CO_2 content is credited.
- **Wood from managed forests** (typical rotation period ~100 years): Only the proportion representing additional sequestration beyond a no-harvest scenario is credited. This comprises: 25% of the harvested wood volume from calamity events (e.g., bark beetle infestation, storm damage), credited at 100% plus 30% of biomass growth from sustainable management practices. → This results in a total credit of 55% of the biogenic CO_2 content for standard harvested wood.
- **Calamity wood**: → 100% credit for the biogenic CO_2 content. Project declaration requires origin verification from pest or storm damage.

In all cases, the maximum biogenic CO_2 storage credit is limited to 100% of the Global Warming Potential (GWP-biogenic) associated with the material. This approach implicitly incorporates two climate-relevant effects:

1. Carbon uptake following harvest due to replanting in sustainably managed forests, compared to unmanaged forest growth.
2. Temporary sequestration of biogenic carbon, which reduces the peak temperature increase when paired with rapid and consistent GHG emission reductions.

2.6. *Operational Conversion Factors/Dynamic Emission Factors*

Operational conversion factors follow time-dependent trajectories per carrier. For electricity and renewable district heating, factors decline to zero by 2050, consistent with national pathways; fossil carriers (natural gas, oil, coal) remain relatively constant until phase-out in 2050; biomass follows the national profile defined in the scenario set; district cooling factors are included analogously and mapped to the respective supply mix. The framework uses hourly factors for analysis and model comparisons, and monthly building-code factors (linearly down-scaled to 2050) for declarations to maximize practical compatibility.

Data sources and projections:

- Hourly CO_2 intensity profiles for electricity are derived from Austrian grid mix data and adjusted to annual benchmarks from the Austrian National Energy and Climate Plan for 2030 and 2040.
- A linear trajectory is applied from current values to a CO_2 intensity of 0 $g \cdot kWh^{-1}$ by 2050, consistent with the 100% renewable scenario [49] adapted in [50], including seasonal balancing via methanation of summer electricity surpluses.
- For biomass and district heating from renewable sources, a separate decarbonization pathway is applied based on sectoral transition strategies, reflecting gradual replacement of fossil inputs with renewable heat sources.
- For natural gas and other fossil fuels, virtually no decarbonization is assumed; in line with their inherent combustion emissions, CO_2 intensities recede by 5% until 2050 and must thereafter be completely phased out or covered by CCS.

Use of hourly vs. monthly factors:

- For analysis and research purposes, the framework uses hourly conversion factors. Multiple factors sets from literature and modeling studies have been created and compared, enabling investigation of the timing effects of energy demand and generation. Section 3.3 presents a case study on these effects.
- For practical usability, regulatory consistency, and comparability with the Austrian building code, the official application of this proposed climate neutrality assessment framework for PEDs uses monthly CO_2 conversion factors to normalize hourly values to compliant annual levels (Appendix A Table A3). These are linearly downscaled from 2023 to 2050 according to the decarbonization path for each energy carrier (Section 2.5.1).

Application in operational assessment (Module B6/D2):

Electricity consumption: Hourly or monthly CO_2 intensities are multiplied with consumption values to determine operational emissions.

On-site PV generation: Electricity consumed on-site replaces grid electricity at the corresponding CO_2 intensity.

PV surplus exports: PV electricity exported to the grid is credited with the same CO_2 intensity, representing its grid substitution potential.

Other on-site renewable heat generation (e.g., solar thermal, biomass): Reductions are credited based on the carbon capture potential of the plant-based materials used, such as the production of biochar, rather than simply accounting for the avoided CO_2 emissions from displaced heat sources (e.g., district heating or fossil boilers). This approach emphasizes the importance of incorporating carbon sequestration into the lifecycle carbon balance. By producing biochar, carbon from biomass is locked away in a stable form, thereby removing it from the atmosphere and contributing to long-term carbon storage. This method is vital for aligning the assessment with the broader goals of carbon neutrality and addressing the full carbon cycle, including both emission reductions and carbon sequestration. Additionally, this approach avoids the common pitfall of solely relying on emission avoidance through reduced fossil fuel use, which does not fully capture the potential of biomass as a tool for climate mitigation.

Treatment of embodied emissions of energy system: Dynamic conversion factors in this framework only include operational emissions from energy generation and supply. Embodied emissions from the construction and maintenance of generation assets (e.g., PV systems, district heating infrastructure) are excluded, as these are already covered within the national sectoral carbon budget for the energy system and deducted when deriving the building sector's carbon budget. In contrast, other methodologies (e.g., IEA lifecycle factors) include both operational and embodied emissions, which would cause double counting here.

Gradual Emission Intensity Reductions 2025 to 2050

The total operational emissions over a 50-year period are calculated by multiplying the simulated annual operational emissions by the integral of the decarbonization pathway, as shown in Figure 5. The annual interpolation of emission reductions is based on established literature sources: a transition to a 100% renewable electricity system by 2030 [51] and the achievement of a fully climate-neutral energy system by 2050 [52]. After this point, no operational emissions are considered (fossil fuels are assumed to be phased out).

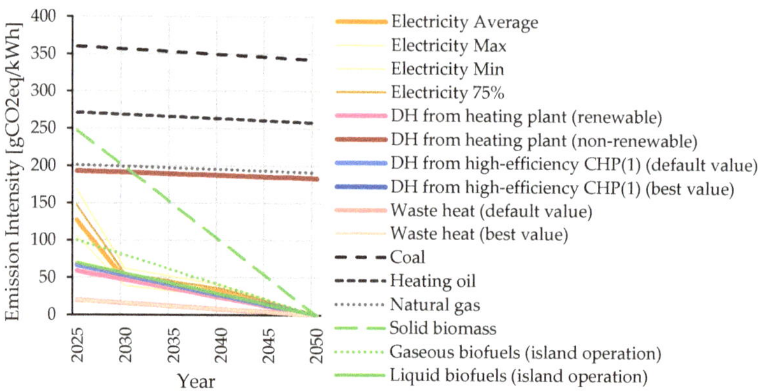

Figure 5. Projected emission intensity reduction paths for GHG balancing 2025–2050. Electricity corridors reflect differences in historic weather and production mix baselines (2016–2022).

2.7. Example Assessments

2.7.1. Selection of Use Cases

Six representative Positive Energy District (PED) projects were chosen to demonstrate the applicability of the proposed climate neutrality PED assessment framework. The selection was guided by three main criteria. First, all projects are located in Austria or in comparable climatic regions, ensuring consistent boundary conditions for energy and emission analyses. Second, sufficient data availability was a prerequisite, covering both life cycle and operational aspects. Third, the cases were selected to reflect a broad diversity of conditions relevant for PED development:

This diversity is expressed in several dimensions: building typologies range from renovation projects to new constructions, from low, rural to high, metropolitan density, and from small-scale interventions to large district developments. Different usage mixes are represented, as well as various energy system configurations, including air and ground source heat pumps and district heating, with and without cooling, and with different ratios of mechanical ventilation and heat recovery. The projects also employ contrasting building materials, spanning from conventional reinforced concrete with EPS insulation to hybrid solutions, wood–concrete compounds, and timber structures with ecological insulation materials such as hemp or straw. Finally, mobility-related conditions differ significantly across the cases, reflecting settings from metropolitan to rural locations, with corresponding variations in annual mileage and electric vehicle penetration levels ranging from 10% to full adoption.

Each of the six investigated projects was characterized by a set of key input parameters relevant for energy and emission balances, as shown in Table 5. These include the gross floor area (GFA), use type (residential, office, education, retail), the district density in terms of floor space index (FSI), and the share of renovation versus new construction and prevalent construction type (conventional, ecological). For energy demand calculations, climate datasets (TRY datasets) were assigned to each location, and hourly outdoor air temperatures and irradiation were considered. In addition, parameters for user electricity, lighting, domestic hot water, heating, cooling, and ventilation demand were included. Table 3 provides an overview of the six projects and their key parameters. Note the specific PV yield's inverse proportionality to the district density, which complicates positive energy balances for higher densities. Operationalization for this effect in the form of target amending context factors is described in [24].

Table 5. Example district baseline parameters.

Project	D1	D2	D3	D4	D5	D6	Unit
Location	Innsbruck	Vienna	Salzburg	Vienna	Vienna	Vienna	
Gross Floor Area (GFA)	5886	13,069	31,799	9761	180,492	33,506	m^2
Net to Gross Floor Area	80	80	80	80	80	80	%
Construction Type	Conv	Eco	Eco	Eco	Conv	Conv	
Refurbishment Share	0	18	0	100	100	0	%
Buildings	5	2	16	2	71	3	
District Plot area	7341	14,095	28,000	4662	48,131	7323	m^2
Floor Space Index (FSI)	0.80	0.93	1.14	2.09	3.75	4.58	–
Context Factor Density [24]	−11.3	−3.7	5.6	26.2	37.9	40.7	$kWh_{PE}·m^{-2}{}_{NFA}·a^{-1}$
Context Factor Mobility [24]	14.1	12.9	14.9	14.1	16.7	25.6	$kWh_{PE}·m^{-2}{}_{NFA}·a^{-1}$
Context Factor Renovation [24]	0	2.2	0	15.0	15.0	0	$kWh_{PE}·m^{-2}{}_{NFA}·a^{-1}$
Residential Space Use	100		96	100	91	52	%
Office/Commercial Space Use			3		9	38	%
Education Space Use		100	1			2	%
Retail Space Use						8	%
Heating System	AS HP	GS HP	DH	GS HP	DH	GS HP	
Cooling System	None	GS HP	None	GS HP	None	GS HP *	
DHW System	AS HP	GS HP	DH	DH	DH	GS HP	
Window Ventilation	0100	0100	0100	0100	90	0100	%
Mechanical Ventilation					10		%
PV installed capacity	196	345	547	91.1	127	560	kW_p
PV specific yield	39.8	32.7	20.0	10.7	0.9	19.5	$kWh·m^{-2}{}_{NFA}·a^{-1}$
EV Share	70	70	70	70	70	70	%
EMIM Reduction	27%	20%			10%	10%	
Flexible TABS	+3K heat	+1K heat	None	+3K heat −2K cool	±2K	+3K heat −2K cool	

* In office and commercial spaces.

2.7.2. Data Collection and Assumptions

For each district, a consistent dataset from project declaration and assessment reports was compiled. To ensure transparency and reproducibility, a detailed catalog of all input fields of the framework (over 250 parameters, including units and default assumptions) is provided in the online data repository [41]. A summary of model categories and parameters can be found in Appendix A Table A4. The data covers building characteristics, energy systems, mobility integration, life cycle inventories, and operational performance and were either taken from project documentation, energy performance certificates, and simulation models or derived from the Austrian LCA database *baubook.at*.

3. Results

3.1. Comparative Analysis

This section compares six baseline districts against the compliance framework using consistent project descriptors and harmonized KPIs. Lifecycle performance is summarized in Table 6 and Figures 6–8: cumulative GHG per m^2_{GFA} is decomposed into A1–A3 (construction incl. TES + EMIM embodied), B-ops (time-resolved electricity and thermal operation), EMIM operation, and PV substitution credits (D2); we also report the primary energy (PE) balance. Compliance bands are shown at L3 = 320, L2 = 196, L1 = 72 $kg_{CO_2e} \cdot m^{-2}_{GFA}$.

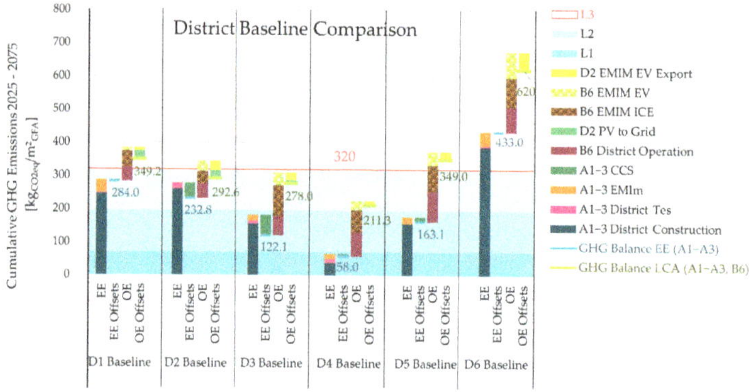

Figure 6. Baseline life cycle GHG balance by district. Cumulative emissions 2025–2075 per m^2_{GFA}, decomposed into A1–A3 (building, TES, and EMIM embodied), B-ops (building/TES with time-resolved grid factors), EMIM operation, and PV substitution credits (D2); CCS shown where applicable. Horizontal bands denote compliance thresholds (L3 = 320, L2 = 196, L1 = 72 $kg_{CO_2e} \cdot m^{-2}_{GFA}$).

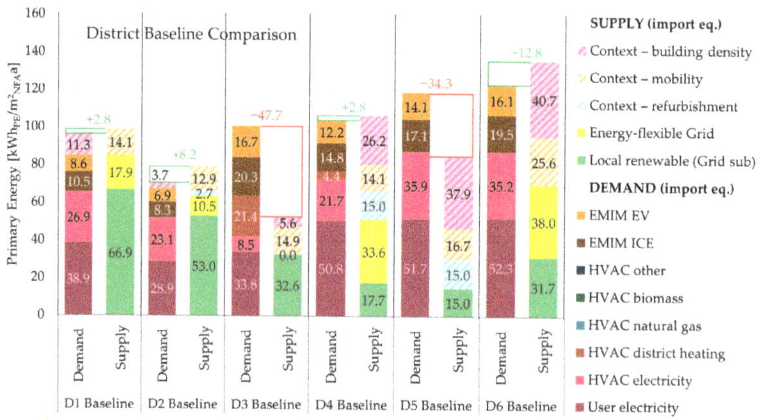

Figure 7. Annual primary energy balance by district. Positive energy criterium is met if supply (PV self-use and export as grid substitution, energy-flexible operation, contextual factors) is greater or equal (=green box) to the demand (import equivalent HVAC heat, electricity, mobility), otherwise (=red box) positive energy criterium is not met. Balancing definition and method is based on [24].

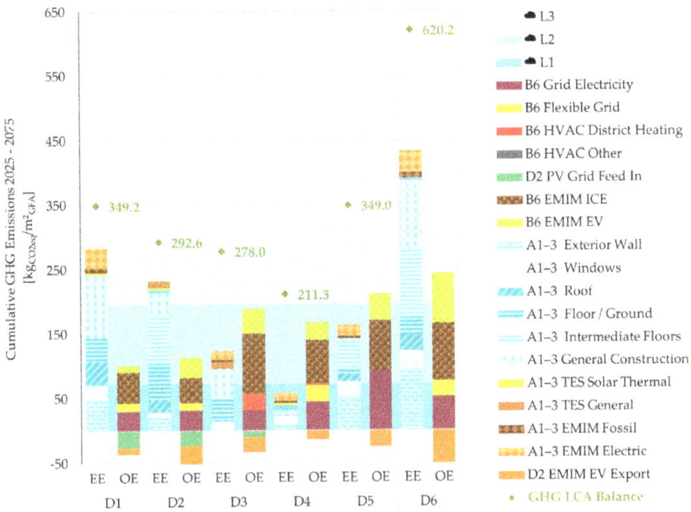

Figure 8. Cumulative GHG balance (2025–2075) with full LCA breakdown. For each district (D1–D6), the left bar (EE) shows embodied impacts (A1–A3, B2–B5) from construction and technical systems (TES, PV, batteries, ground probes) and embodied mobility. The right bar (OE) shows operational impacts (B6) from grid electricity and HVAC carriers (district heating, natural gas, biomass, other) plus EMIM (ICE/EV, B8.1-2) operation and flexible-grid effects; negative segments reflect avoided-grid D2 substitution (e.g., PV feed-in, EV other travel exports). Green diamonds indicate each district's net LCA total, while shaded bands mark according compliance thresholds L3 = 320, L2 = 196, L1 = 72 $kg_{CO_2e}\cdot m^{-2}{}_{GFA}$. Positive stacks add emissions; negative stacks reduce the balance.

Table 6. Baseline district scenarios comparative compliance (✓= yes/✗ = no) summary.

District	Heat Demand $kWh\cdot m^{-2}{}_{NFA}\cdot a^{-1}$	GHG LCA $kg_{CO_2e}\cdot m^{-2}{}_{GFA}$	Δ vs. L3 * $kg_{CO_2e}\cdot m^{-2}{}_{GFA}$	<L3	PE Balance ** $kWh_{PE}\cdot m^{-2}{}_{NFA}\cdot a^{-1}$	>0	CN PED ***
D1	16.0	349.2	+29.2	✗	2.8	✓	✗
D2	21.3	292.6	−27.4	✓	8.2	✓	✓
D3	25.6	278.0	−42.0	✓	−47.7	✗	✗
D4	33.3	211.3	−108.7	✓	2.8	✓	✓
D5	37.9	349.0	+29.0	✗	−34.3	✗	✗
D6	13.8	620.2	+300.2	✗	12.8	✓	✗

* Δ vs. L3 = GHG LCA − 320; positive (red) = overshoot; negative (green) = undershoot/compliance. **, ***: ✓/green = compliance, ✗ /red = non-compliance.

Results are heterogeneous across use types and system choices. Relative to the L3 budget, the modeled GHG balances show: D1 +29.2, D2 −27.4, D3 −42.0, D4 −108.7, D5 +29.0, D6 +300.2 $kg_{CO_2e}\cdot m^{-2}$ (overshoot positive; undershoot negative). PE balance passes in D1, D2, D4, and D6 but not in D3 or D5. District-level drivers include the following:

D1 (greenfield, conventional): Non-compliant at L3 (+29.2). High A1–A3 intensity dominates; EMIM measures reduce operational mobility loads but cannot offset embodied peaks.

D2 (school campus): Benefits from the EMIM context (only commuting staff; high PV self-use + flexible charging), hybrid-eco materials achieving L3 compliance (−27.4) and a positive PE balance.

D3 (urban housing, DH): Meets L3 (−42.0) but fails PE (−47.7 $kWh_{PE}\cdot m^{-2}\cdot a$). Lack of supply side flexibility on DH limits net-positive operation despite acceptable GHG.

D4 (ecologic renovation + TABS/GSHP + mobility measures): Strongest performer across both metrics (−108.7; positive PE). Renovation and density factors reduce embodied intensity; mobility context halves residual EMIM; TABS/GSHP enable flexible, low-carbon operation.

D5 (conventional renovation, gas→DH): Switch to DH alone is insufficient for L3 (+29.0) and fails PE; Section 3.5 explores compliant variants (non-fossil TES + eco materials + flexibility).

D6 (low heat demand, high embodied/EMIM): Despite a positive PE balance, very high embodied and/or mobility-related loads lead to substantial L3 overshoot (+300.2).

Effects of PV TES Embodied Emission (Non-)Inclusion

As argued in Section 2.4.1, embodied emissions from PV systems are not part of this PED system boundary but instead are nationally budgeted in the energy sector. Table 7 shows the omitted PV EE in contrast to the remaining balance: especially for low-density districts that require a significantly positive PE balance and PV export to offset their negative density context factor, and for refurbishment projects with inherently low EE, PV EE can increase overall LCA emissions by more than 10% (assuming 1.16 $t_{CO_2e}\cdot kW_p{}^{-1}$ incl. inverters and cabling).

Table 7. Embodied emissions of example district PV TES EE in total and relative to the remaining EE and total LCA emissions.

Scenario	PV Yield $kWh\cdot m^{-2}a^{-1}$	EE A1–A3 w/o PV $kgCO_2e\cdot m^{-2}{}_{GFA}$	EE A1-3 TES PV $kgCO_2e\cdot m^{-2}{}_{GFA}$	ΔEE [%]	ΔLCA [%]
D1	39.8	284.0	38.6	+13.6	+11.0
D2	32.7	232.8	30.6	+13.1	+10.5
D3	20.0	122.1	19.9	+16.3	+7.1
D4	10.7	58.0	10.8	+18.7	+5.1
D5	0.9	163.1	8.1	+5.0	+2.3
D6	19.5	433.0	19.4	+4.3	+3.1

3.2. Case Study District 3: Redevelopment of Urban Medium Density Residential District: Interplay of Embodied and Operational Emission Reduction Measures

The case study evaluates multiple redevelopment scenarios for an urban medium-density residential district (District 3). Variants differ in construction materials (fossil-intensive vs. ecological), PV system sizing (1×, 2.6×, 2×, 1.7×), mobility assumptions (70% EV, 100% EV, reduced traffic), and heating/DHW supply (district heating vs. ground-source heat pump). The main assessment criteria are the cumulative GHG balance (2025–2075) and the primary energy balance for PED compliance.

The results in Table 8, Figures 9 and 10 show that all scenarios with ecological construction remain within the L3 compliance limit, while only the fossil construction variants exceed it. The combination of 2× PV oversizing with mobility measures (full electrification and reduced traffic) performs particularly well, even reaching the precautionary compliance limit for cumulative GHG emissions. This underlines the necessity of combining renewable supply expansion with systemic demand-side measures.

Table 8. Summary of case study results for District 3—urban medium-density residential redevelopment, reporting GHG LCA emissions ($kg_{CO_2e}/m^2{}_{GFA}$), L3 and L2 compliance, primary energy balance and zero PE-Balance and Climate Neutral PED compliance (✓/green = yes, ✗ /red = no).

Scenario	Construction	PV Size	Mobility	TES	GHG LCA *	<L3	<L2	Primary Energy Balance $kWh_{PE}\cdot m^{-2}{}_{NFA}\cdot a^{-1}$		Climate Neutral PED
Fossil Constr	Fossil	1×	70% EV	DH	378.2	✗	✗	−47.4	✗	✗
Eco Constr	Ecological	1×	70% EV	DH	277.6	✓	✗	−47.4	✗	✗
EC + 2.6 × PV	Ecological	2.6×	70% EV	DH	225.6	✓	✗	+6.9	✓	✓
EC + 2 × PV + Mob	Ecological	2×	90% EV, 80% EMIM	DH	170.1	✓	✓	+5.7	✓	✓
EC + 1.7 × PV GSHP	Ecological	1.7×	70% EV	GS HP	254.7	✓	✗	+0.9	✓	✓

* Unit: $kg_{CO_2e}\cdot m^{-2}{}_{GFA}$.

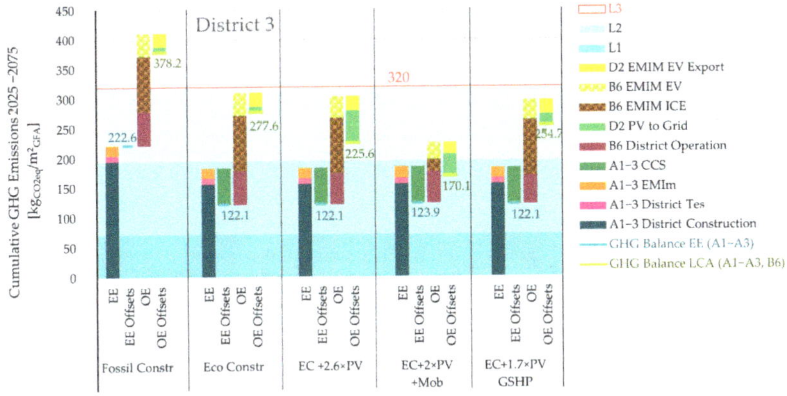

Figure 9. Cumulative GHG emissions (2025–2075) of district scenarios compared to compliance, precautionary, and Paris-compatible limits. Bars show embodied emissions (construction, technical systems, mobility), operational emissions, avoided emissions (grid feed-in), and carbon capture. Results highlight that fossil-based construction exceeds budget limits, whereas scenarios with extended PV and mobility measures approach or meet precautionary compliance thresholds.

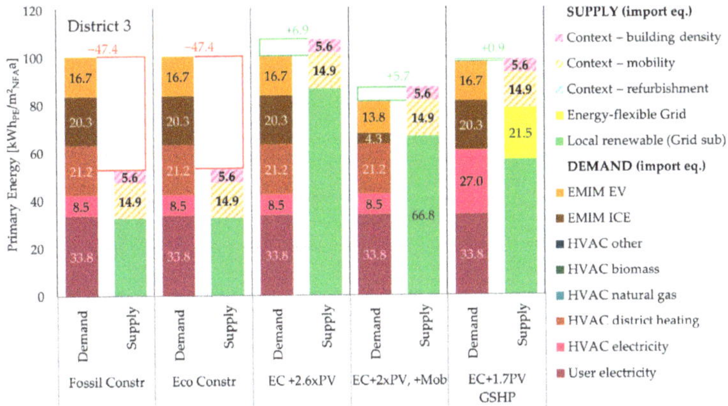

Figure 10. Primary energy balance of district 3 scenarios (demand and supply side contributions as grid equivalents). Stacked bars illustrate energy demand categories (HVAC, electricity, mobility) and supply contributions (PV own consumption and grid feed-in both visualized as grid substitution equivalent, energy-flexible grid demand equivalent, contextual measures). Scenarios compare fossil and eco construction, enhanced PV deployment, additional mobility measures, and renewable heat supply (GSHP, DH). "Positive Energy" criterium is met if supply is at least equal (=green box) to the demand (import equivalent HVAC heat, electricity, mobility), otherwise (=red box) positive energy criterium is not met.

At the same time, the assessment framework requires not only compliance with the finite GHG budget but also a positive primary energy balance. Here, the analysis reveals significant challenges: while PV oversizing improves operational balances, achieving a net-positive balance is difficult in configurations relying on district heating (DH) even when assuming a renewable CHP-based supply with favorable emission factors. In contrast, scenarios with on-site GSHP systems combined with PV come closest to fulfilling both compliance criteria simultaneously, emphasizing the importance of coupling low-carbon construction materials with renewable-based, highly efficient supply systems.

3.3. Case Study District 4: Effect of Dynamic vs. Static Grid Emission Factors and Flexibility Signals

For District 4 we quantify how the temporal resolution and level of grid-GHG signals affect (i) total operational emissions from B6 electricity and EMIM-EV charging, and (ii) the average effective grid-intensity seen by the project. We compare monthly, building-code style factors (OIB '19/'23, currently in effect) with hourly, live-tracking based factors for the same vintages. We first test flexibility via TABS GS HP heating (+3K) and cooling (−2K), then in combination with building battery (BESS) sized = PV installed capacity (kW_p) and EV bidirectional charging (V2B, down to 50% SoC_{RES}). Loads and controls held constant: Building demand, PV generation, HVAC control set-points (22 °C heating, 25 °C cooling), occupancy, and the EMIM travel demand are fixed across scenarios. Only the GHG emission factors and flexibility signals vary:

GHG Emission factors:

- Monthly and annual factors (2019 M, 2023 M, 2019 A) Twelve constant monthly values per vintage or a constant annual value (see Appendix A Figure A2).
- Hourly factors (2023 H). 8760 time steps based on historic live tracking; annual means match the corresponding OIB vintage 2023 (see Appendix A Figure A2)

Flexibility Signals:

- Wind Peaks (for 2020, above 50% nominal Power)
- GHG Emission rolling averages (RA): signal if current GHG EF(t) is y% below Rolling Average (x hour future window) ($x \in \{+24, +48, +72\}$, $y \in \{5\%, 10\%\}$)

Detailed scenario definitions for case study 4 can be found in Appendix A Table A5. The hourly flexibility signals are visualized in Appendix A Figure A2.

This use case highlights why hourly emission accounting is essential. As shown in Table 9, with monthly factors the import-weighted average EI is higher than the annual grid average (≈241→231 $g_{CO_2e} \cdot kWh^{-1}$ for 2019 and 189→175 for 2023) because districts typically import in winter (carbon-intensive) and export PV in summer (cleaner). Annual bookkeeping would therefore "hide" the winter penalty; capturing it is critical—but without flexible dispatch comes at the cost of higher emissions—arguably as it should be.

Table 9. Operational life cycle Emissions (OE), PE Balance, Imported Electricity (ED), relative differences to the Baseline (and Emission Intensities (AVG = Annual Grid Average, Actual = Actual district import average).

Scenario	OE (B6-D2)	ΔBL %	PE Balance $kWh \cdot m^{-2} \cdot a^{-1}$	ED $kWh \cdot m^{-2} \cdot a^{-1}$	ΔBL %	EI_{Grid} $g_{CO_2e} \cdot kWh^{-1}$	$EI_{district}$ $g_{CO_2e} \cdot kWh^{-1}$	$\Delta EI/EI_{Grid}$ %
[%]19M	175.4	+14.4	+6.3	43.2	0.0	231	241.0	+4.2
19A	167.3	+9.1	+6.6	43.2	0.0	227	224.6	−1.1
23M (Baseline)	153.3	-	+2.8	43.2	-	175	188.9	+7.4
23H 24-5	141.3	−7.8	+22.2	44.2	+2.3	175	162.1	−8.0
23H 48-5	141.2	−7.9	+22.5	44.2	+2.3	175	161.9	−8.1
23H 48-10	139.4	−9.0	+14.4	43.6	+0.9	175	160.6	−9.0
23H 72-10	139.8	−8.8	+14.7	43.6	+1.0	175	161.1	−8.6
23H 48-10 BESS	137.3	−10.4	+19.3	44.2	+2.3	175	154.2	−13.5
23H 48-10 BESS V2G	137.9	−10.0	+25.6	45.4	+5.1	175	151.3	−15.7

Switching to hourly factors and applying rolling-average GHG signals enables flexible loads to target low-EI hours. In our runs the import-weighted EI drops from ~189 $g_{CO_2e} \cdot kWh^{-1}$ (2023 monthly) to ~160 $g_{CO_2e} \cdot kWh^{-1}$ with hourly signaling alone, and to ~150 $g_{CO_2e} \cdot kWh^{-1}$ when BESS and V2B are added. The grid EI for inflexible demand remains comparatively high (≈200 $g_{CO_2e} \cdot kWh^{-1}$) because it is drawn during scarcity events when storage is depleted—another effect only visible with hourly accounting.

The results presented in Table 9 and Figure 11 also show that flexibility generally slightly increases total electricity demand (round-trip and conversion losses; more charging activity), but because a larger share becomes flexible and is shifted to low-EI hours, net GHG emissions fall (our KPI), and the average effective EI of imports/charging (second KPI) improves materially. In short: hourly signals + storage/V2B make compliance harder to "paper over" yet provide a real, quantifiable pathway to lower life-cycle emissions.

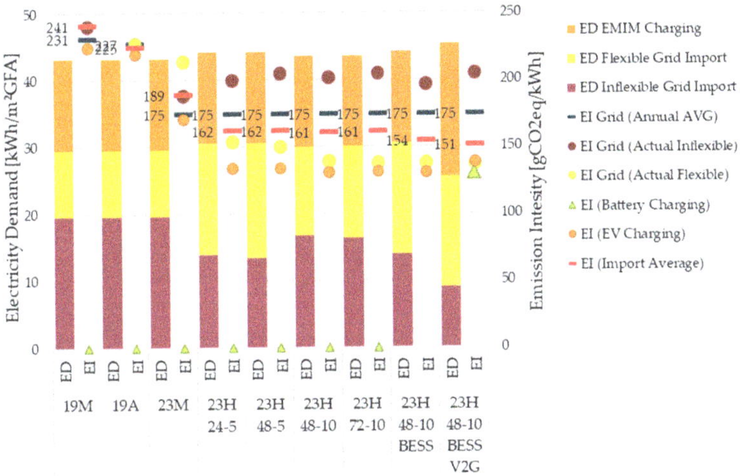

Figure 11. Temporal GHG signals and flexibility for District 4. Stacked bars (left axis) show annual electricity demand split into inflexible grid import, flexible grid import, and EMIM EV charging. Markers (right axis) report effective emission intensities (EI) for imports and charging. Scenarios progress from monthly factors (19M, 19A, 23M) to hourly factors with rolling-average signals (23H 24-5, 48-5, 72-10, 48-10) and three flexibility options (TABS—all, BESS = PV-sized battery; BESS V2G = battery plus EV V2B). The dashed blue line is the annual-average grid EI; filled markers show actual EI of inflexible imports, flexible imports, EV charging, and battery charging; the red bar give the realized import-weighted average EI.

Appendix A Figure A2 shows how energy demand, supplies and flexibility options are dispatched on an hourly basis for Scenario "23H 48-10 BESS V2G".

3.4. Case Study District 5: Mobility and Renovation Sensitivity

The case study illustrates the cumulative GHG emissions of an identical district configuration (70 multistory buildings) across different spatial contexts, with mobility-related emissions added. It shows that EMIM can dominate total emissions, even in not-decarbonized districts such as this gas-heated example. We investigate District 5 with a factorial scenario set spanning four drivers of mobility-related impacts and their interaction with energy supply and envelope choices as shown in Table 10. Figure 12 shows the mileage differentiation of [53] used in this framework as described in [24,54].

Table 10. Case Study District 5 Factor Variations and resulting Scenarios.

Factor	Levels (Examples)	Unit	How Operationalized
Location (EMIM mileage)	34 (Rural-2)/24 (Rural-1)/23/22/12 (Suburban)/11 (Urban)/91 (Metropolitan)	km·cap^{-1}·a^{-1}	Per capita annual motorized mileage profile
EV share	5/30/70/90	% of EMIM	Split km into ICE/EV; apply EI_c, $EF_c(t)$
TES	Gas/DH/GSHP + PV	—	Carrier paths; PV substitution and D2 for feed-in
Renovation	None/Conventional/Ecologic	—	A1–A3 inventories; B6 demand

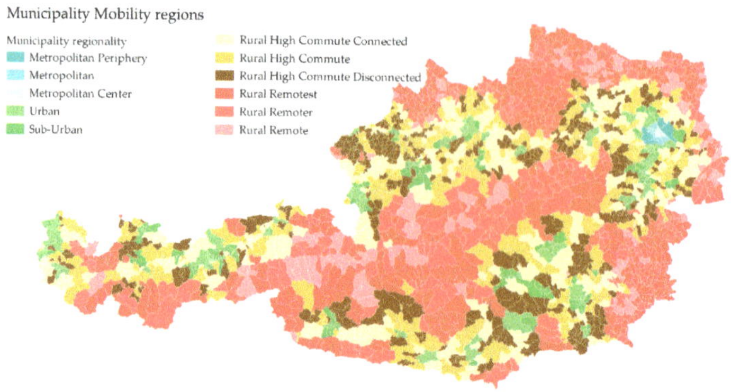

Figure 12. Mobility regions based on [53,55] determining annual EMIM mileage subject to public transport availability and average trip length.

District location and accessibility to public transport strongly influence travel demand, resulting in average annual personal motorized mileage ranging from about 4000 km in metropolitan contexts to 11,831 km in rural areas with high commute and low public transport availability (Region 24). Consequently, cumulative district emissions shown in Figure 13 vary widely—from about 740 $kg_{CO_2e} \cdot m^{-2}{}_{GFA}$ in the fossil-fuel-based, rural commuter case down to 27 $kg_{CO_2e} \cdot m^{-2}{}_{GFA}$ in the fully electrified, metropolitan case with high shares of alternative transport modes. Even at intermediate penetration levels (30–70% EV share), overall district mobility emissions remain dominated by fossil-fuel operation, accounting for roughly 90% of total EMIM-related emissions at 30% EV penetration and still around 60% at 70% EV penetration. These emissions can further be reduced by active mobility measures in the district such as bike and car sharing opportunities to reduce the total annual mileage by EMIM. This shows that L3 compliance (320 $kg_{CO_2e} \cdot m^{-2}{}_{GFA}$ over 2025–2075) is possible in all locations, but complete electrification of vehicles until 2050 (averaging an EV share of 50%+ from 2025 to 2075) is essential. Measures to shift the modal split away from EMIM to public transport, pedestrian and bikes can help also but may not be sufficient by themselves.

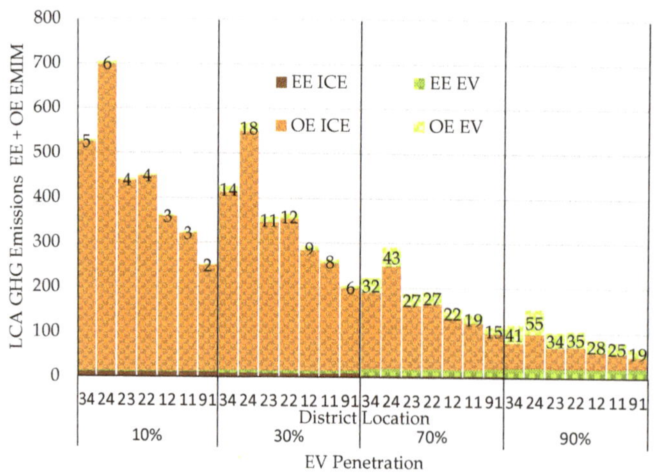

Figure 13. LCA GHG Emissions from EMIM (B8.1-B8.2) in $kg_{CO_2e} \cdot m^{-2}{}_{GFA}$, bars showing embodied emissions (EE) and operational emissions over 50 years (OE) for vehicles with internal combustion engines (ICE) and electric vehicles (EV) in $kg_{CO_2e} \cdot m^{-2}{}_{GFA}$ per district Location (region code in Figure 12) and EV Penetration rate (10–90%) on the x-axis.

3.4.1. Impacts of Refurbishment

Table 11 reports KPIs, and Figures 14 and 15 visualize composition for seven refurbishment options under the metropolitan mobility case with 70% EV: Unrenovated fossil gas (Gas SQ) violates all limits (~753 $kg_{CO_2e}\cdot m^{-2}{}_{GFA}$). Switch to District heating without refurbishment (DH SQ) reaches GHG compliance (~196, L2) but without efficiency measures fails the PED/primary energy target. Conventional renovation is not reliably compliant (DH Conv-Reno ~334; GSHP Conv-Reno ~360), whereas ecologic renovations reduce A1–A3 and are robust (DH Eco-Reno ~246; GSHP Eco-Reno ~245). GSHP Conv-Flex remains within L3 (~293) thanks to higher avoided-grid credits and reduced operation, and—critically—enables a positive primary energy, illustrating that a climate-neutral PED is difficult but feasible when non-fossil TES is combined with flexibility even for an arguable worst case scenario of dense urban areas with little potential for energy efficiency and local renewable energy (Roof PV installations are required on all roofs up to conservative estimates).

Table 11. Sensitivity Analysis of LCA results to TES, renovation and flexibility for District 5. (✓/green = compliance, ✗ /red = non-compliance).

Scenario	TES	Renovation	GHG LCA $kg_{CO_2e}\cdot m^{-2}{}_{GFA}$	L3 Compliance	PE Balance $kWh_{PE}\cdot m^{-2}{}_{GFA}a^{-1}$	PED Compliance
Gas SQ	Gas	None	752.7	✗	−188.3	✗
DH SQ	DH	None	196.0	✓	−64.4	✗
DH Conv (Baseline)	DH	Conventional	349.0	✗	−28.5	✗
GS HP Conv	GS HP	Conventional	359.7	✗	−23.7	✗
DH Eco-Reno	DH	Ecological	245.7	✓	−28.5	✗
GS HP Eco-Reno	GS HP	Ecological	244.9	✓	−23.7	✗
GS HP Conv Flex	GS HP	Conventional	292.9	✓	+0.4	✓

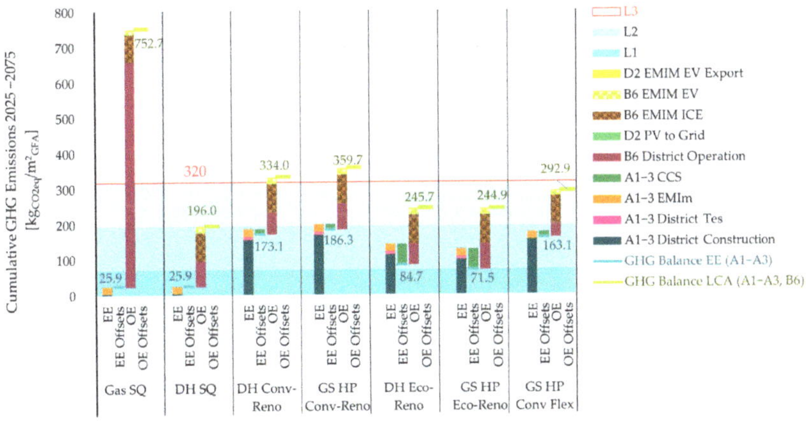

Figure 14. Cumulative GHG LCA emissions (2025–2075) per m^2 GFA for seven refurbishment options—status quo (Gas SQ, district heating DH SQ), renovated conventionally (DH Conv-Reno, GSHP Conv-Reno, GSHP Conv-Flex), and ecologically (DH Eco-Reno, GSHP Eco-Reno) under the metropolitan mobility case with 70% EV. Stacks decompose district-operation (B6), EMIM (ICE/EV), TES-related A1–A3, construction A1–A3, avoided grid emissions (D2), and CCS; shaded bands show compliance thresholds (L3 = 320, L2 = 196, L1 = 72 $kg_{CO_2e}\cdot m^{-2}$).

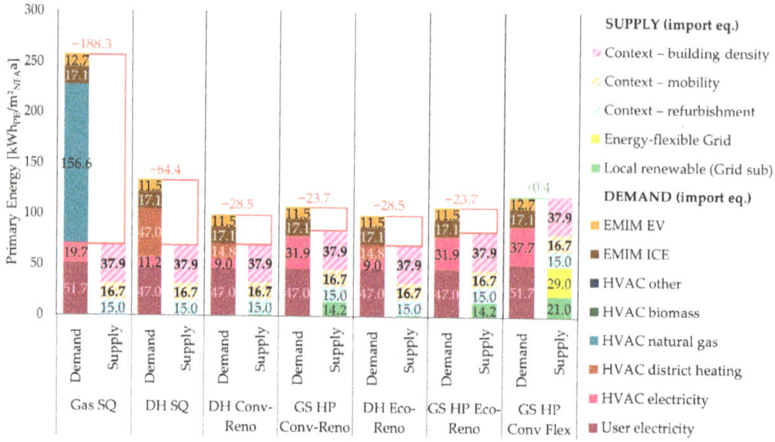

Figure 15. Annual primary energy balance (demand vs. supply) for D5. Demand stacks show user electricity, HVAC electricity/district heating/natural gas/biomass, other HVAC, and passenger-car fuels; supply stacks show local renewable substitution and energy-flexible supply (TABS load shifting) plus contextual credits. "Positive Energy" criterium is met if supply is at least equal (=green box) to the demand (import equivalent HVAC heat, electricity, mobility), otherwise (=red box) positive energy criterium is not met.

3.4.2. Sensitivity to Location and Renovation

Figure 16 plots embodied emissions on the x-axis and operational emissions (including EMIM) on the y-axis for five location types (rural-1, rural-2, suburban, urban, metropolitan). Blue diagonals mark the cumulative-budget limits L3 (320), L2 (196), and L1 (72) $kgCO_2e{\cdot}m^{-2}$GFA to 2050. Marker fill indicates EV share (5–100%), outlines the thermal energy system (blue = gas, red = district heating, yellow = GSHP), and shapes the construction choice (square = no renovation, triangle = code-compliant renovation, circle = ecological renovation).

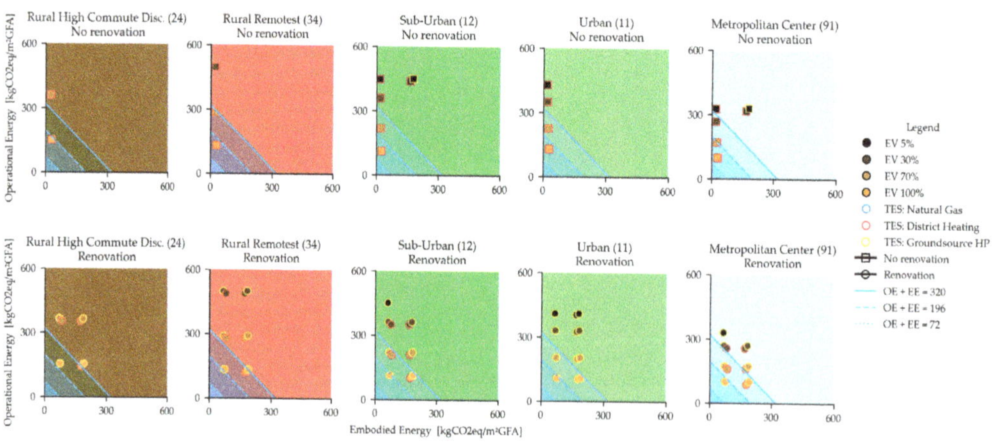

Figure 16. Embodied (x-axis) vs. Operational incl. EMIM (y-axis) emissions by location (background color matching Figure 12) and intervention. Blue lines show cumulative budget thresholds (solid L3, dashed L2, dotted L1). Fill = EV share (5–100%); outline = thermal system (blue gas, red DH, yellow GSHP); shape = construction choice (■ no renovation, • ecological).

In the "no renovation" cases (top row), points cluster far left—reflecting very low embodied emissions—but sit high on the y-axis because the envelope is unchanged. Where district heating is available, several variants fall below L3 on GHG alone; gas heating generally breaches L3, particularly in rural settings where mobility dominates. Raising the EV share pushes points downward across all locations yet seldom achieves L2 or L1 without fabric measures.

With renovation (bottom row), points move right as embodied emissions rise, while y-values drop markedly—most strongly when GS HPs are combined with high EV penetration. In suburban, urban, and metropolitan contexts, many renovated variants approach or meet L2, and a few edge toward L1 when ecological materials are used. Rural districts remain the most challenging, because mobility keeps operational totals elevated.

Overall, if the sole objective were cumulative GHG-budget compliance, a "do-nothing fabric" strategy paired with low-carbon district heating could pass L3. However, it fails PED/primary energy goals: without efficiency upgrades, primary energy demand remains high, and the district cannot achieve a positive PE balance. Meeting both the climate-budget and PED criteria requires renovation (preferably but not necessarily ecological), electrified heat (GSHP or very low-carbon CHP DH), and high EV shares to reduce operational and mobility-related emissions simultaneously.

3.5. Case Study District 6: Differences in Construction Practice and Materials

This case study illustrates the application of the simplified biogenic carbon accounting method to the planned green-field development in District 6 and highlights that construction choices dominate D6's life-cycle GHG outcome (Table 12 and Figure 17):

The baseline structural design of the project is based on conventional building practice, with a reinforced concrete framework, thermal insulation using expanded polystyrene (EPS), and mineral wool in areas requiring enhanced fire resistance. The district benefits from high urban density and mixed-use zoning, supported by excellent public transport accessibility. As a result, the underground parking facility can be designed with a single basement level and a reduced parking ratio, further limiting construction-related GHG emissions. The high building compactness enables comparatively low EE during the construction phase.

Despite these advantages, the total CO_2-equivalent emissions from both construction and operation exceed the allocated climate neutrality budget. Even with relatively low operational EI for building use and EMIM, the project surpasses the allowable threshold by more than 75% over the 50-year assessment period.

Scenario "Eco Skeleton" introduces a more ecologically ambitious system, featuring a reduced structural frame with timber elements and mineral wool insulation. Although this scenario represents an improvement over the traditional concrete structure, the CO_2 emissions remain 30% above the maximum compliance limit due to the use of steel-reinforced concrete in the structural elements.

Table 12. Description of District 6 scenarios and GHG LCA results. (✓/green = compliance, d/yellow = near compliance, ✗/red = non-compliance).

Scenario	EE A1–A3	Δ Baseline	EE Total	GHG LCA	Compliance
Conv-Build (Baseline)	386.5	-	433.0	620.2	✗
Eco-Skeleton	266.7	−31%	313.2	500.3	✗
Mixed-Use CLT	90.1	−77%	136.6	323.8	d (L3, <320)
Max-Eco	−91.0	−124%	−44.5	142.7	✓ (L2, <196)

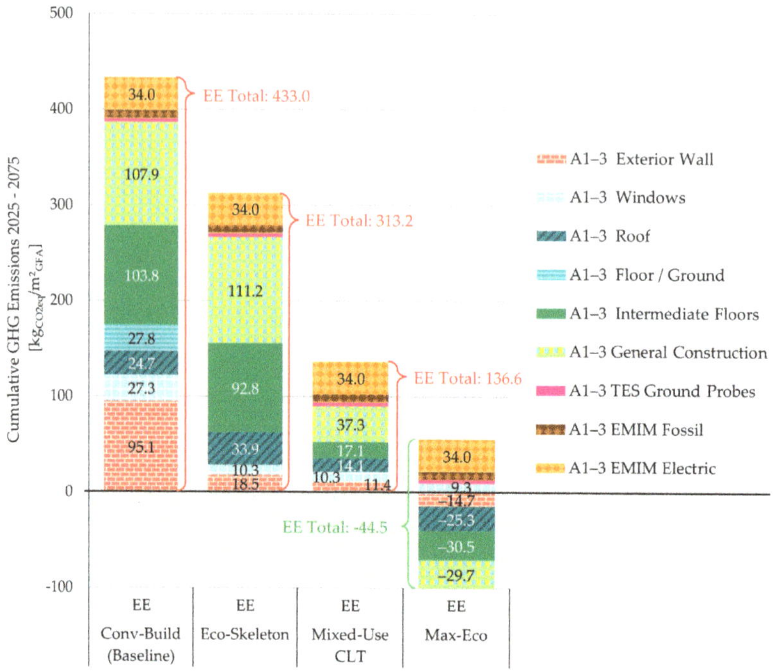

Figure 17. Cumulative A1–A3 GHG emissions by building component for four D6 concepts (EE stacks only). Bars decompose structure, envelope, interior, and TES-related items; negative segments indicate biogenic carbon storage (timber/straw).

Scenario "Mixed-Use CLT" incorporates a combination of concrete and cross-laminated timber (CLT). This mixed approach, including wood–concrete composite decks and CLT for upper floors, leads to a significant reduction in emissions compared to the first two scenarios, while still maintaining structural integrity and near L3 compliance.

Scenario "Max-Eco" represents the most ecologically advanced scenario, pushing the boundaries of climate-neutral construction. It utilizes CLT, straw insulation, wood-fiber insulation, and clay boards, minimizing the use of concrete to only necessary structural components. This results in the lowest emissions, making it the most sustainable option among the four, though it involves compromises, such as relaxing some fire protection regulations. An alternative pathway to compliance would involve further reducing operational and mobility-related emissions. For example:

- Increasing the installed PV Capacity,
- Raising the share of electric vehicles in the mobility mix,
- Implementing more ambitious ecological construction measures (e.g., low-carbon materials).

4. Discussion

The proposed climate neutrality assessment framework for PEDs aims to bridge the gap between research-grade, high-resolution analysis and practical, policy-aligned application. By defining clear system boundaries, integrating dynamic emission factors for multiple energy carriers, and embedding the assessment in a finite CO_2 budget, it offers a robust yet operationally feasible tool for evaluating district-level climate performance. The framework's treatment of local PV—crediting both self-consumption and grid substitution—along with its differentiated decarbonization pathways, supports both design optimization and strategic policymaking. The following subsections discuss the methodological implications, applicability and limitations, and policy relevance of this approach.

4.1. Methodological Implications

Positive Energy Districts are framed here as vehicles for verifiable climate neutrality. The assessment boundary is refined to include operational emissions, EMIM, and selected embodied emissions, while transparently excluding categories allocated to other sectors in the national carbon budget (e.g., PV embodied impacts; end-of-life beyond 2050). This departs from conventional LCA allocations (e.g., attributing the embodied impacts of generation assets to the consuming district) and makes explicit how sectoral budgets are coupled. Many proposed PED assessment frameworks, such as those employed by POCITYF [56] or the SYNIKIA [57] project, use a wide range of KPIs covering many important aspects such as quality of life, comfort, planning processes, waste management [58], and the affordability and accessibility of PED measures. In our proposed declaration framework, in contrast, these aspects are deliberately not included in light of existing well-established national declaration systems covering these aspects in great depth: the klimaaktiv labels "Building" and Districts and Neighborhoods", as well as the TotalQualityBuilding labels all employ methodologically similar points-based approaches, with dozens of assessment aspects across different categories, such as (A) Management, (B) Communication, (C) Urban Development, (D) Building, (E) Supply, (F) Mobility, for the "District and Neighborhoods" declaration [37].

Temporal resolution is central. Hourly emission factors and flexibility (via BESS/V2B/TABS) are shown to significantly reduce operational emissions compared to traditional monthly or annual accounting methods, highlighting the importance of time-resolved emissions in accurate PED assessments. For declarations, monthly factors aligned with Austrian code are used to ensure comparability and administrative simplicity. Carrier-specific decarbonization pathways are differentiated: electricity and district heating follow a declining trajectory to (near) zero by 2050, while gas, biomass, and other fossil carriers retain constant intensities over the horizon. Using static, time-invariant factors would implicitly assume society fails to decarbonize by 2050; the framework instead embeds the policy trajectory, after which the budget is exhausted and historic LCA conventions become less informative.

Local PV is treated on a system basis. Onsite generation offsets grid consumption at the time-varying CO_2 intensity of the avoided imports; exports are credited for grid substitution (D2), recognizing their contribution to system decarbonization. This contrasts with net-zero approaches that ignore export credits, and it materially affects both design optimization and policy interpretation in PED contexts.

Separating measurement effects from control effects is important. Hourly factors expose seasonal import–export asymmetries that annual bookkeeping masks, while dynamic flexibility signals (wind-peak events, GHG rolling averages, price signals, etc.) guide storage, charging, and thermal shifting to lower import-weighted intensities. Declarations remain simple and comparable, while design and operation can be assessed with rigor and temporal fidelity. Crucially, it prevents projects from over-crediting flexibility measures and allows PED compliance to be evaluated transparently against both carbon budgets and energy balance criteria. This clarity improves interpretation but means results are sensitive to the chosen signal and available assets (BESS, TABS, V2B)—an interconnection which would benefit from further research so it can be better designed for. Overall, the approach is auditable and declaration-ready, but data localization and policy-trajectory uncertainty remain the main limits.

A finite-budget perspective tightens alignment with Paris-compatible trajectories. Front-loaded embodied emissions cannot be "repaid" later without crowding out the remaining budget, so low-carbon construction becomes a first-order design variable alongside operational efficiency and mobility measures. Annualized KPIs can conceal this timing problem by spreading construction impacts evenly over service life; the budget lens restores the correct climate signal by tracking accumulation and scarcity over time.

Sensitivity to decarbonization pace and policy timelines. Our baseline assumes electricity and renewable DH factors decline to ~0 $g_{CO_2e}\cdot kWh^{-1}$ by 2050. To reflect slower or partial decarbonization, we examined alternative factor sets: (i) a delayed trajectory achieving near-zero intensities by 2060; and (ii) residual post-2050 intensities (e.g., 30–60 $g_{CO_2e}\cdot kWh^{-1}$ electricity; non-zero DH), with end-of-life and maintenance after 2050 treated accordingly. Under these paths, the import-weighted emission intensity rises, and cumulative totals tighten the available headroom; projects that pass under the baseline may require stronger embodied-carbon reductions (materials, lean design), greater PV and flexibility (BESS/V2B/TABS), or stricter EMIM measures (higher EV shares, demand reduction) to remain below L2/L3. Because the framework is parameter-driven, declarations can be re-issued annually with updated factors and budgets as EU/national trajectories and EPBD/ZEB provisions evolve.

4.2. *Applicability and Limitations*

Applicability. In contrast to other research that found their PED concepts favored certain locations, conditions, and contexts, such as districts with low density [59] or favorable climate conditions [60,61], over others, our framework is applicable to a wide range of PED sizes, usages, and typologies, from small urban districts to large-scale developments, both for renovation and green-field developments in Austrian climates. Its applicability stems from the modular definition of system boundaries and context factors and the flexibility to select conversion factors based on data availability and assessment purpose. In practice, high-resolution operation and mobility inputs are not always available; calibrated defaults allow use at early stages but reduce fidelity.

In contrast to other definitions of PEDs that interpret the positive energy balance without contextual factors [25], the investigated projects show compliance with our assessment targets, even though their energy balance is negative or they have net emissions over the life cycle or both. This indicates their compliance with a net zero emission future, and the strength of the declaration system to map these differences.

Uncertainty and change in underlying assumptions. The calculation of the carbon budget relies on several assumptions about future developments that will need to be revisited as conditions evolve. Key drivers include the total gross floor area, which affects the per-m^2 allocation, and the decarbonization rates of other sectors such as industry. These parameters are outside the direct control of individual projects and require regular updates at national and international levels. A systematic treatment of uncertainties, for example, through Monte Carlo or scenario analysis, lies beyond the scope of this study but should be pursued in future work to provide clearer uncertainty ranges and to link results more closely with established scenario pathways.

In particular, the exclusion of certain embodied emissions, particularly those for PV systems, relies on the integrity of the national budget allocation methodology, which represents a methodological limitation. Future work should investigate and compare other allocation methods.

Regulatory alignment is strong within the Austrian building code, but deviations from international standards, particularly EN 15978, may affect acceptance in cross-border contexts. The focus on a fixed 2050 endpoint aligns with current climate targets but may require revision in the event of accelerated or delayed policy timelines. In fact, current trends indicate that decarbonization might occur slower than mandated, which is why it is crucial for the assessment framework to annually update the remaining carbon budget based on national and international reports. The later sufficient action is taken, the smaller the remaining budget. This periodic update also includes other system developments such as gains in energy efficiency, or potential carbon budget undershoots of realized renovations, which could theoretically lead to a greater carbon budget for future projects if system decarbonization goes better than planned.

Transferability beyond Austria is methodologically possible but requires localizing decarbonization pathways, carrier factors, mobility contexts, and code constraints (e.g., tall-timber fire rules, façade-PV allowances).

The framework embeds a Paris-aligned trajectory to 2050 and assumes (near-)zero-carbon electricity/DH thereafter; if national timelines diverge, factors and budgets must be re-parametrized. Excluding PV embodied impacts at project level relies on coherent sectoral accounting; where this is not in place, a documented alternative (including PV embodied with a compensating sectoral adjustment) may be warranted.

4.3. Policy Implications

Adoption as declaration: The presented framework underpins a national labeling/declaration system in Austria ("klimaaktiv" [37,62]), providing consistent, comparable verification and a practical bridge between research-grade assessment and regulatory adoption. With localized factors, it can harmonize with emerging EU instruments (EPBD revisions, taxonomy), supporting broader uptake while maintaining the rigor needed for credible climate-neutral PED claims. To maintain the claim of "climate neutral districts", the remaining carbon budgets must be routinely monitored and adjusted, preferably on an annual basis.

Adoption in practice: The framework enables municipalities and developers to test whether districts are on Paris-compatible paths by combining two KPIs—a finite cumulative GHG budget and a positive primary energy balance—within transparent boundaries that include user-driven mobility. Moving from annual KPIs to a 50-year cumulative budget refocuses decisions on near-term embodied and operational peaks, closing a key blind spot in conventional PED assessments. Authorities can request framework compliance as part of development agreements and contracts, or set concrete, verifiable targets (e.g., max embodied GHG per m^2, minimum flexible PV/BESS/V2B capacity, DH decarbonization milestones), align procurement and permitting, and offer incentives accordingly. Separating measurement (hourly EF) from control (flexibility signals) also gives a policy lever to reward time-aware operation rather than nominal annual surpluses. This should further be aligned with national adoption plans for Smart Readiness Indicators (SRI).

5. Conclusions

We present a declaration-ready framework for assessing climate-neutral Positive Energy Districts (PEDs) that combines finite CO_2 budgets, explicit system boundaries, and dynamic emission factors. It reconciles research-grade accuracy (hourly CO_2 intensity and differentiated decarbonization paths) with policy usability. Core elements include the following:

- **Two KPIs with targets:** (1) cumulative GHG budget to 2050 and (2) annual primary energy balance for declaring PED status (with/without mobility).
- **LCA scope:** unified accounting of construction, operation, and EMIM (including embodied construction, maintenance, end-of-life) over the next 25 years within a single CO_2 budget.

- **Alignment with national sector targets:** PV embodied impacts remain in the energy sector and are excluded at project level; non-EMIM travel modes (PT, walking, cycling) remain in the transport sector. PV exports are credited with time matching.
- **Time-resolved emissions and pathways:** hourly (or monthly) grid factors; distinct decarbonization trajectories for electricity and renewables; constant intensities for gas/biomass/other fossil carriers (with DH inheriting its supply mix).

Design implications. Case study applications show that even highly efficient PED designs can exceed the available CO_2 budget unless embodied emissions are minimized (through low-carbon/biogenic options, lean structures), renewable energy and flexibility contributions are maximized (through flexible dispatch with BESS/V2B/TABS), and operational emissions from user-driven mobility are effectively reduced (by electrification and demand reduction). The main findings of the case studies indicate that:

- A diverse set of districts, ranging from low-density residential to high-density urban contexts, can achieve compliance with the proposed carbon budget (L3) by adopting low-carbon construction materials, maximizing renewable energy generation (PV oversizing), and electrifying mobility (e.g., EVs).
- Operational mobility emissions (EMIM) remain a significant factor for compliance, particularly in rural areas with lower EV penetration and less access to public transport.
- Hourly emission factors and flexibility (via BESS/V2B/TABS) can significantly reduce operational emissions compared to traditional monthly or annual accounting methods, highlighting the importance of time-resolved emissions in accurate PED assessments.

Limitations. The interpretational claim of "climate neutral" hinges on systemwide decarbonization trajectories and assumptions that remain a source of uncertainty and may require periodic re-adjustment moving forward. Transferability of results depends on localized data. Results obtained in early planning may be subject to change as planning progresses, potentially requiring separate declarations.

Next steps.

- Link to dynamic LCA databases to strengthen embodied-emission estimates.
- Adapt to diverse regulatory contexts and add uncertainty analysis for pathways.
- Automate compliance checks and embed the workflow in planning regulations and municipal decision processes.

Author Contributions: Conceptualization, S.S.; methodology, S.S. and T.Z.; software, S.S. and P.K.; validation, T.Z.; formal analysis, S.S. and T.Z.; investigation, S.S. and R.D.; resources, S.S., T.Z. and R.D.; data curation, S.S., R.D., M.S. and P.K.; writing—original draft preparation, S.S.; writing—review and editing, T.Z., R.D. and J.B.; visualization, S.S.; supervision, J.B. All authors have read and agreed to the published version of the manuscript.

Funding: This research was funded by the City of Vienna MA23, grant number 35-4 "Competence Team for Climate-Fit City Renovation".

Data Availability Statement: Data is available at the GitHub repository [41] under https://github.com/simonschaluppe/peexcel/tree/master/data/publications/2025-Declaration-Ready-Climate-Neutral-PEDs (accessed on 10 October 2025).

Acknowledgments: During the preparation of this manuscript/study, the author(s) used ChatGPT v5 for the purposes of text formulation and review. The authors have reviewed and edited the output and take full responsibility for the content of this publication.

Conflicts of Interest: The authors declare no conflicts of interest.

Abbreviations

The following abbreviations are used in this manuscript:

ACH	Air Changes per Hour
AS HP	Air Source Heat Pump (Air to Water)
BESS	Battery Energy Storage System
CLT	Cross-laminated Timber
COP	Coefficient of Performance
CO_2e	CO_2 Equivalents
DH	District Heating
DHW	Domestic Hot Water
EI	Emission Intensity
EMIM	Everyday motorized individual Mobility
EV	Electric Vehicle
FSI	Floor space index (Gross floor area over plot area)
GHG	Greenhouse Gas
GFA	Gross Floor Area
GS HP	Ground source Heat pump (boreholes)
HVAC	Heating, ventilation, air conditioning and DHW
ICE	Internal combustion engine (vehicle)
LCA	Life Cycle Assessment
NFA	Net Floor Area
PAX	Persons approximate
PEF	Primary Energy Factor
PE tot	Total Primary Energy
PE nre	Non-renewable Primary Energy
PE re	Renewable Primary Energy
PED	Positive Energy District
PV	Photovoltaics/Photovoltaic System
TES	Technical Building Energy Systems
TSD	Time Series Data
V2G	Vehicle to Grid
V2B	Vehicle to Building

Appendix A

Table A1. Input parameters and intermediate calculation values used to derive the floor-area-specific carbon budget limits: global budgets, national allocation, population, floor area statistics. The resulting thresholds are summarized in the main text (Table 2).

Variable	Unit	Timeframe	50%	66%	Source
Paris-compatible CO_2 budget (+1.5 °C, no overshoot)	Mt CO_2	2022–2075	510	280	[27]
Paris-compatible CO_2 budget (+1.5 °C, with overshoot)	Mt CO_2	2022–2075	610	340	[27]
Remaining budget after 2024 (no overshoot)	Mt CO_2	2025–2075	360	130	calc. from [27]
Remaining budget after 2024 (with overshoot)	Mt CO_2	2025–2075	460	190	calc. from [27]
Population (Austria)	million cap	2025	9.0	9.0	[27]
Per capita net budget (no overshoot)	t CO_2 cap^{-1}	2025–2075	40.0	14.5	calc. from [63]
Per capita net budget (with overshoot)	t CO_2 cap^{-1}	2025–2075	51.1	21.1	calc. from [63]
Natural carbon sinks (NCS)	t CO_2 $cap^{-1}a^{-1}$	—	1.0	1.0	[64]
Cumulative NCS added (50 years)	t CO_2 cap^{-1}	2025–2075	50.0	50.0	—
Per capita gross budget (no overshoot + NCS)	t CO_2 cap^{-1}	2025–2075	90.0	64.5	—
Per capita gross budget (with overshoot + NCS)	t CO_2 cap^{-1}	2025–2075	101.1	71.1	—
Buildings + EMIM sector share of per capita budget (A1–A3, B1–B6 + EMIM, C1–C4)	%	—	50	50	[49] based on [26]
Estimated national GFA (all uses)	million m^2	2040	906.7	906.7	[65]
Specific GFA per capita (all uses)	m^2 GFA cap^{-1}	—	101	101	calc from [65]

Table A2. Building LCA balancing boundaries defined in [43].

Boundary	Scope Description
BB1	Complete thermal envelope, including internal floor ceilings
BB3	BB1 plus all interior walls, basement structures, unheated buffer zones; excluding open circulation spaces (stairs, balconies, loggias, galleries)
BB5	BB3 plus shared circulation areas and building services
BB6 (used here)	BB5 plus exterior facilities (carports, bicycle shelters, etc.) and auxiliary buildings

Table A3. Total (PEtot), non renewable (PEnre) and renewable (PEre) primary energy conversion and GHG emission factors for 2024 of selected energy carriers based on [52,66].

Energy Carrier	PEtot	PEnre	PEre	GHG
		$kWh_{PE} \cdot kWh^{-1}{}_{EE}$		$g_{CO_2e} \cdot kWh^{-1}$
Grid Electricity	1.76	0.79	0.97	156
District heating—boiler plant (renewable)	1.72	0.4	1.32	59
District heating—boiler plant (non-renewable)	1.48	1.16	0.32	193
District heating—high-efficiency CHP (default)	0.59	0.41	0.18	67
District heating—high-efficiency CHP (best case)	0.33	–	0.33	20
Waste heat (default)	1.00	–	1.00	20
Waste heat (best case)	0.30	–	0.30	20
Coal	1.46	1.46	0.00	360
Heating oil	1.20	1.20	0.00	271
Natural gas	1.10	1.10	0.00	201
Biomass (solid)	1.13	0.10	1.03	247
Biofuels (gaseous, stand-alone operation)	1.40	0.40	1.00	100
Biofuels (liquid, stand-alone operation)	1.50	0.50	1.00	70
District cooling Vienna [67]	0.42	0.13	0.29	42

Table A4. Assessment model parameter overview.

Category	Description of Data	Parameters
District	Gross floor areas, net to gross ratios, renovation shares per usage	24
Weather	Hourly TSD of ambient temperature and N/E/S/W/H Irradiation	6 × 8760
Heights	Average room heights per usage	8
Building Hulls	Hull areas by type and orientation	8
Building Physics	Thermal transmittances, effective heat capacities, heat bridges, air tightness	10
Windows	Total transmittance, shading offsets summer, winter, mobile shading	32
Ventilation	Per usage: Share of mechanical/window ventilation, night ventilation, ACH, electricity demand, heat recovery rates	44
Usage	Concurrencies, DHW demands, auxiliary and plug loads per usage	35
Lighting	Daylight controls and coefficients, illuminations, light power loads per usage	12
Heating	Heating periods, Heating Powers and COPs of up to four heating systems (2 electric, 2 thermal), setpoint temperatures	33
Cooling	Cooling periods, Cooling Powers and COPs of up to 3 active and 1 passive systems (2 electric, 1 thermal, 1 free cooling), setpoint temperatures	37
DHW	Hot water demand profiles, storage volumes, generation efficiencies, distribution losses	17
PV Systems	Installed PV capacity, module and inverter efficiency, Hourly TSD of PV yield	5 + 8760
BESS	Battery capacity, charging/discharging efficiencies, c-rates, losses	11
Flexibility	Heating, Cooling, DHW, PV, BESS utilization ranges	11
Mobility	Usage intensity, concurrency, motorization, Annual mileage, EV penetration rates, charging infrastructure, minimum SOCs, utilization of bi-directional charging	32
Embodied carbon	Construction materials and components, GHG intensities, life cycles, reference areas	37
PE and Emission factors	Primary energy and emission conversion factors, annual weights, context factors	2 × 8760 13

Table A5. Scenario definition for case study District 4.

Scenario	Grid Emission [66]	Flexibility Signal	Flexibility Option (s)
19M	2019 M	Wind Peaks > 50% P_{nom}	TABS
19A	2019 A	Wind Peaks > 50% P_{nom}	TABS
23M (Baseline)	2023 M	Wind Peaks > 50% P_{nom}	TABS
23H 24-5	2023 H	RA x = 24, y = 5%	TABS
23H 48-5	2023 H	RA x = 48, y = 5%	TABS
23H 48-10	2023 H	RA x = 48, y = 10%	TABS

Table A5. *Cont.*

Scenario	Grid Emission [66]	Flexibility Signal	Flexibility Option (s)
23H 72-10	2023 H	RA x = 72, y = 10%	TABS
23H 48-10 BESS	2023 H	RA x = 48, y = 10%	TABS + BESS
23H 48-10 BESS V2B	2023 H	RA x = 48, y = 10%	TABS + BESS + V2B

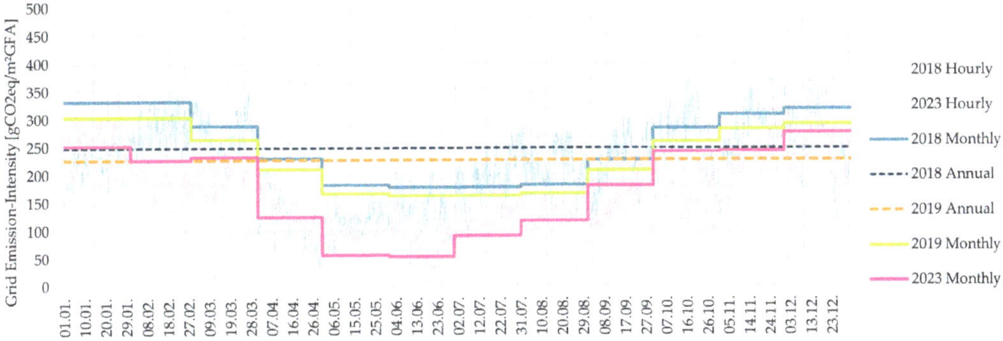

Figure A1. Grid emission intensities of different base years (A—Annual, M—Monthly, H—Hourly).

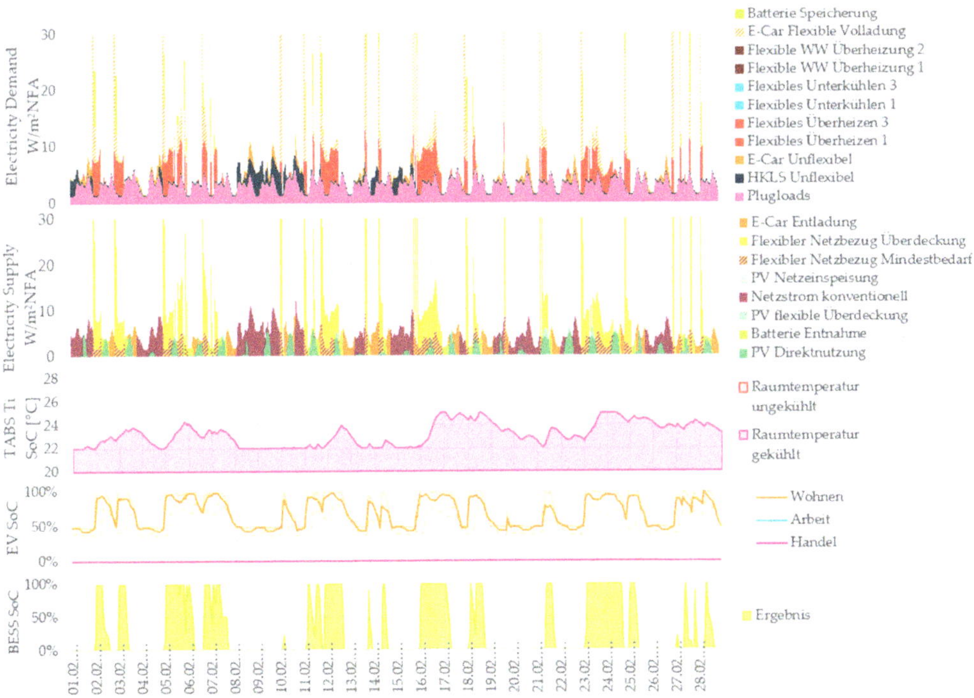

Figure A2. Hourly Flexibility Operation in February Scenario "23H 48-10 BESS V2G" showing electricity demands (Top), electricity supplies (below, note the flexible grid charging in lime and V2B in orange), thermal storage SoCs (TABS indoor Temperature, purple = zones with cooling, red = zones without cooling not present in scenario), EV Battery SoC (dashed = avg offsite SoC, solid = avg SoC, thin solid = avg SoC onsite, orange = residential, purple = retail is overlaying blue = office, both at 0%) and battery energy system SoC (lime green).

References

1. United Nations Framework Convention on Climate Change. Paris Agreement. 2015. Available online: https://unfccc.int/sites/default/files/english_paris_agreement.pdf (accessed on 10 October 2025).
2. Masson-Delmotte, V.; Zhai, P.; Pirani, A.; Connors, S.L.; Péan, C.; Berger, S.; Caud, N.; Chen, Y.; Goldfarb, L.; Gomis, M.I.; et al. (Eds.) *Climate Change 2021: The Physical Science Basis. Contribution of Working Group I to the Sixth Assessment Report of the Intergovernmental Panel on Climate Change*; Cambridge University Press: Cambridge, UK, 2021. Available online: https://www.ipcc.ch/report/ar6/wg1/ (accessed on 10 October 2025).
3. Core Writing Team; Lee, H.; Romero, J.; IPCC (Eds.) *Climate Change 2023: Synthesis Report. Contribution of Working Groups I, II and III to the Sixth Assessment Report of the Intergovernmental Panel on Climate Change*; IPCC: Geneva, Switzerland, 2023. Available online: https://www.ipcc.ch/report/sixth-assessment-report-cycle/ (accessed on 10 October 2025).
4. Pathak, M.; Slade, R.; Shukla, P.R.; Skea, J.; Pichs-Madruga, R.; Ürge-Vorsatz, D. *Climate Change 2022—Mitigation of Climate Change: Working Group III Contribution to the Sixth Assessment Report of the Intergovernmental Panel on Climate Change*; Intergovernmental Panel on Climate Change (IPCC), Ed.; Cambridge University Press: Cambridge, UK, 2023.
5. Rogelj, J.; Shindell, D.; Jiang, K.; Fifita, S.; Forster, P.; Ginzburg, V.; Handa, C.; Kheshgi, H.; Kobayashi, S.; Kriegler, E.; et al. Mitigation Pathways Compatible with 1.5 °C in the Context of Sustainable Development. In *Special Report on Global Warming of 1.5 °C, IPCC*; Cambridge University Press: Cambridge, UK, 2018; pp. 93–174. [CrossRef]
6. Röck, M.; Ruschi Mendes Saade, M.; Balouktsi, M.; Rasmussen, F.N.; Birgisdottir, H.; Frischknecht, R.; Habert, G.; Lützkendorf, T.; Passer, A. Embodied GHG Emissions of Buildings—The Hidden Challenge for Effective Climate Change Mitigation. *Appl. Energy* **2020**, *230*, 115–124. [CrossRef]
7. Brozovsky, J.; Gustavsen, A.; Gaitani, N. Zero Emission Neighbourhoods and Positive Energy Districts—A State-of-the-Art Review. *Sustain. Cities Soc.* **2021**, *72*, 103013. [CrossRef]
8. Knotzer, A.; Geier, S.; Höfler, K.; Venus, D.; Nussmüller, W.; Weiss, T. *IEA SHC Task 40/EBC Annex 52: Internationale Definition von Nullenergiegebäuden*; Bundesministerium für Verkehr, Innovation und Technologie: Wien, Austria, 2014. Available online: https://nachhaltigwirtschaften.at/resources/iea_pdf/endbericht_201417_iea_shc_task40_ebc_annex_52.pdf (accessed on 10 October 2025).
9. Magrini, A.; Lentini, G.; Cuman, S.; Bodrato, A.; Marenco, L. From Nearly Zero Energy Buildings (NZEB) to Positive Energy Buildings (PEB): The next Challenge-The Most Recent European Trends with Some Notes on the Energy Analysis of a Forerunner PEB Example. *Dev. Built Environ.* **2020**, *3*, 100019. [CrossRef]
10. Minelli, F.; Ciriello, I.; Minichiello, F.; D'Agostino, D. From Net Zero Energy Buildings to an Energy Sharing Model—The Role of NZEBs in Renewable Energy Communities. *Renew. Energy* **2024**, *223*, 120110. [CrossRef]
11. Wiik, M.K.; Fufa, S.M.; Andresen, I.; Brattebø, H.; Gustavsen, A. A Norwegian Zero Emission Neighbourhood (ZEN) Definition and a ZEN Key Performance Indicator (KPI) Tool. *IOP Conf. Ser. Earth Environ. Sci.* **2019**, *352*, 012030. [CrossRef]
12. Hedman, Å.; Rehman, H.U.; Gabaldón, A.; Bisello, A.; Albert-Seifried, V.; Zhang, X.; Guarino, F.; Grynning, S.; Eicker, U.; Neumann, H.-M.; et al. IEA EBC Annex83 Positive Energy Districts. *Buildings* **2021**, *11*, 130. [CrossRef]
13. Hinterberger, R.; Gollner, C.; Noll, M.; Meyer, S.; Schwarz, H.G. *JPI Urban Europe/SET Plan Action 3.2. White Paper on PED Reference Framework for Positive Energy Districts and Neighbourhoods*; JPI Urban Europe: Vienna, Austria, 2020. Available online: https://jpi-urbaneurope.eu/app/uploads/2020/04/White-Paper-PED-Framework-Definition-2020323-final.pdf (accessed on 10 October 2025).
14. Shnapp, S.; Paci, D.; Bertoldi, P. *Enabling Positive Energy Districts across Europe: Energy Efficiency Couples Renewable Energy*; Publication Office of the European Union: Luxembourg, 2020. Available online: https://data.europa.eu/doi/10.2760/452028 (accessed on 10 October 2025).
15. European Commission. Directorate General for Research and Innovation. In *100 Climate-Neutral Cities by 2030—By and for the Citizens: Report of the Mission Board for Climate Neutral and Smart Cities*; Publications Office: Luxembourg, 2020. Available online: https://data.europa.eu/doi/10.2777/46063 (accessed on 10 October 2025).
16. Lindholm, O.; Rehman, H.U.; Reda, F. Positioning Positive Energy Districts in European Cities. *Buildings* **2021**, *11*, 19. [CrossRef]
17. Turci, G.; Alpagut, B.; Civiero, P.; Kuzmic, M.; Pagliula, S.; Massa, G.; Albert-Seifried, V.; Seco, O.; Soutullo, S. A Comprehensive PED-Database for Mapping and Comparing Positive Energy Districts Experiences at European Level. *Sustainability* **2022**, *14*, 427. [CrossRef]
18. Guarino, F.; Rincione, R.; Mateu, C.; Teixidó, M.; Cabeza, L.F.; Cellura, M. Renovation Assessment of Building Districts: Case Studies and Implications to the Positive Energy Districts Definition. *Energy Build.* **2023**, *296*, 113414. [CrossRef]
19. Kiviluoma, J.; Pallonetto, F.; Marin, M.; Savolainen, P.T.; Soininen, A.; Vennström, P.; Rinne, E.; Huang, J.; Kouveliotis-Lysikatos, I.; Ihlemann, M.; et al. Spine Toolbox: A Flexible Open-Source Workflow Management System with Scenario and Data Management. *SoftwareX* **2022**, *17*, 100967. [CrossRef]
20. Ala-Juusela, M.; Pozza, C.; Salom, J.; Luque Segura, I.; Tuerk, A.; Lollini, R.; Gaitani, N.; Belleri, A. Workshop on Positive Energy Buildings—Definition. *Environ. Sci. Proc.* **2021**, *11*, 26. [CrossRef]

21. Häkkinen, T.; Kuittinen, M.; Ruuska, A.; Jung, N. Reducing Embodied Carbon during the Design Process of Buildings. *J. Build. Eng.* **2015**, *4*, 1–13. [CrossRef]
22. Chen, X.; Vand, B.; Baldi, S. Challenges and Strategies for Achieving High Energy Efficiency in Building Districts. *Buildings* **2024**, *14*, 1839. [CrossRef]
23. Fawcett, T. Personal Carbon Trading: A Policy Ahead of Its Time? *Energy Policy* **2010**, *38*, 6868–6876. [CrossRef]
24. Schneider, S.; Zelger, T.; Sengl, D.; Baptista, J. A Quantitative Positive Energy District Definition with Contextual Targets. *Buildings* **2023**, *13*, 1210. [CrossRef]
25. Albert-Seifried, V.; Murauskaite, L.; Massa, G.; Aelenei, L.; Baer, D.; Krangsås, S.G.; Alpagut, B.; Mutule, A.; Pokorny, N.; Vandevyvere, H. Definitions of Positive Energy Districts: A Review of the Status Quo and Challenges. In Proceedings of the Sustainability in Energy and Buildings 2021; Littlewood, J.R., Howlett, R.J., Jain, L.C., Eds.; Springer Nature: Singapore, 2022; pp. 493–506. [CrossRef]
26. Steininger, K.W.; Meyer, L.; Nabernegg, S.; Kirchengast, G. Sectoral Carbon Budgets as an Evaluation Framework for the Built Environment. *Build. Cities* **2020**, *1*, 337–360. [CrossRef]
27. Steininger, K.W.; Williges, K.; Meyer, L.H.; Maczek, F.; Riahi, K. Sharing the Effort of the European Green Deal among Countries. *Nat. Commun.* **2022**, *13*, 3673. [CrossRef]
28. Habert, G.; Röck, M.; Steininger, K.; Lupísek, A.; Birgisdottir, H.; Desing, H.; Chandrakumar, C.; Pittau, F.; Passer, A.; Rovers, R.; et al. Carbon Budgets for Buildings: Harmonising Temporal, Spatial and Sectoral Dimensions. *Build. Cities* **2020**, *1*, 429–452. [CrossRef]
29. *prEN 15978:2024*; Sustainability of Construction Works—Assessment of Environmental Performance of Buildings—Requirements and Guidance. European Committee for Standardization (CEN): Brussels, Belgium, 2024. Available online: https://www.cen.eu (accessed on 10 October 2025).
30. Cabeza, L.F.; Rincón, L.; Vilariño, M.V.; Pérez, G.; Castell, A. Life Cycle Assessment (LCA) and Life Cycle Energy Analysis (LCEA) of Buildings and the Building Sector: A Review. *Renew. Sustain. Energy Rev.* **2014**, *29*, 394–416. [CrossRef]
31. Mastrucci, A.; Marvuglia, A.; Leopold, U.; Benetto, E. Life Cycle Assessment of Building Stocks from Urban to Transnational Scales: A Review. *Renew. Sustain. Energy Rev.* **2017**, *74*, 316–332. [CrossRef]
32. Weidner, T.; Guillén-Gosálbez, G. Planetary Boundaries Assessment of Deep Decarbonisation Options for Building Heating in the European Union. *Energy Convers. Manag.* **2023**, *278*, 116602. [CrossRef]
33. Hamels, S.; Himpe, E.; Laverge, J.; Delghust, M.; Van den Brande, K.; Janssens, A.; Albrecht, J. The Use of Primary Energy Factors and CO_2 Intensities for Electricity in the European Context—A Systematic Methodological Review and Critical Evaluation of the Contemporary Literature. *Renew. Sustain. Energy Rev.* **2021**, *146*, 111182. [CrossRef]
34. Pehl, M.; Arvesen, A.; Humpenöder, F.; Popp, A.; Hertwich, E.G.; Luderer, G. Understanding Future Emissions from Low-Carbon Power Systems by Integration of Life-Cycle Assessment and Integrated Energy Modelling. *Nat. Energy* **2017**, *2*, 939–945. [CrossRef]
35. IEA. *CO2 Emissions in 2023*; IEA: Paris, France, 2024. Available online: https://www.iea.org/reports/co2-emissions-in-2023 (accessed on 10 October 2025).
36. Schneider, S.; Drexel, R.; Zelger, T.; Baptista, J. PEExcel: A Fast One-Stop-Shop Assessment and Simulation Framework for Positive Energy Districts. In Proceedings of the 10th Conference of IBPSA-Germany and Austria, Vienna, Austria, 23–26 September 2024; Volume 10, pp. 80–88. [CrossRef]
37. Stadt-Quartiere, Klimaaktiv. Available online: https://www.klimaaktiv.at/erneuerbare/erneuerbarewaerme/stadt-quartiere.html (accessed on 19 August 2024).
38. Schöfmann, P.; Zelger, T.; Bartlmä, N.; Schneider, S.; Leibold, J.; Bell, D. *Zukunftsquartier—Weg zum Plus-Energie-Quartier in Wien*; Berichte aus Energie- und Umweltforschung; Austrian Research Promotion Agency: Vienna, Austria, 2020; p. 203. Available online: https://nachhaltigwirtschaften.at/resources/sdz_pdf/schriftenreihe-2020-11-zukunftsquartier.pdf (accessed on 10 October 2025).
39. Schöfmann, P.; Forstinger, V.; Zelger, T.; Schneider, S.; Leibold, J.; Bell, D.; Schindler, M.; Mlinaric, I.; Hackl, L.; Wimmer, F.; et al. *Zukunftsquartier 2.0 Replizierbare, Thermisch und Elektrisch Netzdienliche Konzeption von (Plus-Energie-) Quartieren im Dichten Urbanen Kontext*; Berichte aus Energie- und Umweltforschung; Bundesministerium für Klimaschutz, Umwelt, Energie, Mobilität, Innovation und Technologie: Vienna, Austria, 2022.
40. Schneider, S. Cities4PEDs Definition of Positive Energy Districts. 2023. Available online: https://energy-cities.eu/wp-content/uploads/2023/07/Cities4PEDs-WP2-Definition-.pdf (accessed on 10 October 2025).
41. Simon Schneider PED Assessment Framework. Available online: https://github.com/simonschaluppe/peexcel (accessed on 21 August 2025).
42. Austrian Standards Austrian Standards International. Nachhaltigkeit von Bauwerken—Umweltdeklarationen für Bauprodukte—Grundregeln für die Produktkategorie Bauprodukte (ÖNORM EN 15804); 91.010.99; Wien. 2022. Available online: https://www.austrian-standards.at/de/shop/onorm-en-15804-2022-02-15~p2614831 (accessed on 10 October 2025).
43. Oekoindex OI3—IBO—Österreichisches Institut Für Bauen Und Ökologie. Available online: https://www.ibo.at/en/building-material-ecology/lifecycle-assessments/oekoindex-oi3 (accessed on 21 August 2025).

44. Hoxha, E.; Passer, A.; Saade, M.R.M.; Trigaux, D.; Shuttleworth, A.; Pittau, F.; Allacker, K.; Habert, G. Biogenic Carbon in Buildings: A Critical Overview of LCA Methods. *Build. Cities* **2020**, *1*, 504–524. [CrossRef]
45. Hawkins, W.; Cooper, S.; Allen, S.; Roynon, J.; Ibell, T. Embodied Carbon Assessment Using a Dynamic Climate Model: Case-Study Comparison of a Concrete, Steel and Timber Building Structure. *Structures* **2021**, *33*, 90–98. [CrossRef]
46. Levasseur, A.; Lesage, P.; Margni, M.; Samson, R. Biogenic Carbon and Temporary Storage Addressed with Dynamic Life Cycle Assessment. *J. Ind. Ecol.* **2013**, *17*, 117–128. [CrossRef]
47. Dobra, T.; Dolezal, F.; Zelger, T. *TimberBioC—Critical Evaluation of the Effect on Climate Change by Biogenic Carbon in Wood Products by Means of Dynamic Models;* IBO Verlag: Vienna, Austria, 2024; p. 2.
48. Zelger, T.; Leibold, J.; Schneider, S.; Bachleitner, F.; Helnwein, J.; Holzer, P.; Stuckey, D.; Koch, A.; Doser, B.; Drexel, C.; et al. *Aspern Klimafit 2.0 Final Report;* Wien 3420 aspern Development AG: Wien, Austria, 2024. Available online: https://www.aspern-seestadt.at/Downloads-Pdfs/aspern%20klimafit.%20Endbericht%20-%20Kurzfassung_aspernKlimafit2_20240917.pdf (accessed on 10 October 2025).
49. Streicher, W.; Schnitzer, H.; Tatzber, F.; Heimrath, R.; Wetz, I.; Titz, M.; Hausberger, S.; Haas, R.; Kalt, G.; Damm, A.; et al. Energieautarkie Für Österreich 2050. 2010. Available online: https://www.uibk.ac.at/media/filer_public/93/94/93941c9a-c2c4-4e68-bede-b275f5309689/energieautarkie_2050.pdf (accessed on 10 October 2025).
50. Schneider, S.; Zelger, T.; Klauda, L. Reflections on the Question of What Share of Renewable Energy Must be Provided Locally in Austria in 2050 (Original German title: ÜBERLEGUNGEN ZUR FRAGE, WELCHER ANTEIL ERNEUERBARER ENERGIE 2050 IN ÖSTERREICH LOKAL AUFGEBRACHT WERDEN MUSS). 2020. Available online: https://simonschaluppe.org/publication/schneider-uberlegungen-2020/schneider-uberlegungen-2020.pdf (accessed on 10 October 2025).
51. BUNDESMINISTERIUM FÜR NACHHALTIGKEIT UND TOURISMUS; BUNDESMINISTERIUM FÜR VERKEHR, I.U.T. #mission 2030 Klima- Und Energiestrategie. 2018. Available online: https://www.ewo-austria.at/wp-content/uploads/2020/03/mission2030_Klima-und-Energiestrategie_Endfassung.pdf (accessed on 10 October 2025).
52. Bundesministerum für Klimaschutz. Integrierter nationaler Energie- und Klimaplan für Österreich. 2023. Available online: https://www.bmluk.gv.at/themen/klima-und-umwelt/klima/nationale-klimapolitik/energie_klimaplan.html (accessed on 10 October 2025).
53. Tomschy, R.; Josef, F.; Follmer, R.; Sammer, G.; Klementschitz, R. *Österreich Unterwegs 2013/2014;* BMFIT: Wien, Austria, 2016. Available online: https://www.bmk.gv.at/dam/jcr:fbe20298-a4cf-46d9-bbee-01ad771a7fda/oeu_2013-2014_Ergebnisbericht.pdf (accessed on 9 October 2025).
54. Schneider, S.; Baptista, J. Annual Hourly E-Mobility Modelling and Assessment in Climate Neutral Positive Energy Districts. In Proceedings of the 2025 IEEE International Conference on Environment and Electrical Engineering and 2025 IEEE Industrial and Commercial Power Systems Europe, Chania, Crete, 15–18 July 2025.
55. Mair am Tinkhof, O.; Strasser, H.; Prinz, T.; Herbst, S.; Schuster, M.; Tomschy, R.; Figl, H.; Fellner, M.; Ploß, M.; Roßkopf, T. *Richt- Und Zielwerte Für Siedlungen Zur Integralen Bewertung Der Klimaverträglichkeit von Gebäuden Und Mobilitätsinfrastruktur in Neubausiedlungen;* Berichte aus Energie- und Umweltforschung; Bundesministerium für Verkehr, Innovation und Technologie: Wien, Austria, 2017; p. 118. Available online: https://nachhaltigwirtschaften.at/resources/sdz_pdf/berichte/schriftenreihe_2017-39_richt-zielwerte-siedlungen.pdf (accessed on 10 October 2025).
56. Angelakoglou, K.; Kourtzanidis, K.; Giourka, P.; Apostolopoulos, V.; Nikolopoulos, N.; Kantorovitch, J. From a Comprehensive Pool to a Project-Specific List of Key Performance Indicators for Monitoring the Positive Energy Transition of Smart Cities—An Experience-Based Approach. *Smart Cities* **2020**, *3*, 705–735. [CrossRef]
57. Salom, J.; Tamm, M.; Andresen, I.; Cali, D.; Magyari, Á.; Bukovszki, V.; Balázs, R.; Dorizas, P.V.; Toth, Z.; Zuhaib, S.; et al. An Evaluation Framework for Sustainable Plus Energy Neighbourhoods: Moving Beyond the Traditional Building Energy Assessment. *Energies* **2021**, *14*, 4314. [CrossRef]
58. Suppa, A.R.; Cavana, G.; Binda, T. Supporting the EU Mission "100 Climate-Neutral Cities by 2030": A Review of Tools to Support Decision-Making for the Built Environment at District or City Scale. In Proceedings of the International Conference on Computational Science and Its Applications, Malaga, Spain, 4–7 July 2022; Springer: Berlin/Heidelberg, Germany, 2022; pp. 151–168.
59. Neumann, H.-M.; Hainoun, A.; Stollnberger, R.; Etminan, G.; Schaffler, V. Analysis and Evaluation of the Feasibility of Positive Energy Districts in Selected Urban Typologies in Vienna Using a Bottom-Up District Energy Modelling Approach. *Energies* **2021**, *14*, 4449. [CrossRef]
60. Bruck, A.; Casamassima, L.; Akhatova, A.; Kranzl, L.; Galanakis, K. Creating Comparability among European Neighbourhoods to Enable the Transition of District Energy Infrastructures towards Positive Energy Districts. *Energies* **2022**, *15*, 4720. [CrossRef]
61. Bruck, A.; Díaz Ruano, S.; Auer, H. A Critical Perspective on Positive Energy Districts in Climatically Favoured Regions: An Open-Source Modelling Approach Disclosing Implications and Possibilities. *Energies* **2021**, *14*, 4864. [CrossRef]
62. Entwicklung Plusenergiequartierzertifizierung inklusive Quickcheck für klimaaktiv Zertifizierung Siedlungen. Available online: https://klimaneutralestadt.at/de/projekte/peq-quickcheck-klimaaktiv.php (accessed on 29 April 2025).
63. Statistics Austria STATatlas. Available online: https://www.statistik.at/atlas/?theme=6 (accessed on 6 June 2022).

64. Agency, I.E. *Net Zero by 2050—A Roadmap for the Global Energy Sector*; IEA: Paris, France, 2021; p. 224. Available online: https://www.iea.org/reports/net-zero-by-2050 (accessed on 10 October 2025).
65. Amann, W.; Goers, S.; Komendantova, N.; Oberhuber, A. *Kapazitätsanpassung der Bauwirtschaft für eine Erhöhte Sanierungsrate*; Berichte aus Energie- und Umweltforschung; Bundesministerium für Klimaschutz, Umwelt, Energie, Mobilität, Innovation und Technologie: Wien, Austria, 2021. Available online: https://nachhaltigwirtschaften.at/resources/sdz_pdf/schriftenreihe-2021-27-kapazitaetsanpassung-bauwirtschaft.pdf (accessed on 10 October 2025).
66. Österreichisches Institut für Bautechnik. OIB-Richtlinie 6. 2023. Available online: https://www.oib.or.at/kernaufgaben/oib-richtlinien/richtlinien/oib-richtlinien-2023/ (accessed on 10 October 2025).
67. Dörn, M.; Fernkälte—Die Kühlung aus der Leitung. ArchiPHYSIK 2020. Available online: https://archiphysik.at/fernkaelte-die-kuehlung-aus-der-leitung/ (accessed on 10 October 2025).

designs

MDPI

Article

Blending Nature with Technology: Integrating NBSs with RESs to Foster Carbon-Neutral Cities

Anastasia Panori [1,2,*], Nicos Komninos [1], Dionysis Latinopoulos [2], Ilektra Papadaki [1] Elisavet Gkitsa [2] and Paraskevi Tarani [3]

1 URENIO Research, Aristotle University of Thessaloniki, 54124 Thessaloniki, Greece; komninos@urenio.org (N.K.); ipapadf@plandevel.auth.gr (I.P.)
2 School of Spatial Planning and Development, Aristotle University of Thessaloniki, 54124 Thessaloniki, Greece; dlatinop@plandevel.auth.gr (D.L.); elisavgc@plandevel.auth.gr (E.G.)
3 Department of Sustainable Development, Major Development Agency Thessaloniki (MDAT S.A.), 54640 Thessaloniki, Greece; tarani@mdat.gr
* Correspondence: apanori@plandevel.auth.gr

Abstract: Nature-based solutions (NBSs) offer a promising framework for addressing urban environmental challenges while also enhancing social and economic resilience. As cities seek to achieve carbon neutrality, the integration of NBSs with renewable energy sources (RESs) presents both an opportunity and a challenge, requiring an interdisciplinary approach and an innovative planning strategy. This study aims to explore potential ways of achieving synergies between NBSs and RESs to contribute to urban resilience and climate neutrality. Focusing on the railway station district in western Thessaloniki (Greece), this research is situated within the ReGenWest project, part of the EU Cities Mission. This study develops a comprehensive, well-structured framework for integrating NBSs and RESs, drawing on principles of urban planning and energy systems to address the area's specific spatial and ecological characteristics. Using the diverse typologies of open spaces in the district as a foundation, this research analyzes the potential for combining NBSs with RESs, such as green roofs with photovoltaic panels, solar-powered lighting, and solar parking shaders, while assessing the resulting impacts on ecosystem services. The findings reveal consistent benefits for cultural and regulatory services across all interventions, with provisioning and supporting services varying according to the specific solution applied. In addition, this study identifies larger-scale opportunities for integration, including the incorporation of NBSs and RESs into green and blue corridors and metropolitan mobility infrastructures and the development of virtual power plants to enable smart, decentralized energy management. A critical component of the proposed strategy is the implementation of an environmental monitoring system that combines hardware installation, real-time data collection and visualization, and citizen participation. Aligning NBS–RES integration with Positive Energy Districts is another aspect that is stressed in this paper, as achieving carbon neutrality demands broader systemic transformations. This approach supports iterative, adaptive planning processes that enhance the efficiency and responsiveness of NBS–RES integration in urban regeneration efforts.

Keywords: nature-based solutions; renewable energy sources; climate adaptation; urban planning

1. Introduction

Integrating renewable energy sources (RESs) into urban environments is a key strategy for achieving the goals of energy transition and climate neutrality [1]. Cities consume the largest proportion of global energy, and, consequently, cities are responsible for 70% of global CO_2 emissions related to energy consumption [2,3]. Therefore, the deployment of RESs in cities is critical to reducing carbon emissions and improving energy autonomy. However, the application of RESs in densely built environments presents challenges, such as limited installation space, interaction with the built environment, and socio-economic parameters.

At the same time, nature-based solutions (NBSs) are increasingly being adopted in urban design to deliver multiple benefits and reduce climate risk in cities. NBS are tools that can enhance the resilience of cities by improving aspects like thermal comfort (urban heat island reduction), air quality, and water and flood management [4]. They are also increasingly promoted to enhance urban sustainability, as they can positively contribute to achieving various Sustainable Development Goals (SDGs), such as tackling poverty, good health and well-being, clean water, life on land, and reducing inequalities (e.g., in access to green spaces) [5,6].

As cities strive to transition toward more sustainable and resilient futures, the integration of NBSs with RESs is emerging as a promising strategy [7,8]. While NBSs primarily focus on enhancing ecosystem services and urban livability, RESs provide a critical pathway for reducing carbon emissions and fostering energy independence. Individually, both approaches contribute to urban sustainability; however, their combined implementation offers untapped potential for maximizing environmental, social, and economic benefits.

Recent literature and empirical applications highlight that combining NBSs with RESs can enhance ecosystem services and contribute to energy efficiency and pollution reduction [9,10]. Specifically, this can be achieved through (i) the creation of synergetic environmental benefits deriving from their combination, such as mutually reinforcing systems where NBSs can enhance RES efficiency; (ii) a comprehensive approach to tackling climate-related challenges; (iii) land use optimization and efficiency; and (iv) the creation of self-sufficient cities that rely less on external energy sources and costly climate-adaptation measures. Therefore, integrating NBSs with RESs offers a holistic sustainability approach, ensuring environmental protection, economic feasibility, and social well-being. However, the challenge remains regarding how we can achieve efficient NBS–RES integration at the urban level.

This study explores planning practices and strategies for integrating NBSs with RESs, emphasizing the need to consider and set interdisciplinary targets/criteria, such as urban environmental enhancement, social well-being, and spatial heterogeneity. Key design interventions analyzed in this paper include permeable pavements, tree canopies, green roofs, and blue infrastructure for water management, alongside RES solutions that are either embedded within NBSs or co-located in shared public spaces. A structured methodology—incorporating best-practice analysis, literature review, and spatial data analysis—guides the development of the proposed integrated NBS–RES framework. This study identifies innovative design typologies that merge NBSs with RESs while ensuring alignment with the unique characteristics of the local urban environment.

The western center of Thessaloniki, Greece, and particularly the railway station district, serves as the pilot area of this study. This district faces escalating challenges related to climate change, including heat island effects, air pollution, and flash flooding. It is also affected by mobility disruptions linked to the prolonged construction of the metro and socio-economic disparities. Given its ongoing transformation, the area seems to be an ideal case for exploring how integrated NBSs (i.e., green infrastructure and green networks) and RES solutions (i.e., photovoltaic panels and solar shading) can enhance urban neutrality. Key areas of intervention include the transformation of public open spaces and the reconfiguration of neglected urban pockets, some of which contain underutilized green spaces.

The structure of this paper is as follows. Section 2 provides an extended literature review on the role of NBS and the ways in which they can be combined with RESs in urban settings. Section 3 describes in detail the case study of the railway station district of Thessaloniki, showing how an integrated NBS–RES framework can be developed. Finally, Section 4 discusses some key findings of this exercise in relation to potential urban projects that could be developed in the intervention area, whereas Section 5 presents a connection to the broader discussion of Positive Energy Districts (PEDS). This paper concludes by discussing how NBS–RES integration could be further developed and implemented to foster carbon-neutral cities, considering potential barriers, limitations, and spatial specificities (Section 6).

2. Literature Review

2.1. Designing NBSs in Urban Settings

According to the European Commission, NBSs are defined as "solutions that are inspired and supported by nature, which are cost-effective, simultaneously provide environmental, social and economic benefits and help build resilience" (https://research-and-innovation.ec.europa.eu/research-area/environment/nature-based-solutions_en (accessed on 23 April 2025)). Moreover, the International Union for Conservation of Nature stresses that these are "actions aimed at protecting, sustainably managing and restoring natural or modified ecosystems and which effectively and adaptively address societal challenges, while providing benefits for human well-being and biodiversity" [11]. Hence, NBSs offer an interdisciplinary and innovative approach to address social, environmental, and economic challenges faced by modern cities.

Their design and implementation integrate the principles of sustainability, biodiversity protection, and enhancement of ecosystem services, making them particularly important, especially in urban areas where adaptation to climate change is essential [12,13]. Specifically, NBSs promote biodiversity by enhancing food security, the health of citizens and ecosystems, and the protection and availability of clean water, as outlined in the Kunming–Montreal Global Biodiversity Framework, which attempts to cease and reverse nature loss by 2030 [14]. Moreover, NBSs encompass urban greenery, such as parks and street trees, which help mitigate high temperatures, regulate air quality, and manage water flows. They also include the allocation of natural habitat spaces in floodplains, which can play a crucial role in controlling and reducing the impacts of flooding [4]. Urban greening through NBSs is considered vital for improving life in cities and climate adaptation efforts, as it contributes to well-being, health, water quality, microclimate, and overall sustainability [15,16].

Even though urban planning has long recognized the importance of green spaces for city residents, the integration of NBSs and their ecosystem services into public policy is a more recent development. Usually, the term "green infrastructure" describes the spatial structure of ecosystems (green spaces) in the urban environment, while the term "nature-based solutions" describes human interventions that utilize the functions of these ecosystems (green infrastructure) to achieve environmental, social, and economic goals. These policies focus on evaluating potential outcomes, analyzing their direct and indirect contributions to human well-being, and assessing the benefits derived from specific ecosystem services [17–19]. However, there is often a lack of focus on how NBSs can effectively address specific vulnerabilities and challenges within urban environments characterized by high spatial heterogeneity, thus limiting the full exploitation of their potential for adaptation and resilience [20]. To achieve a successful implementation, it is essential to consider the local specificities and socio-economic conditions of the area. This ensures that solutions are not only tailored to local needs and characteristics but also embedded within a comprehensive planning framework that fosters sustainable and resilient development [6].

The design challenges of NBSs in urban environments are multidimensional and require an integrated approach. These include, inter alia, the creation of a strong and balanced evidence base capable of assessing the effectiveness of NBSs, especially in terms of their competitive and/or complementary relationship with other more traditional or conventional solutions that are usually technology-based. To the authors' knowledge, even though spatial competition between NBSs and other urban elements has been identified as a potential drawback for their implementation, it has not been investigated in depth by previous studies. In addition, it is necessary to assess the long-term potential impacts of possible alternatives, the choice between which involves a series of trade-offs, such as the scale of application, the ecosystem services selected to be prioritized, and inequalities in benefit/cost distribution [21–23]. Emphasis should, therefore, be placed on effective NBS planning and management to prevent unintended yet foreseeable spatial consequences, such as gentrification.

The complexity of the design and implementation of NBSs calls for an integrated systemic approach and coordinated action. In this context, Dumitru et al. [24] underline that collaboration between scientific, political, and social actors is critical to their success. Adopting an integrated systemic approach means that NBSs should not be designed in isolation but rather complement and strengthen existing local policies related to resilience, sustainability, climate neutrality, and risk management strategies. In terms of its long-term effects, the design of NBSs requires systematic evaluation, both in terms of results (benefits) and the achievement of their objectives, as well as in terms of their ability to adapt to future climate and social changes [25,26].

On the neighborhood scale, challenges related to resilience (e.g., flooding, urban heat island effects, the reduction of air pollution) and climate change mitigation (e.g., achieving a climate-neutral city/neighborhood) should be addressed at the local level [27,28]. These may include measures and solutions in buildings, streets, and open public spaces that should be as extensive as possible to improve the quality of life of residents and to be accepted by the local community [29]. On this scale, it is very important to consider spatial heterogeneity and individual neighborhood characteristics to derive the most appropriate NBSs. To achieve this, a thorough analysis of the study area is essential for identifying a typology of available spaces suitable for green and energy interventions. This typology will serve as a foundation for selecting appropriate NBS categories based on their suitability and effectiveness.

2.2. Integrating NBSs with RES Technologies

As stated at the beginning of this study, integrating RESs in urban environments is critical for achieving energy transition and climate neutrality [1]. The most common RES applications in cities include photovoltaic (PV) systems, urban wind energy, and geothermal energy. Photovoltaics, particularly building-integrated PV (BIPV), are widely used due to their ease of installation and low maintenance requirements [30,31]. Small vertical wind turbines, despite challenges such as noise pollution and visual disturbance, have also emerged as feasible solutions for dense urban areas [32]. However, the efficient integration of these technologies into urban environments still remains complex due to space limitations, regulatory barriers, and socio-economic considerations [33].

Integrating NBSs with RES technologies presents an innovative approach to creating resilient and sustainable urban ecosystems. This integration can contribute significantly to achieving climate neutrality in cities while simultaneously improving the microclimate and enhancing biodiversity. Recent academic literature highlights the potential for creating synergies between these two domains, suggesting that their integration can provide multifaceted benefits, including improved microclimate conditions, enhanced biodiversity, and increased energy efficiency [7,8]. For example, green roofs or green corridors with photovoltaic panels can be used to generate renewable energy but also to improve the visibility and accessibility of green spaces, thus promoting their use by urban residents [8]. Similarly, solar urban gardens incorporate smart solar trees and solar-powered lighting to enhance public spaces while producing renewable energy [8,34]. Another emerging solution is the biomimetic design of wind turbines in the form of trees, which blend aesthetically with urban landscapes while harnessing wind energy with minimal visual disturbance [35]. A study conducted in Zurich, Switzerland, found that green roofs can lead to a higher annual energy yield from PV systems compared to conventional roofs [34]. Additional evidence highlights the potential of community solar gardens to provide equitable access to renewable energy, particularly for low-income communities, by utilizing existing green spaces in urban areas [36]. Similarly, Hakimizad et al. [37] show that solar panels in urban green spaces offer year-round benefits, such as winter heating, summer cooling, and nighttime lighting. Likewise, the use of biomass and geothermal energy in parks and green infrastructure can reduce reliance on fossil fuels, as demonstrated by projects such as the Eden Project in the UK [38].

Below, we present some key applications of NBS and RES integration, indicating a general description, the RES solution, and the NBS applied in each case (Table 1). We can see that there is a diversification when applying synergies between NBSs and RESs in different urban settings, considering the available spaces for interventions as well as the needs for and types of climate mitigation aspects.

Table 1. Examples of combining RES and NBS in different cities around the world.

City	Description	RES Solution	NBS Solution
Gardens by the Bay, Singapore	A 101-hectare urban green space with iconic structures like Supertree Grove and Flower Dome, integrating nature into the city.	Supertrees use solar energy and biomass for power; greenhouses have energy-efficient cooling.	Features 1.5 million plants, vertical gardens, and water management systems for biodiversity and climate resilience.
The Eden Project, Cornwall, UK	An environmental and educational site with massive biomes replicating different climates.	Uses geothermal energy, solar panels, wind turbines, and biomass boilers for power.	Recreates ecosystems, supports biodiversity, and employs rainwater harvesting and wetlands for water management.
The Green Spine, Melbourne, Australia	A sustainable skyscraper development with two towers covered in vertical gardens.	Solar panels, wind power, and smart shading optimize energy use.	Vertical gardens insulate, support biodiversity, and utilize rainwater and greywater for irrigation.
Bosco Verticale, Milan, Italy	Twin residential towers integrating extensive vegetation for urban reforestation.	Solar panels and geothermal heating improve energy efficiency.	Houses 900+ trees and 20,000+ plants for air purification, noise reduction, and biodiversity.
Masdar City, UAE	A low-carbon, zero-waste eco-city near Abu Dhabi, focusing on sustainability research.	Solar farms, rooftop photovoltaics, wind cooling, and waste-to-energy tech.	Green roofs, urban agriculture, and shaded walkways enhance biodiversity and cooling.

Table 1. *Cont.*

City	Description	RES Solution	NBS Solution
Shanghai, China	A city integrating natural elements to boost sustainability and resilience.	Expanding renewable energy and energy-efficient technologies.	Green corridors, planted roofs, and vertical forests improve air quality and provide green spaces.
Freiburg, Germany	A model sustainable city with a focus on renewable energy and eco-friendly architecture.	Photovoltaic roofs, with many energy-positive buildings.	Large green belts, parks, and rainwater collection for irrigation and cooling.
Hammarby Sjöstad, Sweden	A former industrial suburb transformed into a low-impact, sustainable district.	Solar, biogas, and lake water for district heating and cooling.	Canals and green corridors enhance biodiversity and ecosystem connections.
Kalundborg Symbiosis, Denmark	A leading industrial symbiosis model where waste becomes a resource.	Uses biomass and wind power for energy.	Wetlands naturally treat waste and support local biodiversity.

Source: Authors' elaboration.

Despite their potential, integrated NBS and RES solutions face several challenges. The spatial and temporal mismatch between energy production and consumption necessitates efficient energy storage and grid management strategies [39]. Advancements in technology continue to offer innovative solutions, but they also bring new challenges and limitations that need to be addressed for optimal efficiency [40]. For example, Tasneem et al. [41] examine the suitability and efficiency of different urban wind farm models and conclude that commercial wind farms are difficult to replace with urban wind farms due to the chaotic and turbulent nature of city winds, which limits their performance. Additionally, balancing aesthetic considerations with technological advancements remains an ongoing challenge, particularly in densely populated urban areas.

3. Methodological Approach and Case Study Analysis: The Railway Station District of Thessaloniki

The ReGenWest methodology is guided by two contemporary design paradigms. On the one hand, urban sustainability and climate-sensitive planning, as articulated in LEED v4.1 and the forthcoming LEED v5 standards (https://www.usgbc.org/leed/v5), which emphasize the coherence and interdependence of sustainability measures across city, district, and building scales, support an integrated approach that aligns environmental performance with broader urban planning objectives. On the other hand, the smart city and design thinking approach promotes rapid, iterative design cycles and continuous stakeholder and user engagement, which foster continuous feedback and improvement, ensuring that solutions remain adaptable and responsive to both environmental goals and community needs.

The design process of the ReGenWest project includes three subsequent iterations, each building upon and improving the previous one. The first iteration focuses on problem definition, mapping the renovation area, identifying spatial opportunities and potential types of intervention, and conducting an extensive literature review of relevant cases. This phase concludes in the development of a renovation model tailored to the local context. The second iteration presents the selected model to stakeholders, users, and investors, aiming to incorporate considerations of diversity and inclusion, as well as to identify financial opportunities, barriers, and viable business models. The third iteration optimizes the proposed solutions and projects with respect to cost and carbon emissions, codifies the renovation master plan, and estimates the direct environmental impact using an environmental monitoring system.

In the following sections, we present the outcomes of the first iteration process focusing on the mapping of the intervention area, the identification of open space elements, and the potential types of NBS and RES interventions.

3.1. *Analysis of the Intervention Area and Open Space Typologies*

The intervention area, located in the western part of Thessaloniki (Figure 1), retains a neighborhood character with old high-rise social housing and mid-rise apartment buildings (3–4 floors) mixed with commercial and city-center land uses. It is marked by underutilized open spaces, many of which serve as informal parking due to the lack of private parking and the expansion of the metro and the national railway facilities (OSE). Despite the presence of permeable soil, the area lacks well-designed public spaces that meet residents' needs, leaving lower-income communities, often living in cramped housing, to extend private activities into shared spaces. Existing urban squares, parks, and roadside areas require modifications to improve resilience against intensifying climate challenges such as heat waves, floods, and strong winds. Similarly, schoolyards—currently dominated by concrete and minimal greenery—need improvements for climate adaptation. Vulnerable groups, including children, the elderly, a small Chinese community, women, parents, and people with mobility restrictions, face exclusion due to poor accessibility, inadequate amenities, and safety concerns. Those experiencing energy or food poverty would benefit from spaces for socialization, urban agriculture, and renewable energy generation. Additionally, the many unused rooftops and open spaces present opportunities to enhance ecosystem services, create community spaces, and develop sustainable infrastructure. The ongoing design process aims to reclaim and reimagine these neglected spaces, fostering a more inclusive, livable, and climate-resilient urban environment.

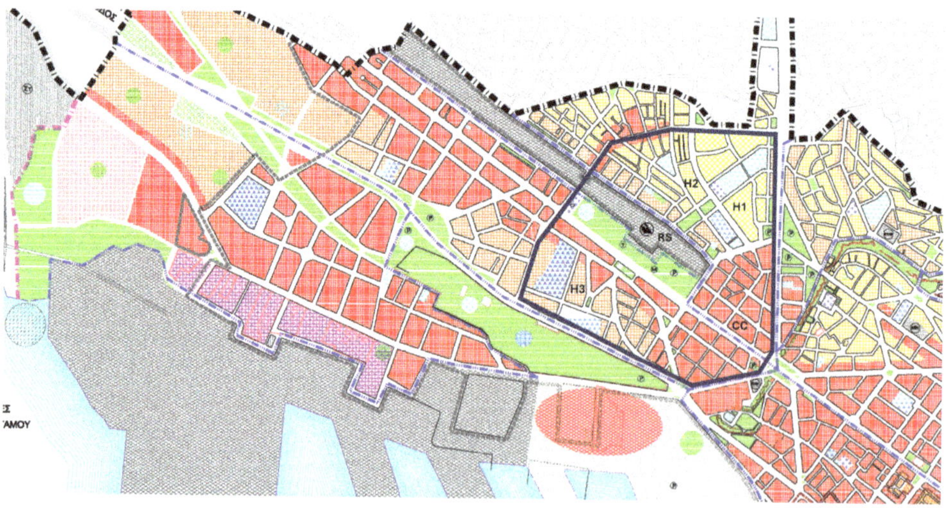

Figure 1. Intervention area. Legend: H1: Pure Residential Uses; H2: Residential and Personal Services Uses; H3: Intermediate-Level Residence, including Commercial, Transport, Hospitality Uses; CC: City-Center Uses.

Our analysis identifies 11 typologies of open spaces suitable for green and renewable energy interventions. These typologies are classified and correspond to specific locations, numbered according to Figure 2. Apart from the identified typologies, throughout the whole intervention area, there are also roadside trees requiring densification and the widening of the natural/perforated roofs of apartment buildings and uncovered apartment buildings/plots.

Figure 2. Identified typologies of open spaces suitable for green and energy interventions. Legend: 1-Empty spaces/plots (including spaces used as parking); 2-Primary and secondary education schools; 3-Higher education facilities; 4-Urban squares designed that require modifications to better behave in extreme weather events (such as heat waves and floods); 5-Small size and activity parks; 6-Uncovered areas around workers' apartment buildings; 7-Roadside trees; 8-Roofs of apartment buildings; 9-Uncovered apartment buildings/plots; 10-OSE Station; 11-Churches with courtyards.

3.2. Identification of Potential NBS Solutions

As a next step, we identified specific NBS categories that are suitable for this intervention area. More specifically, we explored solutions tailored to the urban environment, with a particular focus on the dense urban fabric of the study area. For this purpose, the typology used by the European Commission [4] was chosen as a basis of our analysis, comprising six main general categories of NBSs, each containing several solutions. The six main categories are (i) green solutions in buildings; (ii) urban green areas connected to grey/man-made infrastructure; (iii) parks and (semi-)natural urban green areas; (iv) community land and community gardens; (v) agricultural land; (vi) green areas for water management; and (vii) blue infrastructure/water areas [42–44]. Given the urban nature of our case study, we excluded category (v), which pertains to agricultural land, as it does not apply to the intervention area. Additionally, the last two categories (vi and vii) have been combined into a single category ("Blue and Green Areas for Water Management") with water as a central component. Finally, several subcategories of solutions, primarily related to blue infrastructure, have been removed, as they are not relevant to the study area, which lacks natural water bodies such as lakes or streams.

Beyond identifying the categories mentioned above, another important step was to connect them with specific subcategories acting as solutions tailored to the needs and characteristics of the study area. To perform this correspondence, the World Bank's "NBS Directory for Urban Resilience" [45] and the "Technical guide to NBS" of the UNaLab—

Urban Nature Labs research project [46] have been used as key sources. The identified NBS solutions and their correspondence to the general NBS categories are presented in Table 2, each defined by distinct characteristics and functions. This approach enables the exploration and integration of diverse solutions tailored to the specific needs and characteristics of the Thessaloniki railway station district's urban fabric.

Table 2. Identification of potential NBS solutions adapted to the intervention area needs and specificities.

General NBS Category	NBS Solutions
Green buildings: Include practices and elements designed to create or integrate green features in buildings.	➢ Green walls: Green facades or planted structures on exterior walls, which offer aesthetic value, reduce heat, and improve air quality. These are divided as follows: • Ground-based green walls: Vertical plantings rooted in the ground are often used to shade or reduce heat on exterior walls. • Facade-bound green wall: Plantings directly supported by the building wall, improving thermal insulation and aesthetics. ➢ Green roofs: Surfaces covered with vegetation that help improve thermal insulation, rainwater absorption, and biodiversity. These are divided as follows: • Facade-bound green walls: Lightweight roofs with low-maintenance plants, ideal for large surfaces. • Intensive green roofs: Roofs with the ability to support large plants or even trees, offering greater biodiversity. • Smart roofs: Green roofs that integrate technologies such as sensors, automation, and water management systems to optimize energy efficiency, rainwater management, and temperature. ➢ Green balconies: Plantings on balconies that contribute to the aesthetics and cooling of the building while improving air quality. ➢ Atriums: Interior space with natural light that often includes plantings, creating a green environment within the building. ➢ Green pavements and green parking pavements: Outdoor surfaces of buildings that allow water infiltration, often covered with grass or other plants, contributing to rainwater management. ➢ Green fences and noise barriers: Constructions with plantings that are used as plots/property boundaries or to reduce noise while enhancing aesthetics and biodiversity.
Urban green areas connected to grey infrastructure: Green areas within the urban environment that are directly connected to grey infrastructure, such as roads, railway lines, schools, or parking lots. These areas combine functionality with environmental benefits.	➢ Tree alleys and street trees/hedges: Trees or tall shrubs along roads that provide shading, reduce heat, and improve air quality. In other studies/surveys and classifications, this subcategory is also referred to as street tree canopies or group of trees. ➢ Street green and green verge: Green corridors or small green belts along roads, usually for beautification or environmental management. In other studies/surveys and classifications, this subcategory is also referred to as green avenues, boulevard trees, and urban green corridors. ➢ House gardens: Private gardens around houses offering green spaces. ➢ Green spaces near railway tracks (Railroad banks): Plantings along railway tracks that reduce erosion, enhance biodiversity, and absorb noise. ➢ Green playgrounds/school grounds: Green spaces that enhance children's well-being and contact with nature in their everyday environment. ➢ Green parking lots: Parking spaces designed with green elements, such as permeable surfaces or plantings, to reduce heat and better manage rainwater.

Table 2. *Cont.*

General NBS Category	NBS Solutions
Parks and (semi-)natural urban green areas including urban forests: This category refers to large or small natural areas within cities that include parks, forests, and other green infrastructure, promoting biodiversity, recreation, and environmental management. These areas are critical for enhancing climate resilience and quality of life.	➢ Large urban park: Extensive green areas within cities. ➢ Neighborhood green space: Smaller-scale parks or green areas that serve the needs of local communities/neighborhoods. ➢ Small and pocket parks: Small green areas within densely populated and space-restricted urban centers. ➢ Botanical gardens: Specially designed areas for the conservation and presentation of plants and trees, aiming for education and recreation. ➢ Institutional green space: Green areas belonging to educational or public institutions (beyond school yards). ➢ Church yards. ➢ Sports facilities (Green sport facility): Areas that combine sports activities (e.g., stadiums, fitness equipment, running tartans) with nature and urban greenery. ➢ Shrublands: Natural areas dominated by shrubland. ➢ Abandoned and derelict areas with patches of wilderness: Areas that gradually return to nature, often a refuge for wildlife.
Allotments and community gardens: Describes solutions that focus on participatory cultivation and creation of green spaces, promoting social cohesion, sustainability, and food production.	➢ Community vegetable gardens: Areas for collective cultivation and food production, with actions that promote social participation and sustainable management. ➢ Community (urban) gardens (Horticulture): Describes the systematic cultivation (not necessarily of edible species) for biodiversity, education, and recreation within urban areas.
Blue and Green areas for water management: This category includes nature-based solutions intended for stormwater management (sustainable urban drainage systems) and flood prevention. They combine elements of water (blue infrastructure) and green (green infrastructure). In literature they often appear as sustainable urban drainage systems or as infiltration, filtration and biofiltration structures.	➢ Rain gardens: Landscaped (usually small) green spaces that collect and absorb rainwater, reducing flooding and improving water quality. They are usually planted with aquatic or hydrophilic plants, which withstand wet conditions and contribute to the biological degradation of pollutants. ➢ Swales/Filter strips: The (planted) channels (grooves) collect and guide rainwater, allowing it to be absorbed into the soil and reduce erosion. Similarly, filtration zones are planted areas that are placed along runoff surfaces (e.g., roads, parking lots) to reduce rainwater flow and absorb pollutants before they reach water bodies. They are often referred to as bioswales, i.e., as biological grooves/channels. Soil infiltration and rainwater purification (World Bank, 2021). ➢ Small (artificial) lakes/ponds: Small man-made (or sometimes natural) lakes used to store and manage water while also providing an environment for biodiversity. They contribute to the collection and storage of rainwater, the improvement of local ecology, and the provision of neighborhood recreational areas. They are often divided as follows: (a) Detention ponds capture and temporarily store rainwater during periods of heavy rainfall [47] and evacuate it at a specified time. These lakes can be attractive graphic elements in public spaces, e.g., around playgrounds and sports grounds. (b) Retention ponds are are permanently full of water and have planted ends. They collect rainwater from surrounding areas, add storage capacity, and relieve pressure on surface water treatment and sewage systems while providing urban habitats and enriching the diversity of public green spaces. ➢ Permeable pavements: Surfaces that allow rainwater to penetrate and be absorbed into the soil, reducing surface water concentration and the likelihood of flooding.

3.3. *Matching Open Space Typologies, NBS Solutions, and Ecosystem Services*

As a final step of this exercise, we performed a matching between the identified open space typologies (Figure 2), NBS solutions (Table 2), and the ecosystem services relevant to the railway station district of Thessaloniki. Ecosystem services refer to the ecological characteristics, functions, and processes that contribute directly or indirectly to human well-being through various pathways [48]. The ecosystem services approach highlights nature's role as an active participant in shaping our health and quality of life [49]. In this context, NBS solutions are expected to provide different ecosystem services depending on the characteristics of the urban environment where they are applied [50].

The identified ecosystem services that apply in our case study include cultural, regulatory, provisioning, and habitat and supporting services. More specifically, cultural services relate to entertainment, aesthetic pleasure, cultural heritage, and spiritual and inspirational value. Regulatory services incorporate aspects related to the management of air quality; urban microclimate regulation; the mitigation of environmental hazards, such as flooding and heat stress; and the control of invasive or parasitic species that affect the area's ecological resilience. These services are increasingly supported through NBSs, such as green corridors and permeable surfaces, which not only enhance ecological resilience but also contribute to carbon neutrality when integrated with RESs like solar-powered infrastructure and energy-efficient transit systems. The provisioning type of ecosystem services includes food, water, medicines, and other resources derived directly from nature. In this case, NBSs in urban environments mainly concern community gardens, green terraces, and other micro-farming. Finally, habitat and supporting services include those of habitat provision, biodiversity conservation, nutrient cycle, and soil protection and conservation. The matching results are given in Table 3 below. As we can see, all the identified ecosystem services are affected by the application of NBS solutions, indicating a different mixture each time, depending on the open space typology and its characteristics.

Table 3. Matching intervention site open space typologies, NBS solutions, and ecosystem services.

Open Space Typology	Relevant NBS Solutions	Ecosystem Services	
1-Empty spaces/plots and spaces used as parking	Green playgrounds and yards, green parking Neighborhood parks, pocket parks Permeable pavements	Cultural	●●●
		Regulatory	●●
		Provisioning	●
		Supporting	●●
2-Primary and secondary education schools	Green gardens, extensive and smart green roofs, green walls Community urban gardens Permeable pavements	Cultural	●●●
		Regulatory	●●●
		Provisioning	●●
		Supporting	●●●
3-Higher education facilities	Green walls Rows of trees As the area is not open to visitors, the solutions proposed concern the boundary with the public space	Cultural	●●●
		The current picture of abandonment of the old building of the veterinary school creates an impression of abandonment and insecurity which could be reversed with some green interventions.	
4-Urban squares that require modifications for better performance in extreme weather events	Rows of trees and individual trees near roads, green spaces near roads Permeable pavements	Cultural	●●
		Regulatory	●●●
		Provisioning	●
		Supporting	●●

Table 3. *Cont.*

Open Space Typology	Relevant NBS Solutions	Ecosystem Services	
5-Small/pocket and activity parks	Neighborhood parks and green spaces, small and pocket parks, sports facilities	Cultural	●●●
		Regulatory	●●●
		Provisioning	●
		Supporting	●●●
6-Uncovered areas around old social housing complexes	Community urban gardens and vegetable gardens	Cultural	●●●
		Regulatory	●●
		Provisioning	●●●
		Supporting	●●
7-Roadside tree plantings	Urban green areas connected to grey (artificial) infrastructure Permeable pavements, channels, and filtration zones	Cultural	●
		Regulatory	●●●
		Provisioning	●
		Supporting	●
8-Roofs of apartment buildings	Extensive and smart green roofs	Cultural	●●●
		Regulatory	●●
		Provisioning	●●●
		Supporting	●●
9-Uncovered apartment buildings/plots	Green patios Private gardens, green parking areas	Cultural	●●●
		Regulatory	●●
		Provisioning	●●●
		Supporting	●●
10-OSE Station	Rows of trees and individual trees near roads; green spaces near roads; plantings along railway lines that reduce erosion, enhance biodiversity, and absorb noise; green parking spaces Small and pocket parks Permeable pavements	Cultural	●●●
		Regulatory	●●●
		Provisioning	●
		Supporting	●●●
11-Churches with courtyards	N/A (Churches will not be included in the scope of this project)	N/A	

Source: Authors' elaborations.

4. Integrating NBSs with RES Technologies: Replicable Solutions for Urban Districts

Taking stock of the interventions in the pilot area, we further specify a concrete set of integrated urban projects for combining NBSs with RESs to advance carbon neutrality. A detailed description of the identified project types that can serve as a replicable model for other city districts, thereby increasing urban sustainability and achieving carbon neutrality goals, is provided below.

4.1. Developing Green and Blue Corridors

A holistic approach is necessary to maximize the impact of urban greening. Rather than treating green spaces as isolated features, connecting them through corridors ensures that ecosystem services extend beyond individual parks and squares. These interconnected spaces provide recreational, ecological, and climate resilience benefits [51]. Well-designed green corridors enhance biodiversity, improve air quality, and promote healthier urban environments. They also play a critical role in stormwater management and flood resilience. Green and blue infrastructure, including permeable pavements, bioswales, rain gardens, urban tree cover, small ponds, and wetlands, offers an effective alternative to conventional stormwater systems [52]. These features help retain stormwater, reducing the risk of flooding while simultaneously improving water quality and groundwater recharge.

Another significant benefit is their ability to mitigate urban heat. When designed with adequate length and width and strategically positioned within the urban fabric, green and blue corridors create cooling effects [53]. They provide shade, reduce the urban heat island effect, and promote air circulation, leading to more comfortable urban environments. In addition to climate resilience, green corridors promote sustainable mobility and public health. They encourage walking, cycling, and other forms of micromobility, fostering low-emission transport options [54–56]. These spaces enhance urban livability by creating safe, pleasant routes for pedestrians and cyclists. Improved accessibility and safety contribute to healthier lifestyles and increased social interaction [55,57].

Furthermore, urban green corridors serve as vital habitats for flora and fauna. Research indicates that these spaces support bird populations, which contribute to residents' well-being as part of cultural ecosystem services [58]. The protection of pollinators, such as bees, has also been recognized in urban planning initiatives [59]. Studies further show that fragmented green spaces negatively impact biodiversity, whereas ecological corridors help mitigate habitat loss and species decline [60,61]. Additionally, maintaining healthy ecosystems within cities strengthens resilience against invasive species and pests.

Beyond the environmental benefits, green and blue corridors can enhance economic and social dynamics by linking green spaces with cultural landmarks, commercial areas, and community hubs. Integrating natural elements into urban planning creates multi-functional spaces that improve quality of life, stimulate local economies, and foster social cohesion [55]. Additional research highlights the strengths and challenges of green corridors across Europe and the United States, while recent studies have started to explore their potential in regions such as Malaysia and Saudi Arabia [56].

The ReGenWest intervention area highlights opportunities to implement practices for climate adaptation in cities. The existing network of pedestrian pathways, parks, yards, squares, and vacant plots makes this area particularly well suited for the implementation of green and blue corridors. These corridors will be effectively integrated into the local landscape, enhancing cooling, managing stormwater, and improving urban resilience while also delivering broader environmental and social benefits to the city.

4.2. Enhancing Urban Greenery at the Eye Level

The concept of the "city at the eye level" emphasizes the experiential quality of urban environments from a pedestrian perspective, focusing on the creation of sustainable, visually engaging, and interactive public spaces that foster resilience, social interaction, and comfort [62]. In this context, densifying urban greenery is not simply decorative but also a fundamental aspect of human-centered urban design that enhances sensory richness, psychological well-being, and environmental resilience. This need becomes particularly pressing in densely populated urban environments, like the study area, where creating greener walkable spaces is crucial. Effective urban design methods—such as integrating greenery through street trees, vertical systems, pocket parks, and green street furniture—can transform the urban fabric, making it more pedestrian-friendly and stimulating. These strategies are especially relevant in Thessaloniki's railway station district, where current sidewalk tree conditions, land cover constraints, and urban heat stress phenomena highlight the urgent need for context-sensitive greenery solutions that promote safety, resilience, and quality of life.

In the case study area, sidewalk trees are severely constrained by cement, which limits root growth and makes them vulnerable to collapse during extreme weather events. The removal of numerous trees without proper replacement exacerbates climate-related risks, including urban heat, flooding, and wind damage, and moves the area further away from its climate neutrality goals. This was highlighted in late 2024, when several trees were uprooted by strong winds, posing a danger to pedestrians and causing property damage. Additionally, the lack of exposed soil prevents proper water absorption, contributing to frequent flooding: 208 incidents were recorded between 2012 and 2018, with an increasing frequency in areas like the railway station. Beyond providing cooling and shading, tree roots and permeable surfaces are essential for pedestrian safety and climate resilience. In Thessaloniki, sidewalks are often rendered hazardous by dislodged cement and tiles, a direct result of constrained root systems.

Densifying street trees is, thus, a cornerstone of this strategy by offering shade, aesthetic appeal, and ecological benefits [63] but also encouraging outdoor activity, supporting physical and mental health, enhancing life satisfaction, and generating positive economic impacts on the economy of the nearby streets and blocks [64]. To maximize the benefits of this approach, it is essential to select climate-resilient species that provide dense, effective canopies while also addressing the key challenges associated with implementing street trees as NBSs in dense urban environments. These challenges include the potential obstruction of wind flow, damage to public property from root-induced pavement cracking (a significant issue in the study area that also creates tripping hazards for pedestrians), and the risk of triggering allergies from tree pollen [65]. Thoughtful species selection, strategic placement, and innovative design solutions are, therefore, critical to overcoming these obstacles and ensuring long-term success.

Another prominent strategy is the implementation of vertical green façades and green walls, which are particularly effective in high-density urban environments where horizontal space is constrained. These systems can be integrated into existing building exteriors, maximizing the use of available vertical space and contributing to urban cooling, noise reduction, and air purification while also enhancing the visual appeal of building facades and enriching the sensory experience for pedestrians [66]. Additionally, they provide valuable support for urban biodiversity, fostering a more sustainable urban ecosystem. A similar and very innovative solution to the challenge of limited ground-level green space in high-density areas is green shades. Green shades are a system of tensile sails covered with vegetation, offering an optimal solution on shopping streets, terraces, and squares where, due to a lack of space or the difficulty of intervention, trees or other types of vegetation cannot be installed. Finally, incorporating greenery into street furniture, such as benches and lighting columns, enables the flexible and adaptive greening of public spaces. All these NBS solutions are highly relevant to our case study, as they are specifically tailored to fit highly urban environments with limited space availability.

The densification of trees along urban roads plays a critical role not only in enhancing the pedestrian experience but also in advancing climate neutrality goals through carbon sequestration. Trees act as natural carbon sinks, absorbing CO_2 from the atmosphere and storing it in their biomass and surrounding soil. When strategically planted in dense formations along streets, they increase the cumulative carbon uptake potential of the urban landscape [67]. This process contributes directly to reducing atmospheric CO_2 concentrations, thereby complementing emission-reduction measures. In dense city areas like Thessaloniki's railway station district, where available land for large green spaces is limited, roadside tree densification offers a highly effective and space-efficient solution. Furthermore, long-term studies show that even modest increases in urban tree canopy coverage can significantly enhance the city's overall carbon storage capacity while mitigating the urban heat island effect [68].

Enhancing and densifying urban greenery at the eye level requires a multifaceted approach that considers various factors, including context-specific design, a multi-solution approach, technological advancements, and pedestrian-centric planning. A context-specific design approach emphasizes the need to tailor urban greening strategies to local urban morphology and climate conditions [69]. For example, factors such as urban morphology, shading, and local wind patterns are critical for ensuring that greenery not only thrives but also serves its intended purpose in the urban environment. Incorporating advanced technologies, such as remote sensing and deep learning, can further optimize the placement and performance of greenery interventions [69]. These technologies allow for a more precise understanding of factors like sunlight, shade, temperature, and wind, enabling planners to design spaces that maximize the benefits of greenery.

A multi-solution approach is also essential for densifying greenery at the eye level in dense urban areas, combining various categories of ground-level plantings and vertical greening systems. This strategy maximizes the available space while creating a more engaging urban environment for pedestrians. Ground-level greenery, such as street trees and shrubs, offers direct interaction with nature, while building-integrated green systems like green walls, vertical gardens, and green shades bring nature into architecture and provide more ecosystem services per unit of available space (e.g., contributing to temperature regulation, air quality, aesthetics, etc.). Pedestrian-centric planning is also a crucial element in enhancing urban greenery at the eye level. Hence, by placing plants and green elements where they can be easily interacted with—such as along sidewalks, on building facades, or integrated into street furniture—encourages people to connect with nature in their daily lives as well as promotes movement throughout the city (local neighborhoods), especially during hot summer days.

4.3. Integrating NBSs and Ecosystem Services into Metropolitan Mobility Infrastructures

Metropolitan mobility infrastructures, including central train and bus stations, airports, and other transportation hubs, are critical components of urban development and economic activity. However, these infrastructures often generate significant environmental pressures, such as air and noise pollution, habitat fragmentation, and increased land consumption [70]. As cities face escalating challenges from climate change, biodiversity loss, and urban sprawl, the integration of NBSs and ecosystem services into metropolitan mobility infrastructure presents an innovative and sustainable approach to urban planning and resilience [71]. Rethinking mobility infrastructure in terms of ecological and environmental performance is essential for reducing the negative impacts of urban mobility systems while enhancing the resilience and ecological health of metropolitan areas [72].

Mobility infrastructures in metropolitan areas are closely linked to several environmental and ecological challenges. These include air and noise pollution, land consumption and urban sprawl, habitat fragmentation and biodiversity loss, and a high level of climate vulnerability. Transportation hubs are major sources of greenhouse gas emissions, contributing to climate change and deteriorating urban air quality [73]. The physical footprint of transport infrastructures often results in the fragmentation of natural habitats, reducing landscape connectivity and impeding species migration [74]. This, in connection with land consumption, can disrupt local ecosystems, both natural and human, by creating discontinuous and unfamiliar places. Mobility infrastructures are vulnerable to climate-related events such as heat waves, flooding, and storms.

Integrating NBSs and ecosystem services into metropolitan mobility infrastructures requires a shift from traditional grey infrastructure (e.g., concrete and steel) to hybrid or green infrastructure solutions that incorporate natural elements and processes under sophisticated landscaping strategies. One effective strategy for integrating NBSs into mobility infrastructure is to rethink and redesign them in strong connection with the preexisting landscape. Linear parks, vegetated embankments, and green roofs on transport stations can enhance urban biodiversity while mitigating air pollution and noise. For example, the Thessaloniki Railway Station could be redesigned to include green roofs and vertical gardens, which would enhance biodiversity, provide insulation to reduce energy consumption, and improve air quality. Moreover, planting native vegetation outside the station building or along railway lines and roadways could restore ecological connectivity and create microhabitats for urban natural life.

There are many pioneer examples that follow this design process, such as the Singapore Rail Corridor by MVRDV or the most recent design project for the expansion of Athens International Airport (AIA) by the Anemos Design Consortium, consisting of Grimshaw, Haptic, K-Studio, Arup, Leslie Jones, Triagonal, and Plan A. The Singapore Rail Corridor is more than a park connector or a trail; it is an inspired community space offering the extraordinary experience of the reinvention of hidden spaces within Singapore to inspire movement and connection, intuitive play, and new ways of experiencing the environment [75]. On the other hand, the design approach for AIA's expansion, rooted in a deep respect for the generous Athenian–Mediterranean landscape, prioritizes sustainability and operational excellence, building on AIA's commitment to net-zero operations and low-carbon principles. With a rich landscaping strategy and an infusion of daylight in all parts of the terminal building, the sense of place is always dominant [76]. Other solutions may include integrating bioswales and rain gardens around the Thessaloniki Railway Station, which could reduce surface runoff, improve water quality, and create green spaces for public use. Additionally, restoring natural floodplains along transport corridors would enhance water retention and reduce the vulnerability of transport systems to extreme rainfall events.

Transforming metropolitan mobility infrastructures into multifunctional spaces that integrate ecosystem services can increase urban resilience and adaptability to climate change and provide social and economic benefits. This shift toward ecological urbanism in transport infrastructure requires adaptive governance, innovative financing mechanisms, and stakeholder collaboration.

4.4. Establishing Virtual Power Plants for Smart, Distributed Energy Management

The ReGenWest renovation area presents multiple opportunities for local, distributed renewable energy production, primarily through the installation of solar panels. The most promising locations for integrating RESs are the flat rooftops of apartment buildings. These rooftops offer a valuable opportunity not only to generate clean energy but also to create shared green spaces within a dense urban environment. Their direct exposure to sunlight makes them ideal for capturing solar radiation, while incorporating green roof systems helps mitigate urban heat island effects and improve air quality.

The integration of RESs with NBSs on rooftops yields multiple synergies. Green roofs naturally cool solar panels, enhancing their efficiency, while the vegetation helps filter air pollutants and reduces dust accumulation on the solar cells, further improving performance. When used for micro-agriculture, such as community allotments, these rooftops can also contribute to food security and urban self-sufficiency. Their semi-private nature enables residents to extend their living environments and encourages community interaction. With coordinated rooftop management, apartment buildings can effectively combine photovoltaic panels, solar water heaters, and small-scale gardens or allotments—transforming underused surfaces into resilient, productive urban assets.

Additional opportunities for distributed RESs exist in currently vacant or underutilized plots in the area, such as asphalt-covered car parks, which intensify heat island effects during summer. These spaces can be redesigned to incorporate RESs and NBSs, such as shaded pedestrian paths, solar panel canopies, and EV charging infrastructure. Educational facilities, both primary and secondary, also offer significant potential for rooftop RES installations, which can be paired with appropriate NBSs. Similarly, open areas around social housing complexes could accommodate shaded solar structures for scooters and vehicles, while solar-powered lighting can be deployed using ground-mounted, wall-mounted, or tree-mounted fixtures. This local potential for renewable energy generation provides a strong foundation for a more integrated initiative: the implementation of a virtual power plant (VPP) at the city district level. Such a system would combine solar energy production, smart grid infrastructure, and energy storage solutions—supporting the transition toward self-sufficient net-zero energy districts (S-NZEDs).

To integrate distributed RESs into carbon-neutral smart cities and enhance transparency and control, virtual power plant systems serve as an essential mechanism. A VPP aggregates numerous small-scale Distributed Energy Resources (DERs) into a coordinated virtual unit that functions like a conventional power plant, offering both observability and manageability [77]. Typically, a VPP integrates various technologies such as combined heat and power units, fuel cells, photovoltaic panels, heat pumps, solar thermal collectors, and other thermal or electrical energy systems, all operated via a central control platform [78]. This architecture allows for a blend of generation assets, including both dispatchable units—capable of adjusting output on demand—and intermittent sources dependent on environmental variability, as well as flexible loads and energy storage systems [79].

A notable example of a district-level VPP is the GridFriends initiative, part of the Schoonschip project launched by Amsterdam Smart City [80]. Schoonschip is a sustainable residential community of 46 households located in northern Amsterdam. One of the project's defining features is its intelligent microgrid, which enables residents to act as both energy producers and grid operators. This setup was made possible through a special regulatory exemption granted by the Dutch Ministry of Economic Affairs. The community has a single shared connection to the public electricity grid, while all infrastructure beyond the meter is privately owned and managed by the residents. The smart grid facilitates peer-to-peer energy exchange, with algorithms optimizing energy flows. During summer, the system maximizes local solar energy use, whereas in winter, the priority shifts to minimizing peak loads on the main public grid connection. By coordinating energy storage and distribution, the VPP allows the community to balance supply and demand internally, resulting in an energy-neutral neighborhood powered through platform-based energy management [80]. Schoonschip illustrates how a prosumer-based energy culture can drive sustainable transitions. Since its launch in 2018, the smart grid has expanded across multiple new projects and remains one of the most advanced residential smart grids in the world [81].

The added value of district-level VPPs in the transition toward carbon-neutral smart cities is considerable. Their benefits fall into three primary areas. First, VPPs facilitate a shift in energy culture and behavior—an essential condition for carbon-neutral smart cities. Participants in a VPP, understood as complementors within a platform ecosystem, are no longer passive energy consumers. Instead, they become prosumers, actively influencing both energy supply and demand. As small-scale DERs become more widespread, this cultural shift gains importance. While today's energy production remains largely centralized, the rise of renewable sources is driving a steady move toward decentralization, with VPPs playing a central role in efficiently managing these distributed resources [82].

Second, VPPs help reduce stress on the electricity network. By enabling local energy generation and consumption, they reduce the need for long-distance, high-voltage transmission, thereby minimizing energy losses. More critically, VPPs help overcome network bottlenecks that arise when the capacity of RESs exceeds the distribution system's ability to transfer them. At peak production times, RES units are often curtailed to preserve grid stability. VPPs address this challenge by coordinating generation, storage, and consumption at the local level, thereby increasing the overall efficiency and utilization of renewable energy.

Third, a model we developed to evaluate the transition toward self-sufficient net-zero energy districts in European cities shows that such a transition is feasible through a combination of two energy technologies: energy savings via building refurbishment and smart systems on the one hand, and localized renewable energy generation through photovoltaic panels installed on public and private buildings on the other [83]. In Southern European cities, this combination is sufficient to achieve S-NZEDs. However, in Central and Northern Europe—where energy demand is higher and solar yields are lower—achieving net-zero status requires additional renewable energy imported from external sources.

4.5. Designing an Environmental Monitoring System for Improving Efficiency

The aforementioned analysis has shown that urban planning and design is a critical aspect in this process, as it can optimize the effectiveness of combining theoretical principles and practical applications. However, alongside traditional design practices, the literature also highlights the importance of applying continuous environmental monitoring to maximize the potential outcomes of NBS and RES integration in urban settings [17,84,85].

Environmental monitoring systems (EMSs), by utilizing real-time data collection through remote sensing, IoT sensors, and GIS mapping, provide insights into air quality, temperature regulation, water management, and energy efficiency. Continuous monitoring helps identify potential inefficiencies and areas for improvement, ensuring that implemented solutions adapt to changing environmental conditions [86]. In this context, data-driven approaches facilitate evidence-based decision-making, allowing urban planners to optimize the design, placement, and functionality of NBSs and RESs for enhanced urban resilience and sustainability [84,85].

For the railway station district in Thessaloniki, an initial EMS design has been developed (Figure 3). The key components of the proposed system include a data collection network, consisting of CO_2 sensors, meteorological stations, air quality monitors, and photosynthesis measurement stations, strategically placed in parks and on streets and rooftops. A data management platform connects these sensors through an IoT network, storing data in the cloud for real-time access and historical analysis. Advanced machine learning algorithms help detect trends and anomalies and predict future CO_2 absorption, while GIS mapping provides dynamic visualizations of intervention impacts. Additionally, the EMS incorporates external data sources such as satellite imagery (e.g., Sentinel-2, Landsat), citizen science contributions via mobile apps, and municipal or academic datasets to enrich environmental assessments. To quantify impact, CO_2 absorption indicators compare baseline values with post-intervention data, aligning findings with climate neutrality goals. Finally, interactive digital tools like real-time dashboards, GIS mapping, and automated alerts enhance decision-making and public engagement. Citizen participation is further encouraged through crowdsourced environmental reporting, open data access, and participatory design platforms, ensuring the continuous refinement and optimization of NBS and RES integration in urban settings.

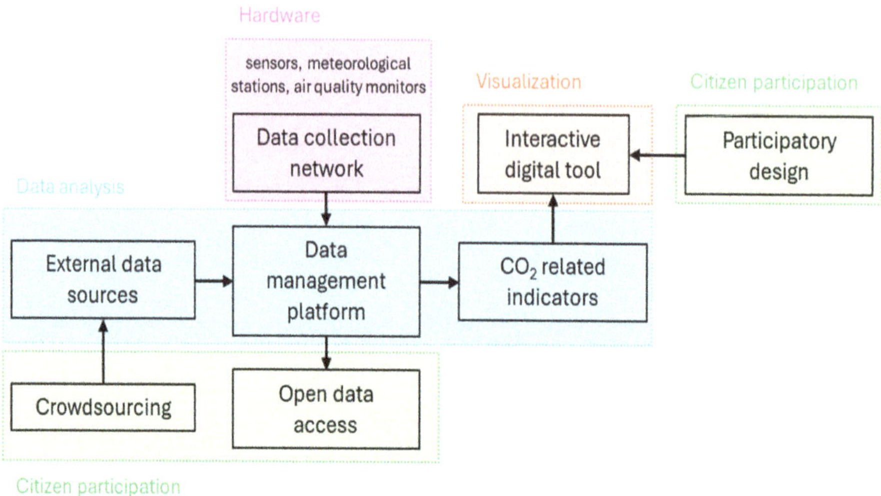

Figure 3. Schematic representation of the EMS designed for the railway station district in Thessaloniki.

Building on this framework, EMS data will enable the assessment of various configurations of NBSs and RESs, evaluating their influence on environmental performance and progress toward carbon neutrality. Agent-Based Modeling can simulate how different public policy interventions in NBS and RES types, as well as actions initiated by citizens and organizations within the district, contribute to advancing closer to or achieving a net-zero state. This iterative modeling supports adaptive design, allowing city planners to test hypotheses, explore trade-offs, and co-create solutions with stakeholders that optimize carbon neutrality impact. Ultimately, the synergy between real-time data, predictive analytics, and community input establishes a dynamic feedback loop, fostering more informed, inclusive, and responsive urban design practices that advance the district toward climate neutrality.

5. Aligning NBS–RES Integration with Positive Energy Districts

The integrated approach of combining nature-based solutions with renewable energy sources, as proposed in this study, provides a critical pathway that must be combined with the development of Positive Energy Districts (PEDs). PEDs represent an advanced urban sustainability model, aiming not only to achieve a net-zero energy balance but also to generate a surplus of renewable energy within clearly defined urban boundaries. While both carbon-neutral city districts and PEDs pursue environmental sustainability, their complementarity lies in the ambition of PEDs to actively contribute renewable energy to the broader grid, thereby reinforcing regional energy resilience and accelerating decarbonization.

The integration of NBSs and RESs supports this transition through three interrelated mechanisms: (1) reducing energy demand via microclimate regulation and enhanced thermal comfort; (2) enabling localized renewable energy generation by embedding energy infrastructure in green urban fabrics and buildings; and (3) offsetting residual emissions by leveraging the carbon sink capacities of urban ecosystems. NBSs—such as tree canopies, vegetated corridors, and green roofs—mitigate the urban heat island effect, lower ambient temperatures, and enhance thermal comfort in both public and private spaces. This localized cooling effect significantly reduces building energy demand, particularly for air conditioning during peak summer months. When co-deployed with RES technologies—such as building-integrated photovoltaics, solar-shaded urban furniture, or hybrid green roofs with PV panels—these multifunctional infrastructures enable on-site renewable energy production while preserving urban biodiversity and spatial quality. In parallel, the ecological services provided by NBSs—such as carbon sequestration through dense vegetation, restored soils, and urban rewilding—play a vital role in compensating for residual emissions, including those embedded in construction materials or resulting from unavoidable transport-related activities.

However, achieving carbon neutrality demands broader systemic transformations. These include the deep energy retrofitting of existing buildings to drastically reduce energy use; smart energy management systems that dynamically balance real-time supply and demand; demand-side flexibility through digital platforms and active community participation; and electrified and integrated mobility solutions, such as Mobility as a Service (MaaS), to reduce transport-related emissions. Smart city technologies further reinforce this transition by enabling adaptive infrastructure governance, real-time monitoring, and participatory planning mechanisms aligned with both environmental goals and community needs.

Crucially, carbon neutrality practices alone are insufficient to address the seasonal mismatch between renewable energy generation and energy consumption. Given the limited capacity for urban energy storage at the city level and the pronounced seasonal variability—characterized by energy surpluses in summer and deficits in winter—districts must function as seasonally positive energy producers. This means exporting surplus renewable energy during periods of high generation (typically summer) and importing equivalent amounts during periods of high demand (typically winter) while maintaining an annual positive energy balance.

The synergetic application of NBSs and RESs, when embedded within broader carbon neutrality strategies, aligns closely with the PED framework's ambitions. The railway station district pilot in Thessaloniki exemplifies how underutilized and typologically diverse urban spaces—such as schoolyards, rooftops, road medians, and vacant plots—can be transformed into multifunctional assets. These interventions simultaneously provide localized climate regulation, generate renewable energy, and function as urban carbon sinks. By combining these design strategies with PED planning, the district can achieve a net-positive energy performance on an annual basis without the need for large-scale seasonal energy storage, thereby accelerating the transition toward low-carbon, regenerative urban futures.

6. Conclusions

As urban areas strive for more sustainable and resilient futures, new approaches to enhancing urban sustainability are being explored. This paper argues that integrating NBSs with RESs presents a promising strategy, as NBSs primarily focus on improving ecosystem services and urban livability, while RESs are essential for reducing carbon emissions and promoting energy independence. Although each approach individually contributes to urban sustainability, their combined implementation holds significant potential to create new opportunities and maximize environmental, social, and economic benefits. To achieve this, it is crucial to develop a comprehensive, well-structured framework for integrating NBSs and RESs that incorporates urban planning and energy principles to effectively address local specificities in the design process.

The railway station district of Thessaloniki has been used as our case study for analyzing how NBSs can be combined with RES solutions, considering the various typologies of the available open spaces as well as the ecosystem services being impacted through NBS solutions. Our results suggest that there are potential windows of opportunity for combining NBSs with RESs, including solutions such as green roofs with photovoltaic panels, solar lighting, and solar parking shaders. The detailed analysis of how the integration of NBSs and RESs can enhance ecosystem services has shown that cultural and regulatory services are consistently impacted across all cases, while provisioning and supporting services are more closely tied to specific interventions and experience a lesser impact from the integration of NBSs and RESs.

We go beyond the immediate context of Thessaloniki by proposing a typology of five major urban project types that offer clear potential for replication and scaling in other city districts. More specifically, replication and scale-up are made possible by developing green and blue corridors to restore ecological connectivity and manage urban microclimates as well as enhancing urban greenery at the eye level to improve walkability, thermal comfort, and social interaction. Moreover, integrating NBSs and ecosystem services into metropolitan mobility infrastructures to support multifunctionality and reduce environmental impacts and establishing virtual power plants for smart, distributed energy management that combines renewable generation with demand-side flexibility offer two additional replicable solutions for boosting NBS–RES integration. In addition, developing an environmental monitoring system (EMS) to guide data-driven decision-making and optimize the efficiency of interventions can be considered a critical factor complementing the previous typologies.

These project typologies provide a flexible yet structured framework that can be adapted to any specific urban morphology, governance structures, and climate priorities of other cities. This strengthens the generalizability of the proposed design approach and supports knowledge transfer for wider implementation in European and international urban contexts. Moreover, by aligning NBSs and RESs within the PED framework, cities can not only enhance energy self-sufficiency and climate resilience but also create replicable models for regenerative urban transformation.

However, the integration of NBSs with RESs faces several limitations, particularly in densely built urban areas, such as the one examined here. These challenges include spatial constraints due to limited available land, socio-economic factors that may hinder adoption and maintenance, and technical difficulties posed by uneven terrain and infrastructure. Moreover, regulatory and governance issues can complicate the coordination of NBS and RES objectives, while the ecological suitability of certain areas, reflected in features such as the lack of natural water bodies or adequate green spaces, limits the types of interventions that can be implemented.

At the same time, transitioning toward integrated NBS and RES solutions requires not only technological advancements but also public awareness and stakeholder engagement. Aligning urban sustainability strategies with financial incentives, participatory governance, and community involvement can further enhance the effectiveness of integrated solutions [87–89]. Energy communities, for instance, foster collective ownership of local energy resources and contribute to urban energy self-sufficiency [33]. Ensuring the long-term sustainability of these solutions requires consistent maintenance and public commitment, as there is a risk of unintended social impacts, such as gentrification, that could displace lower-income residents.

Future research should continue to evaluate the long-term viability of hybrid systems, assess their contribution to urban resilience, and develop standardized metrics for measuring their impact on biodiversity and sustainability [35]. It should also explore context-specific frameworks for optimizing NBS–RES integration across varying urban environments, considering factors such as spatial configuration, energy demands, and local climate conditions. Finally, it should focus on engaging stakeholders—including local authorities, researchers, urban planners, and community representatives –in a co-design process of NBS and RES technologies to allow for adaptive and context-specific planning. This inclusive approach not only enhances social acceptability but also improves the long-term sustainability of implemented solutions by ensuring that they align with community needs.

Author Contributions: Conceptualization, A.P., N.K., D.L., I.P. and P.T.; Data curation, E.G.; Formal analysis, N.K., D.L., I.P., E.G. and P.T.; Methodology, N.K., D.L., I.P. and P.T.; Supervision, A.P.; Writing—original draft, A.P., N.K., D.L., I.P. and P.T. All authors have read and agreed to the published version of the manuscript.

Funding: This research was funded by the European Commission 'NetZeroCities' project, grant number SGA NZC 101121530.

Data Availability Statement: The raw data supporting the conclusions of this article will be made available by the authors on request.

Conflicts of Interest: The authors declare no conflicts of interest.

References

1. Kılkış, Ş.; Krajačić, G.; Duić, N.; Rosen, M.A. Advances in integration of energy, water and environment systems towards climate neutrality for sustainable development. *Energy Convers. Manag.* **2020**, *225*, 113410. [CrossRef]
2. Seto, K.C.; Dhakal, S.; Bigio, A.; Blanco, H.; Carlo Delgado, G.; Dewar, D.; Huang, L.; Inaba, A.; Kansal, A.; Lwasa, S.; et al. Human settlements, infrastructure, and spatial planning. In *Climate Change 2014: Mitigation of Climate Change. Contribution of Working Group III to the Fifth Assessment Report of the Intergovernmental Panel on Climate Change*; Cambridge University Press: Cambridge, UK, 2014.
3. Luqman, M.; Rayner, P.J.; Gurney, K.R. On the impact of urbanisation on CO_2 emissions. *NPJ Urban Sustain.* **2023**, *3*, 6. [CrossRef]
4. Kabisch, N.; Frantzeskaki, N.; Pauleit, S.; Naumann, S.; Davis, M.; Artmann, M.; Haase, D.; Knapp, S.; Korn, H.; Stadler, J.; et al. Nature-based solutions to climate change mitigation and adaptation in urban areas: Perspectives on indicators, knowledge gaps, barriers, and opportunities for action. *Ecol. Soc.* **2016**, *21*, 39. [CrossRef]
5. Dorst, H.; Van der Jagt, A.; Raven, R.; Runhaar, H. Urban greening through nature-based solutions–Key characteristics of an emerging concept. *Sustain. Cities Soc.* **2019**, *49*, 101620. [CrossRef]
6. Cohen-Shacham, E.; Andrade, A.; Dalton, J.; Dudley, N.; Jones, M.; Kumar, C.; Maginnis, S.; Maynard, S.; Nelson, C.R.; Renaud, F.G.; et al. Core principles for successfully implementing and upscaling Nature-based Solutions. *Environ. Sci. Policy* **2019**, *98*, 20–29. [CrossRef]
7. Susca, T.; Zanghirella, F.; Colasuonno, L.; Del Fatto, V. Effect of green wall installation on urban heat island and building energy use: A climate-informed systematic literature review. *Renew. Sustain. Energy Rev.* **2022**, *159*, 112100. [CrossRef]
8. Liu, H.Y.; Skandalos, N.; Braslina, L.; Kapsalis, V.; Karamanis, D. Integrating solar energy and nature-based solutions for climate-neutral urban environments. *Solar* **2023**, *3*, 382–415. [CrossRef]
9. Miguez, M.G.; Raupp, I.P.; Veról, A.P. An integrated quantitative framework to support design of resilient alternatives to manage urban flood risks. *J. Flood Risk Manag.* **2019**, *12*, e12514. [CrossRef]
10. Kumar, P.; Debele, S.E.; Sahani, J.; Aragão, L.; Barisani, F.; Basu, B.; Bucchignani, E.; Charizopoulos, N.; Di Sabatino, S.; Domeneghetti, A.; et al. Towards an operationalisation of nature-based solutions for natural hazards. *Sci. Total Environ.* **2020**, *731*, 138855. [CrossRef]
11. IUCN. *Global Standard for Nature-Based Solutions: A User-Friendly Framework for the Verification, Design and Scaling up of NbS*; IUCN: Gland, Switzerland, 2020.
12. Faivre, N.; Fritz, M.; Freitas, T.; De Boissezon, B.; Vandewoestijne, S. Nature-Based Solutions in the EU: Innovating with nature to address social, economic and environmental challenges. *Environ. Res.* **2017**, *159*, 509–518. [CrossRef]
13. Davies, C.; Chen, W.Y.; Sanesi, G.; Lafortezza, R. The European Union roadmap for implementing nature-based solutions: A review. *Environ. Sci. Policy* **2021**, *121*, 49–67. [CrossRef]
14. Citaristi, I. United Nations Environment Programme—UNEP. In *The Europa Directory of International Organizations*; Routledge: Oxfordshire, UK, 2022.
15. Kleerekoper, L.; Van Esch, M.; Salcedo, T.B. How to make a city climate-proof, addressing the urban heat island effect. *Resour. Conserv. Recycl.* **2012**, *64*, 30–38. [CrossRef]
16. UNFCCC. UN Climate Change Conference—United Arab Emirates. 2023. Available online: https://unfccc.int/cop28 (accessed on 23 March 2025).
17. Raymond, C.M.; Frantzeskaki, N.; Kabisch, N.; Berry, P.; Breil, M.; Nita, M.R.; Geneletti, D.; Calfapietra, C. A framework for assessing and implementing the co-benefits of nature-based solutions in urban areas. *Environ. Sci. Policy* **2017**, *77*, 15–24. [CrossRef]

18. Komninos, N.; Panori, A.; Kakderi, C. Smart cities beyond algorithmic logic: Digital platforms, user engagement and data science. In *Smart Cities in the Post-Algorithmic Era*; Edward Elgar Publishing: Cheltenham, UK, 2019; pp. 1–15.
19. Dumitru, A.; Wendling, L. *Evaluating the Impact of Nature-Based Solutions: A Handbook for Practitioners*; Publications Office of the European Union: Luxembourg, 2021. [CrossRef]
20. Langemeyer, J.; Connolly, J.J.T. Weaving notions of justice into urban ecosystem services research and practice. *Environ. Sci. Policy* **2020**, *109*, 1–14. [CrossRef]
21. Frantzeskaki, N.; McPhearson, T.; Collier, M.J.; Kendal, D.; Bulkeley, H.; Dumitru, A.; Pinter, L. Nature-based solutions for urban climate change adaptation: Linking science, policy, and practice communities for evidence-based decision-making. *BioScience* **2019**, *69*, 455–466. [CrossRef]
22. Chelleri, L.; Waters, J.J.; Olazabal, M.; Minucci, G.J.E. Urbanization Resilience trade-offs: Addressing multiple scales and temporal aspects of urban resilience. *Environ. Urban.* **2015**, *27*, 181–198. [CrossRef]
23. Turkelboom, F.; Leone, M.; Jacobs, S.; Kelemen, E.; Garcia-Llorente, M.; Baro, F.; Rusch, V. When we cannot have it all: Ecosystem services trade-offs in the context of spatial planning. *Ecosyst. Serv.* **2018**, *29*, 566–578. [CrossRef]
24. Dumitru, A.; Frantzeskaki, N.; Collier, M. Identifying principles for the design of robust impact evaluation frameworks for nature-based solutions in cities. *Environ. Sci. Policy* **2020**, *112*, 107–116. [CrossRef]
25. Johnson, B.A.; Kumar, P.; Okano, N.; Dasgupta, R.; Shivakoti, B.R. Nature-based solutions for climate change adaptation: A systematic review of systematic reviews. *Nat.-Based Solut.* **2022**, *2*, 100042. [CrossRef]
26. Goodwin, S.; Olazabal, M.; Castro, A.J.; Pascual, U. Global mapping of urban nature-based solutions for climate change adaptation. *Nat. Sustain.* **2023**, *6*, 458–469. [CrossRef]
27. Palermo, V.; Walsh, C.L.; Dawson, R.J.; Fichera, A.; Inturri, G. Multi-sector mitigation strategies at the neighbourhood scale. *J. Clean. Prod.* **2018**, *187*, 893–902. [CrossRef]
28. Joshi, N.; Agrawal, S.; Lie, S. What does neighbourhood climate action look like? A scoping literature review. *Clim. Action* **2022**, *1*, 10. [CrossRef]
29. Marquez-Ballesteros, M.J.; Navas-Carrillo, D.; Mora-López, L.; Sidrach-de-Cardona, M.; Enciso-Martínez, E.; Rico-Pinazo, P. Citizens and local administration in climate change mitigation. Urban strategies and local actions applicable to neighborhoods. In *Resilient and Sustainable Cities*; Elsevier: Amsterdam, The Netherlands, 2023; pp. 495–517.
30. Adam, K.; Hoolohan, V.; Gooding, J.; Knowland, T.; Bale, C.; Tomlin, A. Methodologies for city-scale assessment of renewable energy generation potential to inform strategic energy infrastructure investment. *Cities* **2016**, *54*, 45–56. [CrossRef]
31. Delponte, I.; Schenone, C. RES Implementation in urban areas: An updated overview. *Sustainability* **2020**, *12*, 382. [CrossRef]
32. Cooney, C.; Byrne, R.; Lyons, W.; O'Rourke, F. Performance characterisation of a commercial-scale wind turbine operating in an urban environment, using real data. *Energy Sustain. Dev.* **2017**, *36*, 44–54. [CrossRef]
33. Niemi, R.; Mikkola, J.; Lund, P. Urban energy systems with smart multi-carrier energy networks and renewable energy generation. *Renew. Energy* **2012**, *48*, 524–536. [CrossRef]
34. Cavadini, G.B.; Cook, L.M. Green and cool roof choices integrated into rooftop solar energy modelling. *Appl. Energy* **2021**, *296*, 117082. [CrossRef]
35. Fang, X.; Li, J.; Ma, Q. Integrating green infrastructure, ecosystem services and nature-based solutions for urban sustainability: A comprehensive literature review. *Sustain. Cities Soc.* **2023**, *98*, 104843. [CrossRef]
36. O'Connor, M. All in the community: Using community solar gardens to bring the benefits of renewable energy to low-income communities. *Georgetown Environ. Law Rev.* **2019**, *31*, 391–407.
37. Hakimizad, S.; Asl, S.R.; Ghiai, M.M. A review on the design approaches using renewable energies in urban parks. *Int. J. Renew. Energy Res.* **2015**, *5*, 686–693.
38. Upton, R. Learning for Change During the Global Climate Crisis: Exploring the (Noneconomic) Impact of Education for Sustainability Using the Eden Project as a Case Study. In *Handbook of Climate Change Management: Research, Leadership, Transformation*; Springer International Publishing: Berlin/Heidelberg, Germany, 2021; pp. 4615–4637.
39. Porse, E.; Fournier, E.; Cheng, D.; Hirashiki, C.; Gustafson, H.; Federico, F.; Pincetl, S. Net solar generation potential from urban rooftops in Los Angeles. *Energy Policy* **2020**, *142*, 111461. [CrossRef]
40. Razzaghi Asl, S. Re-powering the nature-intensive systems: Insights from linking nature-based solutions and energy transition. *Front. Sustain. Cities* **2022**, *4*, 860914. [CrossRef]
41. Tasneem, Z.; Al Noman, A.; Das, S.K.; Saha, D.K.; Islam, M.R.; Ali, M.F.; Badal, M.F.R.; Ahamed, M.H.; Moyeen, S.I.; Alam, F. An analytical review on the evaluation of wind resource and wind turbine for urban application: Prospect and challenges. *Dev. Built Environ.* **2020**, *4*, 100033. [CrossRef]
42. Xing, Y.; Jones, P.; Donnison, I. Characterisation of nature-based solutions for the built environment. *Sustainability* **2017**, *9*, 149. [CrossRef]

43. Ecologic Institute. Assessment of the Potential of Ecosystem-Based Approaches to Climate Change Adaptation and Mitigation in Europe. 2011. Available online: https://www.ecologic.eu/sites/default/files/publication/2022/2345_EbA_EBM_CC_FinalReport_23Nov2011.pdf (accessed on 13 March 2025).
44. Ndubisi, F.; DeMeo, T.; Ditto, N.D. Environmentally sensitive areas: A template for developing greenway corridors. *Landsc. Urban Plan.* **1995**, *33*, 159–177. [CrossRef]
45. World Bank. *A Catalogue of Nature-Based Solutions for Urban Resilience*; World Bank Group: Washington, DC, USA, 2021.
46. Eisenberg, B.; Chiesa, C.; Fischer, L.; Jakstis, K.; Polcher, H.G. *Nature-Based Solutions—Technical Handbook Factsheets*; UNaLab Urban Nature Labs: Stuttgart, Germany, 2022.
47. Eisenberg, B.; Polcher, V. Nature-Based Solutions Technical Handbook. UNaLab Horizon: Brussels, Belgium, 2020. Available online: https://unalab.eu/system/files/2020-02/unalab-technical-handbook-nature-based-solutions2020-02-17.pdf (accessed on 28 February 2025).
48. World Resources Institute. Ecosystem Services—A Guide for Decision Makers, W.R.I. 2008. Available online: https://www.wri.org/research/ecosystem-services (accessed on 2 February 2025).
49. Barbosa, V.R.F.; Damasceno, R.M.; Dias, M.A.; Castelhano, F.J.; Roig, H.L.; Requia, W.J. Ecosystem services provided by green areas and their implications for human health in Brazil. *Ecol. Indic.* **2024**, *161*, 111975. [CrossRef]
50. Almassy, D.; Pinter, L.; Rocha, S.; Naumann, S.; Davis, M.; Abhold, K.; Bulkeley, H. Urban Nature Atlas: A Database of Nature-Based Solutions Across 100 EUROPEAN Cities. Report of H2020 Project Naturvation. 2018. Available online: https://naturvation.eu/sites/default/files/result/files/urban_nature_atlas_a_database_of_nature-based_solutions_across_100_european_cities.pdf (accessed on 22 February 2025).
51. Aziz, H.A.; Rasidi, M.H. The role of green corridors for wildlife conservation in urban landscape: A literature review. In *IOP Conference Series: Earth and Environmental Science*; IOP Publishing: Bristol, UK, 2014; Volume 18, p. 012093.
52. Kaur, R.; Gupta, K. Blue-Green Infrastructure (BGI) network in urban areas for sustainable storm water management: A geospatial approach. *City Environ. Interact.* **2022**, *16*, 100087. [CrossRef]
53. Jiang, Y.; Huang, J.; Shi, T.; Li, X. Cooling island effect of blue-green corridors: Quantitative comparison of morphological impacts. *Int. J. Environ. Res. Public Health* **2021**, *18*, 11917. [CrossRef]
54. Rasli, F.N.; Kanniah, K.D. Green corridors for liveable and walkable city: A case of Kuala Lumpur. *Chem. Eng. Trans.* **2018**, *63*, 391–396.
55. Aloshan, M.; Gharieb, M.; Heba, K.M.; Khalil, R.; Alhumaid, M.H.; Ezz, M.S. Promoting Urban Corridors in Saudi City Center to Enhance Walkability Using Multi-Criteria Decision-Analysis Methods. *Sustainability* **2024**, *16*, 9255. [CrossRef]
56. Balsas, C.J.; Blair, N. Walkable and Sustainable City Centre Greenway Planning. *Sustainability* **2025**, *17*, 2897. [CrossRef]
57. Telega, A.; Telega, I.; Bieda, A. Measuring walkability with GIS—Methods overview and new approach proposal. *Sustainability* **2021**, *13*, 1883. [CrossRef]
58. Turo, K.J.; Gardiner, M.M. From potential to practical: Conserving bees in urban public green spaces. *Front. Ecol. Environ.* **2019**, *17*, 167–175. [CrossRef]
59. Threlfall, C.G.; Walker, K.; Williams, N.S.; Hahs, A.K.; Mata, L.; Stork, N.; Livesley, S.J. The conservation value of urban green space habitats for Australian native bee communities. *Biol. Conserv.* **2015**, *187*, 240–248. [CrossRef]
60. Bailey, S. Increasing connectivity in fragmented landscapes: An investigation of evidence for biodiversity gain in woodlands. *For. Ecol. Manag.* **2007**, *238*, 7–23. [CrossRef]
61. Gilbert-Norton, L.Y.N.N.E.; Wilson, R.; Stevens, J.R.; Beard, K.H. A meta-analytic review of corridor effectiveness. *Conserv. Biol.* **2010**, *24*, 660–668. [CrossRef]
62. Karssenberg, H.; Laven, J.; Glaser, M.; Van't Hoff, M. *The City at Eye Level: Lessons for Street Plinths*; Glaser, M., Ed.; Eburon Uitgeverij BV: Utrecht, The Netherlands, 2016; pp. 14–25.
63. Seamans, S.G. Mainstreaming the environmental benefits of street trees. *Urban For. Urban Green.* **2013**, *12*, 2–11. [CrossRef]
64. Lu, Y. Using Google Street View to investigate the association between street greenery and physical activity. *Landsc. Urban Plan.* **2019**, *191*, 103435. [CrossRef]
65. Weidmüller, N.; Knopp, J.M.; Beber, J.; Krnjaja, G.M.; Banzhaf, E. Local planning scenario for shading from trees as an urban nature-based solution. *City Environ. Interact.* **2024**, *23*, 100154. [CrossRef]
66. Hamling, A.; Van Dam, C.; Dubbeldee, V. Vertical Green Landscape Facades as a Reducer of Pollution in High Density Cities: A Review. *Urban Plan. Constr.* **2024**, *2*, 128. [CrossRef]
67. Nowak, D.J.; Crane, D.E.; Stevens, J.C. Air pollution removal by urban trees and shrubs in the United States. *Urban For. Urban Green.* **2013**, *4*, 115–123. [CrossRef]
68. Strohbach, M.W.; Haase, D. Above-ground carbon storage by urban trees in Leipzig, Germany: Analysis of patterns in a European city. *Landsc. Urban Plan.* **2012**, *104*, 95–104. [CrossRef]
69. Hua, J.; Cai, M.; Shi, Y.; Ren, C.; Xie, J.; Chung, L.C.H.; Webster, C. Investigating pedestrian-level greenery in urban forms in a high-density city for urban planning. *Sustain. Cities Soc.* **2022**, *80*, 103755. [CrossRef]

70. Caba, J.; Berra-Sandín, M.; Krebs, R.; Brian, R. *Rethink Mobility Infrastructure for Better Metropolises*; European Commission & URBACT: Brussels, Belgium, 2022. Available online: https://urbact.eu/sites/default/files/2023-01/RiConnect_Final_Report.pdf (accessed on 29 March 2025).
71. Kabisch, N.; Korn, H.; Stadler, J.; Bonn, A. *Nature-Based Solutions to Climate Change Adaptation in Urban Areas, Theory and Practice of Urban Sustainability Transitions*; Springer International Publishing: Berlin/Heidelberg, Germany, 2017. [CrossRef]
72. Tosics, I.; Krebs, R.; Köllinger, C.; Kézy, B.; Caba, J.; Argelich, E. *Walk'n'Roll Cities Guidebook*; European Commission & URBACT: Brussels, Belgium, 2022. Available online: https://urbact.eu/sites/default/files/2023-01/walknroll.pdf (accessed on 29 March 2025).
73. Dobers, K.; Perotti, S.; Wilmsmeier, G.; Mauer, G.; Jarmer, J.-P.; Spaggiari, L.; Hering, M.; Romano, S.; Skalski, M. Sustainable logistics hubs: Greenhouse gas emissions as one key sustainability performance indicator. *Transp. Res. Procedia* **2023**, *72*, 1153–1160. [CrossRef]
74. Iuell, B.; Bekker, G.J.; Cuperus, R.; Dufek, J.; Fry, G.; Hicks, C.; Hlavác, V.; Keller, V.B.; Rosell, C.; Sangwine, T.; et al. *Wildlife and Traffic: A European Handbook for Identifying Conflicts and Designing Solutions (COST 341-Habitat Fragmentation due to Transportation Infrastructure)*; Office for Official Publications of the European Communities: Luxembourg, 2003. Available online: https://www.miteco.gob.es/content/dam/miteco/es/biodiversidad/publicaciones/cost341-handbook_tcm30-195789.pdf (accessed on 29 March 2025).
75. MVRDV. Green Rail Corridor. 2015. Available online: https://www.mvrdv.com/projects/267/green-rail-corridor- (accessed on 29 March 2025).
76. K-Studio. Athens International Airport. 2025. Available online: https://k-studio.gr/project/athens-international-airport/ (accessed on 29 March 2025).
77. Saboori, H.; Mohammadi, M.; Taghe, R. Virtual power plant (VPP), definition, concept, components and types. In Proceedings of the 2011 Asia-Pacific Power and Energy Engineering Conference, Wuhan, China, 25–28 March 2011; IEEE: Piscataway, NJ, USA, 2011; pp. 1–4.
78. Nikonowicz, Ł.; Milewski, J. Virtual Power Plants-general review: Structure, application and optimization. *J. Power Technol.* **2012**, *92*.
79. Ghavidel, S.; Li, L.; Aghaei, J.; Yu, T.; Zhu, J. A review on the virtual power plant: Components and operation systems. In Proceedings of the 2016 IEEE International Conference on Power System Technology (POWERCON), Wollongong, Australia, 28 September–1 October 2016; pp. 1–6.
80. Amsterdam Smart City. GridFriends at Schoonschip. 2016. Available online: https://amsterdamsmartcity.com/projects/grid-friends (accessed on 26 March 2025).
81. SPECTRAL. Schoonschip: One of the World's Most Advanced Residential Smart Grids (2019–2024). 2024. Available online: https://spectral.energy/project/schoonschip/ (accessed on 26 March 2025).
82. Yu, S.; Fang, F.; Liu, Y.; Liu, J. Uncertainties of virtual power plant: Problems and countermeasures. *Appl. Energy* **2019**, *239*, 454–470. [CrossRef]
83. Komninos, N. Net zero energy districts: Connected intelligence for carbon-neutral cities. *Land* **2022**, *11*, 210. [CrossRef]
84. Chrysoulakis, N.; Somarakis, G.; Stagakis, S.; Mitraka, Z.; Wong, M.S.; Ho, H.C. Monitoring and evaluating nature-based solutions implementation in urban areas by means of earth observation. *Remote Sens.* **2021**, *13*, 1503. [CrossRef]
85. Kumar, P.; Debele, S.E.; Sahani, J.; Rawat, N.; Marti-Cardona, B.; Alfieri, S.M.; Basu, B.; Basu, A.S.; Bowyer, P.; Charizopoulos, N.; et al. An overview of monitoring methods for assessing the performance of nature-based solutions against natural hazards. *Earth-Sci. Rev.* **2021**, *217*, 103603. [CrossRef]
86. Salam, A. Internet of things for sustainable community development: Introduction and overview. In *Internet of Things for Sustainable Community Development: Wireless Communications, Sensing, and Systems*; Springer International Publishing: Berlin/Heidelberg, Germany, 2024; pp. 1–31.
87. Panori, A. Fostering regional intelligence from the ground up: A selection of case studies from EU projects. *Discov. Sustain.* **2024**, *5*, 407. [CrossRef]
88. Komninos, N.; Panori, A. Smart carbon–neutral development: Embracing complexity with multi-level governance and convolution. *World Dev.* **2025**, *192*, 107026. [CrossRef]
89. Panori, A. Digitally Disrupted Space: Proximity and New Development Opportunities for Regions and Cities. Elsevier: Amsterdam, The Netherlands, 2024.

designs

MDPI

Review

Review of Dynamic Building Envelope Systems and Technologies Utilizing Renewable Energy Resources

Mohammad Almesbah and Julian Wang *

Department of Architectural Engineering, Penn State University, University Park, PA 16802, USA
* Correspondence: jqw5965@psu.edu

Abstract: Dynamic building envelopes integrated with renewable energy sources, termed Dynamic and Renewable Source Building Envelopes (DREBE), provide an innovative approach to optimizing building envelope designs. Yet, these systems are not mature enough and not widely adopted in the industry and few literature resources are employed to understand them. These systems dynamically respond and adapt to various environmental, energy, and occupancy demands for higher energy efficiency and comfort levels compared to traditional building envelopes while simultaneously producing energy. Their potential in climate change mitigation and fostering sustainable urban development warrants great attention from industry and urban planners. Especially in positive energy districts, which aim to reach net-positive energy goals through utilizing smart energy efficient building systems on the district level. This paper reviews innovative systems like dynamic photovoltaic shading devices and phase change materials and evaluates their performance by answering two research questions, what are the current DBE trends and are they feasible in achieving net-positive energy consumption? The analysis conducted reveals the dominance of solar-based dynamic renewable energy systems and a great need for alternatives. The study suggests that alternatives like wind as a renewable energy source should be studied with dynamic systems. Moreover, the study highlights current research gaps including insufficient data on long-term application and economic costs associated with such systems. To address this gap, the study suggests exploring in depth some of these systems and then branching into various combinations of dynamic envelope systems with multiple renewable or adaptive components to further enhance the overall building performance. By synthesizing the current body of literature, this paper gives insights into advancing the application of the dynamic building envelope systems and highlights their crucial role in the future of sustainable urban environments.

Keywords: dynamic building envelope; renewable energy source; positive energy districts; phase-change material; building-integrated photovoltaic; review

1. Introduction

The future of building envelopes lies in utilizing renewable energy sources. Non-renewable sources are no longer sustainable due to their limited resources and rapid consumption in the last few decades [1]. Buildings are significant contributors to energy consumption and greenhouse gas (GHG) emissions [2] accounting for 30% of global energy consumption and 26% of energy-related emissions in 2024 [3]. Inefficient building envelope designs can also contribute to higher energy consumption due to higher energy and heat transfer [4]. Traditional building envelope systems mainly utilize static layers that do not respond to environmental changes throughout the year. Outdated inefficient designs

 https://doi.org/10.3390/designs9020041

may not work in modern times considering the increase in energy consumption and high density of urban environments, which contribute to significant climate change effects [5]. Therefore, solutions to reduce energy consumption and GHG emissions in buildings are crucial for a sustainable future.

Positive Energy Districts (PEDs) have been introduced recently as a novel concept in response to the increasing challenges caused by climate change. PEDs consist of interconnected buildings that share an energy exchange system, enhancing the energy efficiency of the connected buildings [6]. In addition, they incorporate smart building systems to improve the building efficiency of the whole PED community. PEDs also utilize Renewable Energy Sources (RES) for generating electricity to provide surplus energy to reach net-positive energy consumption. The incorporation of RES into the building envelope can provide energy production potential while also utilizing large amounts of façade surface areas, especially in dense urban areas.

The building envelope is the exterior protective layer of the building that acts as an environmental barrier between the outdoor and indoor environments. In recent years, the focus of the envelope has shifted towards harnessing the environmental elements rather than deterring them as a barrier [7]. Dynamic building envelopes (DBE) are systems that use some sort of dynamic component that can respond to external factors such as environmental, energy, or occupational changes [8]. These systems can adapt and change their geometric or physical properties to enhance the building energy or thermal performances by harnessing the environment. Some examples of these systems are dynamic shading devices that shade the building façade based on the time of day and seasonal factors to enhance the daylight and thermal performance of the building [9].

The objective of this review paper is to synthesize findings from recent studies and infer the trends of research related to DBE utilizing RES, and discuss the resulting meta-analysis of the studied systems. Two main questions are answered in this review: (1) What are the current trends of DBE integrated with RES; (2) how feasible are they in achieving net-positive energy?

The use of Building Integrated Photovoltaics (BIPV) that is applied to the building envelope is a widely researched topic that can bring benefits by incorporating it into PED design by utilizing roof and vertical surfaces to reach net-zero potential [10] and allow other systems to provide the necessary surplus to reach positive energy consumption. The use of BIPV has been well established and significant research has been conducted that proved the effectiveness of these systems in decarbonization by integrating with the building energy system through RES utilization [11]. Building-integrated photovoltaic-thermal (BIPV/T) systems are hybrid systems consisting of PV and thermal energy collectors that can provide heating to the building while simultaneously generating electricity [12]. The authors in ref. [13] did a systematic review of BIPV/T building systems and highlighted the key parameters that affect its integration for achieving near-zero energy potential, such as PV tilt angle.

Shading devices are one solution for PED integration with BIPV technology. Furthermore, it has been proven that shading systems have greater energy efficiency and savings when made dynamic compared to traditional building envelope systems, especially in south-facing façades in the northern hemisphere [14]. For example, the researchers in ref. [15] reported in their paper multiple types of kinetic shading devices and reported that dynamic shading can provide energy savings of up to 30% of the energy in commercial buildings. Researchers in ref. [16] introduced a parametric model to simulate the two-motion effect of rotation and folding of dynamic façades on energy and daylight performances in high-rise offices and reported a decrease in energy consumption of up to 21%, while researchers in ref. [17] developed a method to analyze dynamic envelopes by using parametric models to assess their energy performances. Another paper investigated the improvements of dynamic building envelopes on energy performance compared to static façades [18].

Another solution for reducing building energy heating and cooling demands is to incorporate Phase-Change Material (PCM) into the building envelope. PCM offers significant energy savings when implemented in the building envelope [19]. PCM is a widely researched technology that has been studied and tested to increase the thermal efficiency of building envelopes for decades and has been increasing in use since then [20]. With the incorporation of PCM in walls, energy savings of up to 30% and comfortable thermal levels can be achieved through a thermal battery-like process that reduces thermal fluctuations between day and night cycles [20]. PCM can be used in different locations on the building envelope in windows and walls or installed alongside other dynamic envelope systems to enhance the building's thermal efficiency. The authors in refs. [21–23] investigated building envelope systems integrated with PCM in terms of its heat storage and release property and thermal energy storage efficiencies. Another review considered the climate conditions regarding the thermal performance of buildings incorporating PCM [24].

To achieve a net-positive energy building, buildings can make use of DBE systems utilizing RES termed Dynamic and Renewable Energy Building Envelopes (DREBE). Envelope components like BIPV, PCM, and dynamic shading devices are efficient solutions to make energy-efficient buildings through the building envelope. However, these individual solutions are not always enough to reduce building energy consumption to net-zero levels. DREBEs can provide enough energy savings for energy generation to cover the rest of the building's demands to achieve net-positive energy. DREBEs combine systems such as BIPV with dynamic shading devices and PCM to improve the building energy and thermal efficiencies while producing energy.

The current body of research regarding the combination of dynamic systems and RES integrated into the building envelope is scarce. The effect of DBE systems on achieving net-positive energy buildings is not well researched. To understand what makes a DBE reach sustainable goals and achieve net-zero/positive energy goals, we need to review different DBE system configurations and assess their performance. Then, we must synthesize their result findings with things such as building typology, location, and system characteristics and configurations. A wide range of topics were discussed regarding the integration of active/dynamic/adaptive building envelopes and many terminologies are used to define the building components [25–27]. However, articles summarizing dynamic building envelopes with renewable energy sources are limited, with very few review papers written on them in the past 11 years.

Current literature reviews do not answer the question of how to achieve net-zero or net-positive goals using dynamic, adaptive façades utilizing renewable energy sources. This paper bridges this research gap by compiling articles regarding DBE systems and providing a meta-analysis of the reviewed papers based on their study methodology, system types, and building typology, so one can be informed about the current trends or gaps that are inferred from the analysis.

This review paper conducts a systematic review on the topic of renewable energy sources incorporated in DBE systems and analyzes the trends of such systems in recent years and synthesizes their configuration with building typology information to highlight what makes such systems achieve higher energy savings and reach net-positive potential. Using systematic search and analysis methods to investigate the recent trends in dynamic envelope systems, classifying each technology by type, and highlighting their performance metrics. The meta-analysis will highlight the primary factors, such as building typology, climate, seasonal testing conditions, and system configurations that contribute to reaching net-positive potential. The contribution of this systematic review will offer insights for future researchers, field specialists, and policymakers on the sustainability of these systems and it will offer a summary of the system objective results for understanding the potential of integrating DBE system designs. Compiling the details of dynamic envelope systems and their energy-saving results will also help highlight current research gaps and future research pathways that can be undertaken by future researchers.

This paper is structured as follows: Section 2 details the review methodology on how the literature body was searched and collected. Section 3 provides an overview of the reviewed systems and conducts a detailed meta-analysis of the papers focusing on the combination of DBE systems and technologies integrated with RES. Section 4 discusses the results and suggests what future research can focus on. Finally, Section 5 concludes the findings from this paper.

2. Review Methodology

This section provides a systematic approach to gathering research articles focusing on DBE systems that incorporate RES using precise inclusion and exclusion criteria and established key terms. The literature search was conducted using the search databases of Web of Science, Engineering Village Compendex, and Google Scholar. Web of Science and Engineering Village Compendex were used as the search medium to find relevant literature on the Dynamic + Renewable topic using a comprehensive search of terms and keywords, while Google Scholar was used to find articles through manual search or through article references.

The criterion of search included a combination of terms and keywords. Terms such as dynamic building envelope, energy generation, adaptive façade, net positive energy buildings, active building envelope façade, renewable energy source, and phase change material PCM were used to narrow down the body of literature to the relevant topic. In addition, the keywords included provide further specifics for the search, these keywords include: BIPV, PCM, piezoelectric, positive energy, thermoelectric materials, renewable energy source, adaptive building, and positive energy district. The search query is (dynamic building envelope) AND (energy generation) OR (adaptive façade) AND (net positive energy buildings) AND (renewable energy source) AND (phase change material PCM) followed by the keywords that were used set to INCLUDE or set to OR operators.

The inclusion criteria include any paper with a DBE system that utilizes some sort of energy generation through renewable energy sources. Articles mentioning climate-responsive building elements studies are included. Studies with a focus on achieving energy savings and energy efficiency through individual components of the system or the combination of dynamic + renewable are included in the literature review. The range of literature searches includes peer-reviewed journal articles published within the past 11 years from 2013–2024.

The exclusion criteria for paper screening are based on whether the studies address dynamic envelopes only with no renewable energy generation present like those with dynamic shading elements. Excluded are articles that study both dynamic envelopes and renewable energy generation as separate non-integrated case studies, and papers presenting a static envelope system and no renewable energy generation. Articles purely focusing on theoretical models or hypothetical materials were excluded.

Figure 1 shows the methodology flowchart used in this review. The reasons for excluding some articles at the end of the screening process is because after accessing and reading the full article it was found that some articles include multiple case studies that are studied separately, and some do not include energy generation or dynamic components.

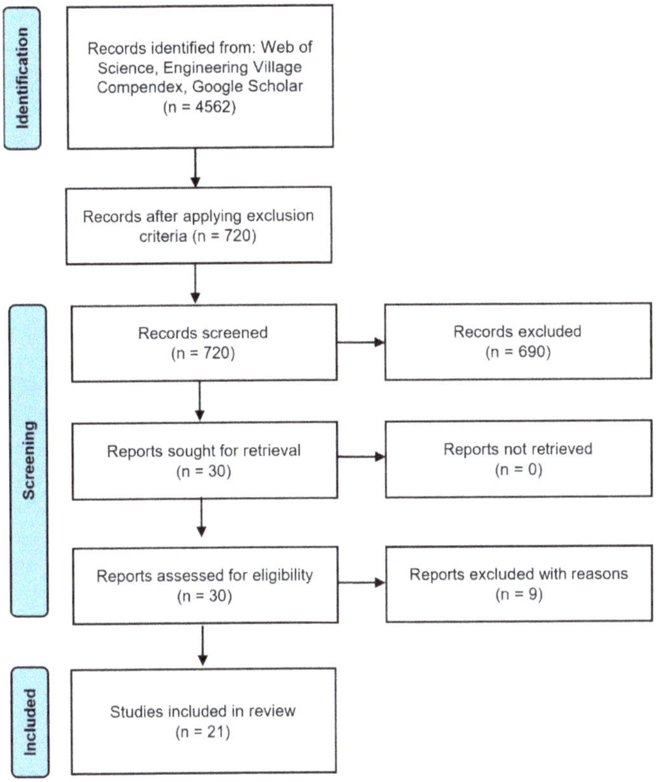

Figure 1. PRISMA flowchart [28].

A challenge in searching the literature is the many different terminologies and keywords used to define the same building envelope system or component. To address this, refs. [25,26] papers' terminology review was used to obtain a general idea of the different naming conventions used for the many building envelope terminologies.

For the purposes of this review, a dynamic envelope is defined as a system where the components act to adapt, respond, or change to sustain an environment at efficient energy and comfort thermal levels. Systems such as adaptive PV panels dynamically adjust to sustain the maximum energy generation possible while also reducing building loads through increased or reduced solar heat gains into the building's glazed façade. A PCM system in this case transitions the phase of the material from solid to liquid and vice versa when absorbing or releasing heat to prolong the heat transfer and reduce thermal fluctuations. After collecting a total of 21 papers, the systems were categorized into two primary types: Kinetic dynamic envelopes and Passive dynamic envelopes. Kinetic dynamic envelopes are characterized by moveable components such as adjustable or motorized PV modules like those used in shading devices. Passive dynamic envelopes include systems incorporating PCM. A meta-analysis of the reviewed papers will outline their system's configuration and objective in terms of what type of envelope component is being studied (e.g., wall, glazed façade, double-skin façade (DSF), windows, etc.), what type of RES is utilized, location and climate characteristics, does the paper mention net-zero or net-positive goals, and how the study is conducted as well as the objective results.

3. Results

3.1. Overview of Dynamic Building Envelopes

For ease of identifying the systems, the dynamic envelope systems studied were classified into two categories: Kinetic Dynamic envelope systems and Passive Dynamic envelope systems. Kinetic dynamic envelope systems include technologies such as dynamic PV shading devices, movable components with solar tracking, and piezoelectric systems. In contrast, Passive dynamic systems include PCM, thermoelectric materials, and rooftop PV units with PCM in walls or paired with PV systems. Dynamic envelope systems in this study can be broadly categorized into paired systems or unpaired. Paired systems are usually located in the same location or are one combined component. Unpaired systems are usually located elsewhere and are not one combined component. Both of those system types share the characteristic of being linked and affecting the building system jointly.

The structure of this section is as follows: First, the configurations of the studied systems and their objectives are detailed. Second, the methods of study employed to achieve the objectives are presented. Third, the system performances and the objective results are summarized. This review paper analyzed a total of 21 papers and conducted a meta-analysis, which is detailed in this section. Table 1 summarizes the reviewed papers in terms of which dynamic envelope component was used and what renewable energy sources are present. The main points detailed in this review are the type of envelope component, renewable energy source, study methodology, building typology, season, and climate. The types of envelope components include dynamic PV and PCM, and renewable energy sources include solar energy harvesting PV, thermoelectric generators, and piezoelectric generators. The methodology of the studies includes experimental, simulation, or both, and building typology includes commercial or residential buildings. Finally, included are the season and climate of which the study investigates. The results of the envelope systems studied are reported in the subsequent sections focusing on their energy, thermal, and visual performance metrics.

Table 1. Summary of referenced studies.

Reference	Envelope Type	Envelope Location	Renewable Energy Source	Objectives	Publication Date
[29]	Dynamic PV	Glazed façade	Solar	Energy saving, generation, and daylight	2024
[30]	Dynamic PV	Glazed façade	Solar	Optimal geometry for energy saving and generation	2024
[31]	Dynamic PV	Glazed façade	Solar	Energy generation and visual comfort	2024
[32]	PCM	Wall	Solar	Improving thermal and electrical efficiency	2024
[33]	Dynamic PV	Window	Solar	Enhance thermal and energy efficiencies	2024
[34]	Dynamic PV	Window	Solar	Shade and energy generation	2024
[35]	Dynamic PV	Window/Roof	Solar	Shade and energy generation	2023
[36]	Dynamic PV	Glazed façade	Solar	Maximize irradiance on panels and minimize daylighting values	2023
[37]	PCM	Wall	Solar	Reduce rooftop PV size	2022
[38]	PCM	Wall	Solar	Electrical cost reductions in PV system	2022

Table 1. *Cont.*

Reference	Envelope Type	Envelope Location	Renewable Energy Source	Objectives	Publication Date
[39]	PCM	Wall	Solar	Thermal and electrical performance analysis	2022
[40]	PCM	Trombe Wall	Solar	Thermal and electrical performance analysis	2022
[41]	Dynamic PV	Window	Solar	Shade and energy generation	2021
[42]	Dynamic sunscreen	Glazed façade	Piezoelectric	Shade and energy generation	2020
[43]	PCM	Wall	Thermoelectric	Energy generation	2020
[44]	Dynamic PV	Glazed façade	Solar	Shade and energy generation	2019
[45]	PCM	DSF	Solar	Energy performance analysis	2019
[46]	PCM	Wall	Solar	Enhance PV thermal performance	2019
[47]	Dynamic PV	Glazed façade	Solar	Shade and energy generation	2017
[48]	PCM	DSF	Solar	Energy and thermal pefromance analysis	2017
[49]	PCM	DSF (on PV)	Solar	Energy and thermal pefromance analysis	2016

3.2. Renewable Energy Sources

Three renewable energy sources are utilized in the reviewed papers: solar energy through PV generators, waste heat harvesting through thermoelectric generation, and wind-induced vibrations captured by piezoelectric generators. Solar energy harvesting was the most studied RES, as has been mentioned in refs. [29–41,44–49], which accounted for 90% of the reviewed papers from the time interval between 2013 and 2024, highlighting the predominant use of PV technologies. PV technologies used include dynamic PV shading devices, solar tracking PV modules, and rooftop PV installation. Thermoelectric generation has been discussed in one study in ref. [43]. Thermoelectric generation uses blocks embedded in the walls to collect heat that is transferred through the wall throughout the day and night cycles. And only ref. [42] mentioned using piezoelectric energy generation through wind-induced actuation and vibrations. Figure 2 visualizes the distribution of each envelope system for the studied papers.

Most of the studied envelopes are located on the walls, specifically glazed façades. Some of these system's components are located at different locations and others are located in the same shared space. For example, in ref. [37] PCM was used to reduce the size of the rooftop PV system. For dynamic PV systems, a general method used is multi-objective optimization due to the conflicting nature of having both the dynamic shading and energy generation in one shared location in the envelope. For most PCM systems, the energy saving from PCM contributes to the overall building energy system while being located in a different space from the PV modules.

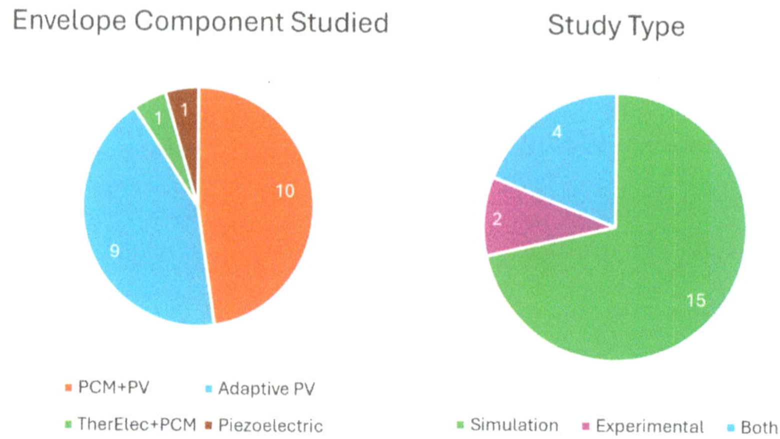

Figure 2. Type of envelope components (**Left**) and method of study (**Right**).

3.3. Dynamic Envelopes

About 47% of the papers mention PCM in their study [32,37–40,43,45,46,48,49]. The topic of dynamic façades has slowly gained traction in the last few years, specifically in regard to dynamic facades with integrated RES (Figure 3). Adaptive PV has been the most widely used dynamic envelope system in recent years.

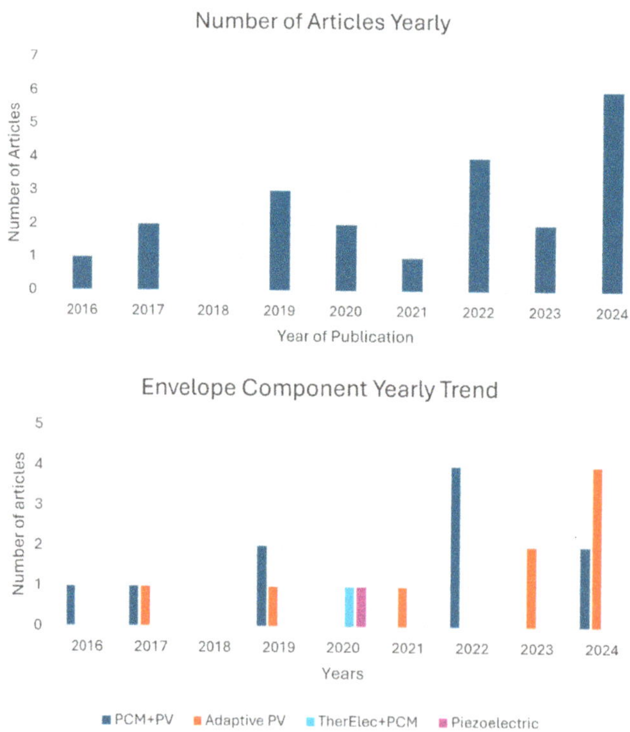

Figure 3. Yearly trends of the reviewed papers.

3.3.1. Kinetic Dynamic Envelopes

About 43% of the studied papers are about dynamic or adaptive PV systems with some sort of movement control [29–31,33–36,41,44,46]. Their system configurations and characteristics as well as objectives are described in this section.

The authors in ref. [29] proposed a dynamic, adaptive, and vertical photovoltaic integrated building envelope (dvPVBE) that consists of 24 aluminum alloy blind slats integrated with PV cells on a glazed façade. The slats act as a weather and occupant-responsive dynamic envelope system that provides thermal and visual comfort, and energy-saving potential. The slats can be controlled manually by indoor occupants or automatically by responding to weather conditions and occupants' demands with three prioritization strategies: power generation priority (PGP), energy saving priority (ESP), and natural daylight priority (NDP). Researchers in ref. [44] introduced dynamic thin-film PV panels utilizing soft-robotic solar tracking technology to enhance the energy generation, heating, and shading performances of the building façade. In a similar case study, researchers in ref. [47] used the same dynamic envelope prototype as a prior research study. Researchers in ref. [36] implemented dynamic BIPV façade modules in a case study. The BIPV modules consist of 72 PV panels spread over two stories of glazed façade that each change their respective orientations to provide the maximum amount of solar irradiance on the cells while also minimizing the daylight illuminance values below 300 and above 3000 lux. Researchers in ref. [31] studied a dynamic PV façade employing a biomimetic design that copies the adaptive characteristics of nature. Each PV is an individual component, and the components are motorized for precise controls and for optimizing harvesting and saving energy. The study also employs a lighting control system to ensure adequate lighting levels are present in the room and sensors for occupancy and energy saving when the room is not used. The objective of the study is to optimize the energy generation potential with visual comfort in glazed office buildings using the dynamic PV façade.

Researchers in [30] studied the geometry and operation of a dynamic PV shading device (PVSD) in terms of reducing building energy loads and increasing PV energy generation to find the optimal geometry and operational configurations. Researchers in ref. [35] considered two Dynamic PV-integrated moveable shading devices (PVIMSD) and overhangs (PVIO). These dynamic envelopes are termed under the umbrella of Adaptive envelope Technologies (AET). The overhangs are situated atop windows to provide shading for energy saving and PV generation through dynamic position and angle changes throughout the year. They can change from vertical to horizontal states during winter and summer conditions, respectively. The PVIMSDs are deployed atop roofs to allow the cool roof surface to be shaded or exposed depending on the seasonal conditions. In the summer, the PVIMSDs are moved to shade the south façade of the building and expose the cool roof, while in the winter the PVIMSD is set to cover the cool roof and keep the façades exposed. Researchers in ref. [41] Introduced new PV-integrated dynamic overhangs in windows that have the ability to slide and rotate to provide shade and energy generation. Researchers in ref. [33] introduced a novel approach to sliding windows with PV and PCM to enhance the energy and thermal efficiency of the buildings. Researchers in ref. [34] introduced a similar concept of PVSDs to ref. [41].

Piezoelectric generators use vibrations to generate electricity. Only one paper mentioned the piezoelectric system in their study [42] that introduced a smart origami sunscreen system as a redesign of the dynamic sunscreen system that was developed for the Al Bahar towers in Abu-Dhabi. This new design utilizes an actuation and wind harvester that also acts as a dynamic sunshade for the building façade using piezoelectric generators termed Wind and Actuation Energy Harvesting Tensegrity Adaptive Building Screens

(WTABS). The design uses 42 piezoelectric cables to generate energy in the sunscreens through wind-induced vibrations.

3.3.2. Passive Dynamic Envelopes

Passive dynamic envelopes primarily use Phase-Change Material (PCM) for their energy-saving potential and thermal performance enhancements by incorporating it inside walls, roofs, or PV modules. These systems effectively stabilize the thermal levels of the envelope and reduce cooling and heating demands. Researchers in ref. [37] used PCM to reduce the system size and costs of a hybrid solar system. The hybrid solar system consists of solar- and fossil-based systems to power the building. The Photovoltaic-Thermal (PVT) system is installed on the rooftop and is used to collect heat and generate electricity for the building, while the PCM is installed in the building façade to provide energy savings. Researchers in ref. [38] introduced a planet-based PCM to investigate the effect of electricity cost reduction using the PCM as a thermal battery instead of normal PV storage batteries. Their envelope system consists of PCM in the building insulation and a rooftop PV array. Their research studies the use of PCM in the wall on increasing the self-consumption of the rooftop PV modules. The study uses an optimization approach using a home energy management system (HEMS) to minimize building energy costs and maximize PV self-consumption.

Researchers in ref. [45] studied a Double Skin Façade (DSF) system energy performance with the addition of BIPV and PCM in the outer glazed façade. In ref. [49], a PV-PCM DSF envelope system was analyzed as part of the study with the PCM layer behind the PV layer. Researchers in ref. [48] introduced a PV-PCM glazed double skin façade building envelope similar to their previous study in ref. [49] with the PCM integrated behind the PV modules to absorb excess heat to improve thermal levels and reduce cooling loads by diverting and using that excess heat elsewhere. Researchers in ref. [32] introduced a double PCM building envelope with PV to improve the thermal and electrical efficiency of the building. The objectives of the study were to enhance PV efficiency by incorporating PCM, therefore reducing overheating and enhancing thermal insulation to minimize indoor energy use. Researchers in ref. [40] studied a PV Trombe wall (PV/TW) incorporating PCM for the thermal and electrical performances of the PV. Researchers in ref. [39] studied the thermal and electrical performance of a building envelope with PCM and a distributed PV system and compared it to a building with PV only. Researchers in ref. [46] introduced a BIPV with PCM behind the modules to enhance the thermal performance of the building and PV panels and termed it BIPV/PCM. Researchers in ref. [43] Utilized in their study a thermoelectric generator to utilize the potential of waste heat in the building walls to generate electricity to power sensors and circuits. A thermoelectric energy harvesting block was employed with PCM in building walls to harvest waste heat from the temperature difference and heat flow between building wall surfaces to generate electricity.

3.4. Method of Study

The methods used in the papers were also collected. These methods include experiments, simulations, or both performed in the same study. Experiments may include on-site, laboratory, or similar on-site experimentation. The experimental work has been performed with different types of dynamic components and systems and no experimental work is the same. The simulations conducted by the reviewed studies differ, some are numerical calculations and others perform the simulations using tools and software like Rhino3D and the Grasshopper Plugin, the TRNSYS tool, and other tools.

In total, 16 out of 21 papers conducted only simulations in refs. [29–32,34–38,41,42,45–49], 2 out of 21 conducted only experimental work [34,37], and 3 out of 21 did both experimental and simulations [33,39,44]. Figure 2 (right) shows the distribution of methods of the studies for all 21 papers. Researchers in ref. [29] used EnergyPlus simulations for the energy performance of the dvPVBE to calculate the heating and cooling loads, energy generation, and daylighting performance. Only the PGP and ESP control strategies were employed and the NDP strategy was left for future studies due to its dependency on occupancy which was not included in this study. Researchers in ref. [35] used EnergyPlus simulations to assess the energy efficiency of a prototype building under three combinations of dynamic shading configurations. The AETs are controlled via an Energy Management System (EMS) that provides high-level supervisory controls that are like those used in real buildings. The control strategy was set to maximize annual energy savings by adjusting the AET's positioning for the dynamic roof and overhangs. Researchers in ref. [37] used the TRNSYS tool and numerical simulations to assess the three Es, energy, exergy, and economic impacts of the envelope system.

Researchers in ref. [38] conducted a numerical simulation with the use of HEMS to optimize the operation of the building HVAC system to reduce operation costs and maximize PV-self consumption. Using four simulation scenarios of fixed HVAC operation using deadband control without PCM, deadband control with PCM, HEMS optimized control without PCM, and HEMS control with PCM. The researchers in ref. [42] conducted a numerical simulation of the dynamic WTABS to simulate the actuation motions and wind-induced vibrations. Researchers in ref. [46] did a numerical simulation for correlating various parameters such as PCM and air gap thickness, PV height, and air mass flow rate to the energy production of the PV units. Using a Taguchi L9 method to optimize the objectives of maximizing the PV energy production and heat extraction and minimizing PV panel temperature. Researchers in ref. [32] conducted a parametric numerical simulations study using TRNSYS for the double PCM layer envelope and compared it to three system configurations: one with PV and no PCM, one with a single PCM layer near the PV, and a single PCM layer near the interior layer.

Researchers in ref. [45] considered climate conditions, orientation of the building, cavity width, and glazing type as the parameters studied for the DSF. A simulation was performed in the DesignBuilder software with and without PCM integration into the DSF for three cases without PCM, with PCM and PV integration, and with PCM but without PV integration. Researchers in ref. [47] did a similar case study to [44]; they developed a simulation framework to provide results that can help in the control strategies of the dynamic PV orientations. Researchers in ref. [49] did a numerical analysis to investigate the thermal and electrical performance of a PV-PCM integrated DSF system. The integration of PCM was used to provide the PV modules with a higher thermal efficiency. The primary goal of the paper was to develop a physical-mathematical modeling of the proposed PV-PCM DSF.

Researchers in ref. [36] did a multi-objective optimization (MOO) approach to obtain maximum irradiance on the dynamic BIPV modules while also minimizing both internal values of illuminance below 300 lux and above 3000 lux. The MOO approach was performed in the Rhino3D and Grasshopper software using a non-dominated sorting genetic algorithm (NSGA-2) using the Wallacei Grasshopper plugin. Using Honeybee (Grasshopper plugin inside Ladybug tools), the illuminance and irradiance analysis were conducted. Researchers in ref. [31] conducted a simulation using Rhino/Grasshopper to model the problem parametrically and using the Honeybee plugin for the environmental and energy simulations with Radiance and Daysim to optimize between visual comfort and energy savings. Using Galapagos to do a MOO approach to optimize PV layer orientation and lighting system controls. The daylight factor (DF) and Daylight Glare Probability (DGP) are the two main indexes used in this paper for assessing the simulations. DF is used to provide an idea about the availability of natural sunlight in a space. The higher the DF means the better lit the room is. DGP is used to provide a metric for the probability of glare happening in space due to brightness contrast. Researchers in ref. [34] conducted parametric simulations using Ladybug tools in the Grasshopper ecosystem for simulating energy consumption and daylighting. Similar to ref. [41], they employed three dynamic control strategies: rotation, sliding, and a hybrid combination of sliding and rotation. Researchers in ref. [48] used a MOO approach to investigate the use of PV-PCM technology in a glazed DSF building envelope using MATLAB's genetic algorithm framework with the goals of minimizing thermal loads and indoor temperature variations and used the results in the TRNSYS software. The goal of the optimization is to assess the effect of the parameters of ventilation flow rates and operation on the thermal and energy performance of the building.

Researchers in ref. [41] conducted a simulation and optimization to find the optimal configurations for the dynamic PV overhang to maximize energy production and minimize energy use through sliding and rotating overhangs located above windows. The authors investigated parameters such as depth and tilt angle for the overhang. Utilizing DOE-2.2 for the building thermal loads and the System Advisor Model (SAM) for simulating the PV energy production. The dynamic PV shading devices were compared with their static counterparts to assess their energy-saving potential. Three configurations were investigated: static, sliding only, and sliding-rotating overhangs with three options of no PV, load-tracking for maximum reduction in building loads, and PV-tracking overhangs for maximum energy generation. Researchers in ref. [30] utilized a parametric approach using Rhino3D and Grasshopper to optimize the geometric patterns of the PVSD and determined four case patterns for simulating. The four scenarios simulated included scenario A with a 4-panel configuration with limited motion control, B with 36 panels with limited motion controls, C with 36 panels with extended motion control, and D with 4 panels and precise and individual motion controls. Using Galapagos to optimize the objective functions.

The following studies used experimental work to conduct their research. Researchers in ref. [44] developed a real-scale dynamic envelope prototype to study the objectives of maximizing PV energy generation and reduction in building energy usage for their prototype dynamic envelope and simulated the energy production and saving performances for three other locations. Researchers in ref. [43] did experimental work to find the annual energy generation of the thermoelectric energy-harvesting block. They used the TRNSYS tool to simulate the outdoor environment and provided the resulting annual data to conduct the experimental work and simulate the day-night cycles. Researchers in ref. [39] did experimental work by building two small-scale building models, one without PCM and one incorporating PCM. They gathered data on-site and used it for the EnergyPlus input for the building model validation. Researchers in ref. [40] did experimental work to assess the impacts of PCM on the electrical and thermal performance of the PV Trombe wall with PCM integrated. They built two small-scale experimental models with a south-facing Trombe wall, one integrated with PCM and one without. Using a combination of temperature, voltage, current, and airflow sensors to assess the performance of the envelope system. Researchers in ref. [33] conducted experimental work and numerical simulations using ANSYS 2020 R2.

Site and Building Typology

This part summarizes the typology of the buildings and sites being studied. Each location and building type, commercial or residential, is summarized in this part. Table 2 summarizes the articles' study types, building typologies, locations, and seasons. The experimental work or simulations were conducted in different seasonal or climate conditions. With how certain renewable energy source availability changes based on climate, geographical location, and seasonal changes affecting solar radiance intensity, the studies' location and climate are detailed in this section. By providing this section, we can analyze what systems might work better in what climate or seasonal conditions compared with what might happen when it is somewhere else.

Table 2. Study type, building typology, and location of the reviewed papers.

Reference	Study Type	Building Type	Location	Season
[29]	Simulation	Commercial, Office	Beijing, China	Yearly
[30]	Simulation	Commercial, Office	Houston, Texas, US	Summer
[31]	Simulation	Commercial, Office	Montreal, Canada	NS
[32]	Simulation	NS	Guangzhou, China	Yearly
[33]	Both	NS	Egypt	Summer
[34]	Simulation	NS	Qingdao, China	Yearly
[35]	Simulation	Commercial, Office	Varies	NS
[36]	Simulation	Educational	Sydney, Australia	Summer
[37]	Simulation	Residential	Tabriz, Iran	Yearly
[38]	Simulation	Residential	Australia, Varied cities	Yearly
[39]	Both	Residential	Changsha, China	Summer and winter
[40]	Experimentation	NS	Hawija, Iraq	Summer
[41]	Simulation	Residential	US, varied cities	NS
[42]	Simulation	Commercial, Office	Abu Dhabi	NS
[43]	Experimentation	Commercial, Office	Seoul, South Korea	NS
[44]	Both	Office, Residential	Zurich, Switzerland, other cities	Summer
[45]	Simulation	Residential	Iran, six cities	Yearly
[46]	Simulation	Commercial, Office	Arlington, Virginia, US	NS
[47]	Simulation	Commercial, Office	Zurich, Switzerland	Summer
[48]	Simulation	Commercial, Office	Venice, Italy	Summer and winter
[49]	Simulation	Commercial, Office	Varies	NS

NS: Not Specified.

The authors in ref. [29] simulated their proposed dynamic vertical façade in the city of Beijing, China for a typical office building in the autumn, vernal equinox, and winter and summer solstices. In ref. [37], the researchers used a two-story residential building in Tabriz, Iran as an annual case study. And in ref. [38], the authors did an annual case study in Australia for five different cities including Syndey, Brisbane, Melbourne, Adelaide, and Perth. They studied fifty residential homes in each city (10 for Perth) with real energy demand and PV generation data. Researchers in ref. [30] studied their envelope in a hot and humid climate on a typical sunny summer day condition in Houston, Texas. Researchers in ref. [44] used a case study for a prototype in Zurich, Switzerland in the summer conditions while also conducting simulations for Helsinki, Zurich, and Cairo. Similarly, the authors in ref. [47] conducted a simulation for the same case study in Zurich. Researchers in ref. [42] studied the dynamic sunscreen system as a redesign of an existing envelope of Al-Bahar towers in Abu Dhabi. Researchers in ref. [45] utilized three design scenarios and simulated them in six cities of Iran, Tehran, Tabriz, BandarAbas, Esfahan, Shiraz, and Yazd with different climates for a year-round simulation. In ref. [49] three climate conditions of Venice (Italy, warm), Helsinki (Finland, snow, fully humid), and Abu Dhabi (UAE, hot arid) were studied. Researchers in ref. [35] simulated a commercial office building in Denver as the baseline and compared it to the performance of buildings located in Tucson, AZ, El Paso, TX, Denver, CO, and Rochester, MN in the United States. Researchers in ref. [36] used the University of Technology Sydney in the city of Sydney, Australia as the case study for its dynamic BIPV envelope system. The times simulated are 8AM, 12PM, and 4PM for the solar analysis for the March equinox (Summer season in Australia). Researchers in ref. [31] simulated a south-facing glazed façade office space in Montreal, Canada. Researchers in ref. [48] did their simulation for the temperate climate of Venice, Italy in summer and winter conditions. The simulations were performed for the whole year. Researchers in ref. [46] used the weather of Arlington, Virginia for their numerical calculation simulations. Researchers in ref. [41] studied their dynamic PV overhang for the four types of climates of Chicago, IL; Boulder, CO; San Francisco, CA; and Phoenix, AZ in the US. Researchers in ref. [32] simulated their study for a south-facing room of a building in Guangzhou, China in the summer, winter, spring, and autumn seasons. Researchers in ref. [34] did their simulations in Qingdao, China for an office building room with a south-facing window façade. Their simulation was conducted in the whole year conditions. Most of these studies were conducted on singular buildings with no interaction with neighboring buildings.

Researchers in ref. [43] did their experimental and simulation analysis for the climate of Seoul, South Korea. Researchers in ref. [39] did their experimental work in Changsha, China for the summer and winter seasons. Researchers in ref. [40] did the experimental work in the city of Hawija, Iraq in the summer conditions with two Trombe wall systems with a south-facing façade. Researchers in ref. [33] did the experimental work in Egypt in summer conditions.

3.5. Building Performance Metrics and Outcomes

The energy-saving and generation results are summarized in this part. The performances include thermal, visual, and energy. Economic and environmental performance results are also included and summarized in this section. The type of results expected and used as an objective is mentioned for each study to obtain a general understanding of what type of dynamic envelope system is used for what performance objectives or goals. In the case of this review paper, it was important to understand whether each reviewed paper mentions net-zero or net-positive goals explicitly or not.

3.5.1. Energy Performance

The energy saving resulting from incorporating dynamic envelope systems with RES using various systems combinations and configurations are detailed in this section. Researchers in ref. [29] reported 131% of the annual energy for office rooms has been met, and there has been an increase of 226% in annual net energy output when using the dvPVBE compared to static PV blinds. The optimal slat angles ranged between 45° and 60° for the case in Beijing, China. Researchers in ref. [36] state the difference in energy generated potential was reported to be a 21.53% increase compared to the static envelope. An increase of 1.16% was also achieved by minimizing indoor illumination values above 3000 lux for the dynamic envelope compared to its static counterpart. Researchers in ref. [31] reported that the dynamic PV façade reduced the lighting and cooling loads in the summer months by more than 50%. Researchers in ref. [41] reported annual energy savings for the sliding-rotating overhangs relative to the no-overhang cases with savings of up to 105% for San Francisco and 92% for Phoenix, which are relatively hot climates, and savings of up to 40% and 56% for Chicago and Boulder, respectively, which are colder climates. Researchers in ref. [34] reported their hybrid control strategy reduced energy consumption by 50.38% while rotation and sliding only reduced it by 32.13% and 47.22%, respectively. The PV generation in the months of March and April exceeded building energy demands providing net-positive results. Researchers in ref. [35] reported annual site energy savings of 99% in Tucson, 109% in El Paso, 76% in Denver, and 47% in Rochester when combining dynamic shading devices with integrated PV energy generation. The paper reports on the tradeoff of the dynamic windows, in which they can reduce total energy use but affect the daylighting performance of the building.

Researchers in ref. [37] reported a 27.4% energy reduction for the heating and cooling loads was achieved when introducing PCM, a peak demand reduction of 28.57%, and an improvement of 18.4%, 13%, and 28.5% in fuel consumption, exergy efficiency, and unit cost of the system, respectively. The study reported a net-positive energy consumption in the 6 months from April to September (not explicitly mentioning net-positive), while the other remaining months did not produce enough energy to meet the building demand due to low solar intensity. It also reported a cost reduction of 20% with the addition of PCM. The overall plant cost was reduced by 39.8% with the introduction of PCM in the building envelope. Researchers in ref. [38] mention the results of their research and highlight that adding PCM does not increase the self-consumption of the rooftop PV system, rather it reduces it by about 1–3%. However, HVAC consumption was reported with a 30% decrease in all the cities. Researchers in ref. [48] reported that the addition of PCM reduced the cooling loads by up to 48% in the summer conditions. In winter, minimal ventilation provided a higher reduction in heating loads by keeping PCM in its solid state. While in summer, the extended ventilation improved the PCM performance and reduced the cooling loads. Researchers in ref. [39] reported peak cooling load reduction of 47% and a delay of 1 h in the summer season in August, and annual energy consumption reduction averages of 2.8%, 4.56%, and 0.53% in the summer and winter seasons, respectively, using only 5.2 vol% of PCM compared with the model without PCM. Researchers in ref. [40] presented results showcasing the impact PCM has on PV/TW systems with an increase in electrical and thermal efficiency of 4.8% and 0.3%, respectively. Researchers in ref. [32] mention that their proposed double PCM envelope results in reduced cooling loads of 7.94% compared to their first comparative case. PV had 8.5% higher power outputs in the summer compared to the first case. Researchers in ref. [33] reported a delayed peak temperature of 75 min and 2 h for heat flux.

Researchers in ref. [30] reported that the energy generation potential scenario A generated two times more solar energy than the static panels. Researchers in ref. [44] reported adding a static PV shading device improved the net energy consumption with a decrease of 37–73% and a further 6–19% if the PV shading device is made dynamic. As a retrofit solution, this PV shading device can provide up to 45–88% reduction in annual energy consumption for commercial office or residential buildings. Researchers in ref. [47] reported a net energy saving of 20–80% for the dynamic PV façade compared to static ones.

Researchers in ref. [42] reported the energy produced by the WTABS is 20 Wh/m^2 and 0.263 mWh/m^2 for the actuation and wind, respectively. Al Bahar Towers consist of 1000 WTAB modules, each having an area of 8.75 m^2 that produces equal energy to that provided by 233 PV panels or 87 rooftop wind turbines with 1 m^2 surface area. Researchers in ref. [43] reported passive energy generation of 2.1 kWh/m^2 annually, with the highest for May generating 3.54 Wh/m^2 daily, which is enough to power small-scale modern digital and radio circuits.

Researchers in ref. [45] investigated DSF energy performance with the addition of BIPV in the outer glazed façade integrated with PCM. The results indicate that the heating load is not reduced when applying the PCM into the DSF, but the cooling loads were significantly reduced by 10.38%, 5.38%, 9.62%, 6.48%, 9.2%, and 9.66% when applied in the western wall for the six cities of Tehran, Tabriz, Shiraz, Esfahan, Bandar Abbas, and Yazd. Researchers in ref. [49] reported that the addition of a PCM layer in the DSF provided a reduction of 20–30% of the monthly cooling loads. The PCM reduced the surface temperature of the PV, but this did not have major effects on the electrical efficiency. Table 3 below summarizes the key findings of the papers with climate zones of each study included based on the Köppen climate classification [50].

Table 3. Summary of key result findings for each studied envelope.

Reference	Envelope Type	Envelope Location	Renewable Energy Source	Climate	Key Findings
[29]	Dynamic PV	Glazed façade	Solar	Monsoon-influenced Humid Continental	131% of office annual energy demand met
[30]	Dynamic PV	Glazed façade	Solar	Humid Subtropical	2x times energy generation and up to 30% higher daylight performance
[31]	Dynamic PV	Glazed façade	Solar	Warm-Summer Humid Continental	More than 50% reduction in lighting and cooling loads
[32]	PCM	Wall	Solar	Humid Subtropical, monsoon-influenced	7.94% reduced cooling loads
[33]	Dynamic PV	Window	Solar	Hot Desert	Significant delay in peak temperature and heat flux
[34]	Dynamic PV	Window	Solar	Monsoon-Influenced Humid Subtropical	Up to 50.38% energy consumption reduction
[35]	Dynamic PV	Window/Roof	Solar	Varies	47–109% energy savings reported for different cities
[36]	Dynamic PV	Glazed façade	Solar	Humid Subtropical	21.53% increase in energy generation
[37]	PCM	Wall	Solar	Cold Semi-Arid	27.4% heating and cooling load reduction
[38]	PCM	Wall	Solar	Varies	30% HVAC consumption reduction
[39]	PCM	Wall	Solar	Humid Subtropical	Up to 47% peak cooling load reduction
[40]	PCM	Trombe Wall	Solar	Hot Semi-Arid	Electrical and thermal efficiency increase of 4.8% and 0.3%
[41]	Dynamic PV	Window	Solar	Varies	40–105% energy savings for different US cities

Table 3. *Cont.*

Reference	Envelope Type	Envelope Location	Renewable Energy Source	Climate	Key Findings
[42]	Dynamic sunscreen	Glazed façade	Piezoelectric	Hot Desert	1000 WTABS provide comparable generation to 233 PV
[43]	PCM	Wall	Thermoelectric	Monsoon-influenced Humid Continental	2.1 kWh/m^2 annual passive generation
[44]	Dynamic PV	Glazed façade	Solar	Oceanic Climate, Varies	45–88% reduction in energy consumption
[45]	PCM	DSF	Solar	Varies	Cooling load reduction between 5 and 10% in different cities in Iran
[46]	PCM	Wall	Solar	Humid Subtropical	Optimum BIPV design configurations
[47]	Dynamic PV	Glazed façade	Solar	Oceanic Climate	20–80% energy savings
[48]	PCM	DSF	Solar	Humid Subtropical	48% cooling load reduction
[49]	PCM	DSF (on PV)	Solar	Varies	20–30% cooling load reduction

Table 3 above shows that studies with dynamic PV components report significant energy saving, reaching more than 50% in most cases and up to 100%, and considerable improvement in daylight performance with dynamic shading. PCM also provided significant savings in the range of 20–50% for the cooling loads, mostly in the summer conditions. Significant energy generation comes from using solar-based renewable energy sources. Thermoelectric and piezoelectric generators currently can only provide minimal energy generation to power things like sensors and circuits when installed on a small scale. Thermoelectric generators have low energy conversion efficiencies and are not suggested to replace PV power generation but only to utilize otherwise untapped potential to convert wasted heat into energy [43].

Net-Zero or Net-Positive Energy

Some articles mention reaching net-zero goals explicitly, while the results of others imply net-positive energy potential. Researchers in ref. [35] report annual positive net energy demand up to 109% when using the PV integrated with the roof and overhangs in El Paso, while the authors in ref. [41] proved that dynamic PV envelopes can provide net-positive results, particularly for apartment buildings in the San Francisco case. However, significant saving can also occur when using the sliding-only overhangs with angles closer to the latitude of the locations, which can result in a similar saving to the sliding-rotating overhangs with less complex operation. In addition, the authors in ref. [34] reported net-positive building energy for the dynamic PVSD in the months of March and April.

3.5.2. Non-Energy Metrics

Visual comfort and economic and environmental performance are reported in this section. The authors in ref. [44] suggest the use of the retrofit solution of their prototype due to its lightweight design and ease of application on any glazed façade. The researchers in ref. [35] investigated the cost-effectiveness of the AETs with PV integration using a breakeven cost analysis for the office buildings in the different US cities and reporting cost savings of up to $47,047 in El Paso. Researchers in ref. [31] provided the results for the visual comfort performance metrics and found that the dynamic PV façade improved the daylight distribution and reduced glare in the space. Their lighting control system contributed to energy savings of 10.7% to 20.9%, compared with using conventional occupancy sensors. Researchers in ref. [34] reported that their dynamic PVSD hybrid control strategy provides enhanced daylighting performance and visual comfort levels. Researchers in ref. [30] reported better daylight performance for the four scenarios with an increase of 25–30%, and 36% in Spatial Daylight Autonomy (sDA300/50%) for scenarios A and C, and a reduction in glare by 20% for A.

4. Discussion

4.1. Key Findings, Research Gaps, and Future Directions

Dynamic and renewable energy envelopes are a fairly new concept that emerged in recent years. This review highlights a few limitations in the current body of literature regarding these dynamic envelopes. A limited focus on achieving net-positive energy buildings and a lack of comprehensive analysis exploring differing combinations of dynamic system envelopes utilizing different renewable energy sources. Most of the studies lean heavily on Solar energy generation reflecting its natural abundance and availability year round. However, dynamic PVs are solar-based energy generation meaning they are inactive during the night emphasizing the need to incorporate other renewable energy sources like wind. This trend underscores the need to explore new avenues for RES integration other than solar. Further studies should consider this in their design of dynamic systems. Also, developing robust frameworks for simulating, analyzing, and implementing these systems will significantly advance future research endeavors. Establishing strategies or guidelines for envelope system selection based on climate conditions or energy needs is also essential. For example, locations without adequate solar intensity for PV generation to reach net-positive may incorporate additional wind energy harvesting ability.

Both solar energy utilization and phase change material in envelopes have been well established; however, dynamic PVSDs in theory can benefit from further integration with PCM. Utilizing dynamic PVSDs provided the highest energy savings for all the systems studied in this study. The inclusion of PCM with dynamic PVSD modules can theoretically further improve the performance of the building and PV module efficiency. Studies such as ref. [49] showcase the integration of PCM in static PV modules to enhance energy conversion efficiency by reducing overheating. However, this has not been studied cohesively with dynamic PVSDs, for example. Unfortunately, no studies talked about this integration of dynamic PVSDs and PCM to enhance both thermal and electrical efficiencies of the dynamic building envelope. The Authors in ref. [51] introduced a new concept of using PCM as a movable dynamic layer; otherwise, there is a lack of movable PCM integration studied with dynamic PV components. In addition to that, there is limited focus on other envelope avenues, such as the roof envelopes and opaque walls that can also benefit from RES integration. Currently, there is a heavy focus on the glazed façade for Dynamic PV and opaque walls for PCM. Moreover, there is no literature research conducted on the performance of PEDs incorporating DREBE on an urban cluster level. Combining such concepts in the early stages paves the way for better compatibility in installations before considering retrofit solutions for future PEDs.

The exploitation of natural greenery and vegetation can also help in reducing overall building energy consumption through the reduction in cooling loads in the summer conditions through natural shade. In addition, green building technologies offer significant CO_2 reduction when applied extensively in dense urban areas to mitigate climate change [52]. Researchers in ref. [53] reviewed green building technology implemented in building envelopes and reported significant savings in the energy consumption of the building. In ref. [54], new green building practices on how to incorporate renewable energy sources in the building industry were investigated. These technologies can be combined with green spaces and green systems to increase the energy and thermal energy of the buildings. Furthermore, if applied in the urban context, they can help mitigate climate change and thermal comfort challenges like Urban Heat Island (UHI) effects that may be associated with dense urban environments.

Research articles that talk about how to achieve PEDs are scarce. This review paper provides insight into some technologies that can help in achieving the net-positive energy of PEDs, but future studies should implement frameworks for studying the combined effect of multiple buildings with dynamic façades on the overall efficiency of the PED and the electrical grid. This paper reviews building systems with RES integrations at the urban cluster level [55]. Another paper also reviews the urban cluster level incorporating RES but focuses on polishing the cluster methodology [56]. More studies need to focus on the implementation of existing technologies and optimizing placement with more smart system controls for further energy efficiency enhancement of existing infrastructure. Retrofit solutions of existing technologies must be studied and improved by making systems simpler to install and cost-effective while providing maximum energy savings and generation potential. The authors in ref. [57] provided a review of the modular construction of building envelopes. These modular envelopes can be used for future PED construction for easily installed envelope systems. The authors in ref. [58] conducted a systematic review of papers to determine differences in terminologies of adaptive facades and designs. On a similar note, the definition of PEDs shares the same function as some widely accepted terminologies such as near Zero Energy Buildings (nZEB), and there is still no consensus or specific analysis frameworks given for such a novel concept [59].

Utilizing RES in building envelopes to offset building operational energy demands can contribute to better space utilization when efficiently placed or designed on the building envelope or façade. Better utilization of shared space such as a roof or vertical façade can be further studied. With how limited space is in dense urban environments and how urbanization is exponentially increasing, the deployment of RES on-site contributes to efficient space utilization by employing RES locally. However, for certain contexts, space is very limited. In this case, high-rise buildings have very limited roof space, while large areas of vertical façade are primarily made of glazing material. Therefore, consideration of space optimization and system configuration must be implemented carefully. A paper presented an optimization model for integrating renewable energy into existing buildings as a retrofit solution to achieve near-zero energy buildings [60]. BIPV parametric models using Rhino3D+Grasshopper were used to assess their benefits of installing in existing buildings [61]. The visual and aesthetic characteristics of the envelopes should also be studied for more pleasantly looking building façades. User satisfaction and comfort with integrating dynamic facades in buildings were addressed in [62]. In addition, indoor user satisfaction studies regarding the dynamic PVSDs should be studied.

Future research can investigate the combination of multiple renewable energy sources and dynamic envelope systems and assess the viability and efficiency of such combinations and can conduct an economical assessment of the systems to provide the industry and policymakers with general cost-effective measurements of implementing these systems. Since these types of systems have not reached maturity in the industry, there is a general lack of long-term data and life cycle costs and maintenance that might be associated with keeping them up to date and working. To expedite the process of implementing these technologies, studies must first define their system to a standard set of terminologies agreed upon. Second, their implementation in both specific and general conditions that can be applied to more than one case that is different in climate or location and assess their viability should be studied. Third, compile and develop a general standard for dynamic envelope systems that can later incorporate climate/location and objective-specific components or types. Then, expand on the experimental work starting with simpler dynamic envelope systems and branching to variations in building typologies and envelope components other than glazed façades or rooftops. Finally, develop a standard for designing and implementing dynamic envelopes and introduce it to the general engineering industry.

4.2. Implications for Urban Sustainability

Retrofitting existing buildings with dynamic envelope systems can have an immense impact on urban sustainability. Such retrofits can potentially transform existing urban communities into positive energy districts, reducing dependency on the electrical grid and making communities self-sustained. Authors in ref. [37] used their hybrid-solar system as a two-way grid-connected system where surplus energy can be sold and electricity can be bought from the grid when the energy generated is not enough to meet the building demand. In ref. [36], the authors proposed their system as a climate-conscious retrofit solution that can provide greater energy efficiency for existing buildings.

The potential to save immense amounts of energy when implementing dynamic envelope systems should not be ignored. Each system provides significant energy savings by providing efficient systems that utilize natural environmental elements to improve the thermal, visual, and energy performance of the buildings. Utilizing RES on-site also reduces the dependency on the electrical grid, making each building/community independent in terms of energy consumption. The widespread implementation of such systems will yield greater energy efficiency for the buildings cluster/districts/communities and contribute to the reduction in GHG emissions. Using such technological innovations with efficient urban planning, newer emerging communities, and existing ones can transform from energy-inefficient environments to sustainable and greener environments. Continued research, development, and interest in this matter can help drive the transition to higher adoption of PEDs and sustainable cities.

Achieving sustainable goals for the future of the built environment requires gradual PED's development across the globe or in this case the European union for the sustainable goals of 2050. The authors in ref. [63] mentioned the additional benefits PEDs can provide in the European cities of Evora, Amsterdam, and Espoo. Using PEDs with the integration of DREBE to reach net-positive potential is a great solution. DREBEs can be used to improve the overall efficiency of the building by providing high-efficiency systems and buildings through higher energy saving and renewable energy source (RES) utilization.

5. Conclusions

This review paper provides comprehensive summarization and meta-analyses of dynamic building envelope systems integrated with renewable energy sources, revealing insights into their significant potential to enhance overall building energy performance. The findings highlight the significant energy savings from adaptive systems such as dynamic PV shading devices and PCM for improving energy and thermal efficiency. Furthermore, systems like thermoelectric and wind-induced piezoelectric energy generation offer a newer alternative to solar-based energy production albeit a less efficient one.

While those systems can potentially allow buildings to reach net-positive energy, there are still challenges in implementing them in new or existing buildings due to the lack of economic analysis or standardization for general industry applications. Due to that, the study is limited in terms of discussing the DBE system's applicability and economic concerns. Another limitation is that the review does not address the effects of DBEs on the outdoor urban environment in terms of Urban Heat Island effects. Addressing these challenges will pave the way for general application and provide considerable benefits to the building infrastructure. Therefore, future research can focus on developing strategies for implementing such systems on a wider basis with careful considerations of climate and location conditions.

Retrofitting dynamic envelope systems into existing buildings can help transform them into sustainable urban environments, and with wider application can potentially transition cities into positive energy districts. Visual and user satisfaction is also a necessary factor to consider when retrofitting. Immense efforts by researchers, policymakers, and industry specialists are needed to realize the completion of sustainable goals for the future. This review highlights the importance of advancing DREBE into higher efficiency systems while also gradually employing them in the real world for climate change mitigation and a sustainable future for the urban environment.

Funding: This research was funded by U.S. National Science Foundation grant number [1953004] and [2001207].

Conflicts of Interest: The authors declare no conflict of interest.

Abbreviations

GHG	Greenhouse Gas
PED	Positive Energy District
RES	Renewable Energy Source
DBE	Dynamic Building Envelope
PV	Photovoltaic
BIPV	Building Integrated Photovoltaics
BIPV/T	Building-integrated Photovoltaic-thermal
PCM	Phase-Change Material
DREBE	Dynamic and Renewable Energy Building Envelope
DSF	Double-skin façade
dvPVBE	dynamic and vertical photovoltaic integrated building envelope
PVSD	Photovoltaic shading device
PVIMSD	Photovoltaic integrated moveable shading devices
PVIO	Photovoltaic integrated overhangs
AET	Adaptive envelope Technologies
WTABS	Wind and Actuation Energy Harvesting Tensegrity Adaptive Building Screens
PV/TW	Photovoltaic Trombe wall
PVT	Photovoltaic-Thermal
HEMS	Home energy management system
EMS	Energy Management System
DF	Daylight factor
DGP	Daylight Glare Probability
MOO	Multi-Objective Optimization
SAM	System Advisor Model
UHI	Urban Heat Island

References

1. Hannah Ritchie and Pablo Rosado. "Fossil Fuels" Published online at OurWorldinData.org. 2017. Available online: https://ourworldindata.org/fossil-fuels (accessed on 9 November 2024).
2. Bosu, I.; Mahmoud, H.; Ookawara, S.; Hassan, H. Applied single and hybrid solar energy techniques for building energy consumption and thermal comfort: A comprehensive review. *Solar Energy* **2023**, *259*, 188–228. [CrossRef]

3. IEA. Buildings—Energy System. Available online: www.iea.org/energy-system/buildings (accessed on 9 November 2024).
4. Gupta, V.; Deb, C. Envelope design for low-energy buildings in the tropics: A review. *Renew. Sustain. Energy Rev.* **2023**, *186*, 113650.
5. Resch, E.; Bohne, R.A.; Kvamsdal, T.; Lohne, J. Impact of urban density and building height on energy use in cities. *Energy Procedia* **2016**, *96*, 800–814.
6. Derkenbaeva, E.; Vega, S.H.; Hofstede, G.J.; Van Leeuwen, E. Positive energy districts: Mainstreaming energy transition in urban areas. *Renew. Sustain. Energy Rev.* **2022**, *153*, 111782.
7. Badarnah, L. Form follows environment: Biomimetic approaches to building envelope design for environmental adaptation. *Buildings* **2017**, *7*, 40. [CrossRef]
8. Wang, G.; Fang, J.; Yan, C.; Huang, D.; Hu, K.; Zhou, K. Advancements in smart building envelopes: A comprehensive review. *Energy Build.* **2024**, *312*, 114190. [CrossRef]
9. Kunwar, N.; Cetin, K.S.; Passe, U. Dynamic shading in buildings: A review of testing methods and recent research findings. *Curr. Sustain./Renew. Energy Rep.* **2018**, *5*, 93–100.
10. Boccalatte, A.; Fossa, M.; Ménézo, C. Best arrangement of BIPV surfaces for future NZEB districts while considering urban heat island effects and the reduction of reflected radiation from solar façades. *Renew. Energy* **2020**, *160*, 686–697.
11. D'Ambrosio, V.; Losasso, M.; Tersigni, E. Towards the energy transition of the building stock with BIPV: Innovations, gaps and potential steps for a widespread use of multifunctional PV components in the building envelope. *Sustainability* **2021**, *13*, 12609. [CrossRef]
12. Rajoria, C.S.; Kumar, R.; Sharma, A.; Singh, D.; Suhag, S. Development of flat-plate building integrated photovoltaic/thermal (BIPV/T) system: A review. *Mater. Today Proc.* **2021**, *46*, 5342–5352.
13. Kazem, H.A.; Chaichan, M.T.; Al-Waeli, A.H.; Sopian, K.; Alnaser, N.W.; Alnaser, W.E. Energy Enhancement of Building-Integrated Photovoltaic/Thermal Systems: A Systematic Review. *Solar Compass* **2024**, *12*, 100093.
14. De Luca, F.; Voll, H.; Thalfeldt, M. Comparison of static and dynamic shading systems for office building energy consumption and cooling load assessment. *Manag. Environ. Qual. Int. J.* **2018**, *29*, 978–998. [CrossRef]
15. Hosseini, S.M.; Mohammadi, M.; Rosemann, A.; Schröder, T.; Lichtenberg, J. A morphological approach for kinetic façade design process to improve visual and thermal comfort. *Build. Environ.* **2019**, *153*, 186–204. [CrossRef]
16. Shi, X.; Abel, T.; Wang, L. Influence of two motion types on solar transmittance and daylight performance of dynamic façades. *Solar Energy* **2020**, *201*, 561–580. [CrossRef]
17. Wang, J.J.; Beltran, L. A method of energy simulation for dynamic building envelopes. *Proc. SimBuild* **2016**, *6*, 298–303.
18. Goia, F.; Cascone, Y. The impact of an ideal dynamic building envelope on the energy performance of Low Energy Office Buildings. *Energy Procedia* **2014**, *58*, 185–192. [CrossRef]
19. Reddy, V.J.; Ghazali, M.F.; Kumarasamy, S. Advancements in phase change materials for energy-efficient building construction: A comprehensive review. *J. Energy Storage* **2024**, *81*, 110494.
20. Al-Yasiri, Q.; Szabó, M. Incorporation of phase change materials into building envelope for thermal comfort and energy saving: A comprehensive analysis. *J. Build. Eng.* **2021**, *36*, 102122.
21. Wang, P.; Liu, Z.; Zhang, X.; Hu, M.; Zhang, L.; Fan, J. Adaptive dynamic building envelope integrated with phase change material to enhance the heat storage and release efficiency: A state-of-the-art review. *Energy Build.* **2023**, *286*, 112928. [CrossRef]
22. Rathore, P.K.S.; Gupta, N.K.; Yadav, D.; Shukla, S.K.; Kaul, S. Thermal performance of the building envelope integrated with phase change material for thermal energy storage: An updated review. *Sustain. Cities Soc.* **2022**, *79*, 103690. [CrossRef]
23. Narbuts, J.; Vanaga, R. Revolutionizing the Building Envelope: A Comprehensive Scientific Review of Innovative Technologies for Reduced Emissions. *Environ. Clim. Technol.* **2023**, *27*, 724–737. [CrossRef]
24. Tripathi, B.M.; Shukla, S.K. A comprehensive review of the thermal performance in energy efficient building envelope incorporated with phase change materials. *J. Energy Storage* **2024**, *79*, 110128.
25. Mohtashami, N.; Fuchs, N.; Fotopoulou, M.; Drosatos, P.; Streblow, R.; Osterhage, T.; Müller, D. State of the art of technologies in adaptive dynamic building envelopes (adbes). *Energies* **2022**, *15*, 829. [CrossRef]
26. Luo, Y.; Zhang, L.; Bozlar, M.; Liu, Z.; Guo, H.; Meggers, F. Active building envelope systems toward renewable and Sustainable Energy. *Renew. Sustain. Energy Rev.* **2019**, *104*, 470–491. [CrossRef]
27. Harry, S. Dynamic adaptive building envelopes—An innovative and state-of-the-art technology. *Creat. Space* **2016**, *3*, 167–184. [CrossRef]
28. Page, M.J.; McKenzie, J.E.; Bossuyt, P.M.; Boutron, I.; Hoffmann, T.C.; Mulrow, C.D.; Shamseer, L.; Tetzlaff, J.M.; Akl, E.A.; Brennan, S.E.; et al. The PRISMA 2020 statement: An updated guideline for reporting systematic reviews. *BMJ* **2021**, *372*, 71.
29. Zou, W.; Wang, Y.; Tian, E.; Wei, J.; Peng, J.; Mo, J. A new dynamic and vertical photovoltaic integrated building envelope for high-rise glaze-facade buildings. *Engineering* **2024**, *39*, 194–203.
30. Oh, S.; Choi, G.S.; Kim, H. Climate-Adaptive Building Envelope Controls: Assessing the Impact on Building Performance. *Sustainability* **2023**, *16*, 288. [CrossRef]

31. Valinejadshoubi, M.; Athienitis, A.K.; Bagchi, A.; Abtahi, M. Integrated Dynamic Photovoltaic Facade for Enhanced Building Comfort and Energy Efficiency. *Biomimetics* **2024**, *9*, 463. [CrossRef]
32. Cai, Y.; Huang, Y.; Shu, Z.; Liu, Z.; Zhong, H.; Zhao, F. Investigation of double-PCM based PV composite wall for power-generation and building insulation: Thermal characteristics and energy consumption prediction. *Energy Built Environ.* 2024; *in press*.
33. Soliman, A.S.; Radwan, A.; Fouda, M.S.; Sultan, A.A.; Abdelrehim, O. Energy assessment of a sliding window integrated with PV cell and multiple PCMs. *J. Energy Storage* **2024**, *86*, 111341.
34. Jiang, Y.; Qi, Z.; Ran, S.; Ma, Q. A study on the effect of dynamic photovoltaic shading devices on energy consumption and daylighting of an office building. *Buildings* **2024**, *14*, 596. [CrossRef]
35. Dehwah, A.H.; Krarti, M. Energy performance of integrated adaptive envelope technologies for commercial buildings. *J. Build. Eng.* **2023**, *63*, 105535. [CrossRef]
36. Biloria, N.; Makki, M.; Abdollahzadeh, N. Multi-performative façade systems: The case of real-time adaptive BIPV shading systems to enhance energy generation potential and visual comfort. *Front. Built Environ.* **2023**, *9*, 1119696. [CrossRef]
37. Gholamian, E.; Barmas, R.B.; Zare, V.; Ranjbar, S.F. The effect of Incorporating phase change materials in building envelope on reducing the cost and size of the integrated hybrid-solar energy system: An application of 3E dynamic simulation with reliability consideration. *Sustain. Energy Technol. Assess.* **2022**, *52*, 102067.
38. Rahimpour, Z.; Verbič, G.; Chapman, A.C. Can phase change materials in building insulation improve self-consumption of residential rooftop solar? An Australian case study. *Renew. Energy* **2022**, *192*, 24–34.
39. Sun, X.; Lin, Y.; Zhu, Z.; Li, J. Optimized design of a distributed photovoltaic system in a building with phase change materials. *Appl. Energy* **2022**, *306*, 118010.
40. Abdullah, A.A.; Atallah, F.S.; Ahmed, O.K.; Daoud, R.W. Performance improvement of photovoltaic/Trombe wall by using phase change material: Experimental assessment. *J. Energy Storage* **2022**, *55*, 105596.
41. Krarti, M. Evaluation of PV integrated sliding-rotating overhangs for US apartment buildings. *Appl. Energy* **2021**, *293*, 116942.
42. Miranda, R.; Babilio, E.; Singh, N.; Santos, F.; Fraternali, F. Mechanics of smart origami sunscreens with energy harvesting ability. *Mech. Res. Commun.* **2020**, *105*, 103503. [CrossRef]
43. Byon, Y.S.; Jeong, J.W. Annual energy harvesting performance of a phase change material-integrated thermoelectric power generation block in building walls. *Energy Build.* **2020**, *228*, 110470.
44. Svetozarevic, B.; Begle, M.; Jayathissa, P.; Caranovic, S.; Shepherd, R.F.; Nagy, Z.; Hischier, I.; Hofer, J.; Schlueter, A. Dynamic photovoltaic building envelopes for Adaptive Energy and Comfort Management. *Nat. Energy* **2019**, *4*, 671–682. [CrossRef]
45. Ziasistani, N.; Fazelpour, F. Comparative study of DSF, PV-DSF and PV-DSF/PCM building energy performance considering multiple parameters. *Solar Energy* **2019**, *187*, 115–128. [CrossRef]
46. Kant, K.; Pitchumani, R.; Shukla, A.; Sharma, A. Analysis and design of air ventilated building integrated photovoltaic (BIPV) system incorporating phase change materials. *Energy Convers. Manag.* **2019**, *196*, 149–164. [CrossRef]
47. Jayathissa, P.; Luzzatto, M.; Schmidli, J.; Hofer, J.; Nagy, Z.; Schlueter, A. Optimising building net energy demand with dynamic BIPV shading. *Appl. Energy* **2017**, *202*, 726–735. [CrossRef]
48. Elarga, H.; Dal Monte, A.; Andersen, R.K.; Benini, E. PV-PCM integration in glazed building. Co-simulation and genetic optimization study. *Build. Environ.* **2017**, *126*, 161–175. [CrossRef]
49. Elarga, H.; Goia, F.; Zarrella, A.; Dal Monte, A.; Benini, E. Thermal and electrical performance of an integrated PV-PCM system in double skin façades: A numerical study. *Solar Energy* **2016**, *136*, 112–124.
50. Beck, H.E.; Zimmermann, N.E.; McVicar, T.R.; Vergopolan, N.; Berg, A.; Wood, E.F. Present and future Köppen-Geiger climate classification maps at 1-km resolution. *Sci. Data* **2018**, *5*, 180214. [CrossRef]
51. de Gracia, A. Dynamic building envelope with PCM for cooling purposes—Proof of concept. *Appl. Energy* **2019**, *235*, 1245–1253. [CrossRef]
52. Dimitrijević-Jovanović, D.; Živković, P.; Stevanović, Ž. The impact of the building envelope with the green living systems on the built environment. *Therm. Sci.* **2018**, *22* (Suppl. 4), 1033–1045.
53. Lai, F.; Zhou, J.; Lu, L.; Hasanuzzaman, M.; Yuan, Y. Green building technologies in Southeast Asia: A review. *Sustain. Energy Technol. Assess.* **2023**, *55*, 102946.
54. Chen, L.; Hu, Y.; Wang, R.; Li, X.; Chen, Z.; Hua, J.; Osman, A.I.; Farghali, M.; Huang, L.; Li, J.; et al. Green building practices to integrate renewable energy in the construction sector: A review. *Environ. Chem. Lett.* **2024**, *22*, 751–784.
55. Zhang, X.; Lovati, M.; Vigna, I.; Widén, J.; Han, M.; Gal, C.; Feng, T. A review of urban energy systems at building cluster level incorporating renewable-energy-source (RES) envelope solutions. *Appl. Energy* **2018**, *230*, 1034–1056. [CrossRef]
56. Wilberforce, T.; Olabi, A.G.; Sayed, E.T.; Mahmoud, M.; Alami, A.H.; Abdelkareem, M.A. The state of renewable energy source envelopes in urban areas. *Int. J. Thermofluids* **2024**, *21*, 100581.
57. Mohammed, M.A.; Budaiwi, I.M.; Al-Osta, M.A.; Abdou, A.A. Thermo-Environmental Performance of Modular Building Envelope Panel Technologies: A Focused Review. *Buildings* **2024**, *14*, 917. [CrossRef]

58. Tabadkani, A.; Roetzel, A.; Li, H.X.; Tsangrassoulis, A. Design approaches and typologies of adaptive facades: A review. *Autom. Constr.* **2021**, *121*, 103450.
59. Gondeck, M.; Triebel, M.A.; Steingrube, A.; Albert-Seifried, V.; Stryi-Hipp, G. Recommendations for a Positive Energy District Framework–Application and Evaluation of Different Energetic Assessment Methodologies. *Smart Energy* **2024**, *15*, 100147.
60. Hassan, A.A.; El-Rayes, K. Optimizing the integration of renewable energy in existing buildings. *Energy Build.* **2021**, *238*, 110851. [CrossRef]
61. De Freitas, J.; Cronemberger, J.; Soares, R.M.; Amorim, C.N. Modeling and assessing BIPV envelopes using parametric rhinoceros plugins grasshopper and ladybug. *Renew. Energy* **2020**, *160*, 1468–1479. [CrossRef]
62. Bakker, L.G.; Hoes-van Oeffelen, E.C.M.; Loonen, R.C.G.M.; Hensen, J.L.M. User satisfaction and interaction with automated dynamic facades: A pilot study. *Build. Environ.* **2014**, *78*, 44–52. [CrossRef]
63. Ntafalias, A.; Papadopoulos, P.; van Wees, M.; Šijačić, D.; Shafqat, O.; Hukkalainen, M.; Kantorovitch, J.; Lage, M. The Benefits of Positive Energy Districts: Introducing Additionality Assessment in Évora, Amsterdam and Espoo. *Designs* **2024**, *8*, 94. [CrossRef]

designs

MDPI

Article

Implementing Optimal Operation of Multi-Energy Districts with Thermal Demand Response

Martina Capone and Elisa Guelpa *

Energy Department, Politecnico di Torino, 10129 Torino, Italy; martina.capone@polito.it
* Correspondence: elisa.guelpa@polito.it

Abstract: The combination of different energy vectors in the context of multi-energy systems is a crucial opportunity to reach CO_2 reduction goals. In the case of urban areas, multi-energy districts can be connected with district heating networks to efficiently supply heat to the buildings. In this framework, the inclusion of the thermal demand response allows for significantly improve the performance of multi-energy districts by smartly modifying the heat loads. Operation optimization of such systems provides excellent results but requires significant computational efforts. In this work, a novel approach is proposed for the fast optimization of multi-energy district operations, enabling real-time demand response strategies. A 3-step optimization method based on mixed integer linear programming is proposed aimed at minimizing the cost operation of multi-energy districts. The approach is applied to a test case characterized by strongly unsteady heat/electricity and cooling demands. Results show that (a) the total operation cost of a multi-energy district can be reduced by order of 3% with respect to optimized operation without demand side management; (b) with respect to a full optimization approach, the computational cost decreases from 45 min to 1 s, while the accuracy reduces from 3.6% to 3.0%.

Keywords: multi-energy systems; district heating; demand side management

1. Introduction

1.1. Context: Multi-Energy District

For several years, fossil fuels have been exploited to produce electricity in large power plants and heat in the boilers installed in the buildings. Therefore, traditionally, the electric, chemical, and thermal energy vectors were integrated into a simple and predefined connection. During decades, the exploitation of different technologies has radically increased:

- New renewable energy technologies, with highly variable and unpredictable generation profiles, e.g., wind and solar, along with a more advanced version of traditional renewable technologies, e.g., hydro, geothermic, and biomass.
- Different types of storage systems, e.g., batteries, thermal/chemical storage.
- Components for the energy vector conversion, e.g., high-efficiency cogeneration plants (gas into electricity and heat), heat pumps (electricity into heat), and power-to-gas systems (power into gas).
- Low exergy heat, e.g., waste heat [1].

In this new framework, various benefits can be achieved by the adoption of connected and combinedly managed energy infrastructures (as energy infrastructures are intended conversion technologies, grids, and storages [2]). Examples of potentials of the combined management are (a) the possibility to transform energy in a different form when a special request occurs or in case of high purchase costs, and (b) to convert energy in a different form in order to store it [3]. A practical example consists in storing excess renewable electricity in thermal storage or thermal network (very cheap storage options) after being converted by heat pumps. Therefore, the energy vectors and infrastructures that were originally developed mainly independently are now operated combinedly to obtain further flexibility [4], also adjusting operation according to the fluctuating CO_2/electricity pricing [5]. Multi-energy systems in the case of urban areas are also called multi-energy districts; these are districts where the various loads (e.g., electrical, thermal) are supplied by different technologies that can interact with each other.

1.2. Multi-Energy District Operation Optimization

Multi-energy systems are typically characterized by the availability of various technologies, which can be combined in different ways to cover the demand for electricity, heating, cooling, etc. The main design parameters are the kind and the sizes of the technologies installed, their connections, and the characteristics of the energy infrastructure. The various combinations can provide different outcomes in terms of total energy cost, primary energy consumed, emission, etc. Among the various combinations, it is possible to find an optimal solution depending on the goal expressed by the objective function (e.g., minimum cost, carbon emission, primary energy). The best operation continuously changes in time since this depends on the various demands (electricity, heat, and cold), the ambient conditions, the constraints on the available technologies, etc. The optimization becomes much more complicated in case the system is connected to a district heating or/and one or multiple thermal energy storages. In these cases, the analysis is time-dependent. This means that the mathematical optimization cannot be solved separately for each time step, but this requires the solution to be obtained over a representative period (hours or days). This makes the number of independent variables much larger, as the set of variables is repeated within each time step. Various models have been proposed in the literature to face this kind of problem [6]. Linear programming is used when the technology efficiencies do not vary with the thermal load [7]. Most of the studies have been performed using mixed integer linear programming, both operation [8,9] and design [10] optimization. Non-linear programming or mixed integer non-linear programming is usually reserved for problems that are highly non-linear [11]. Results of the literature analysis show that the optimization of energy systems guarantees cost savings of the order of 5–25% [12,13] and CO_2 emission reductions of 5–9% [14,15]. A graphical representation of the problem is provided in Figure 1.

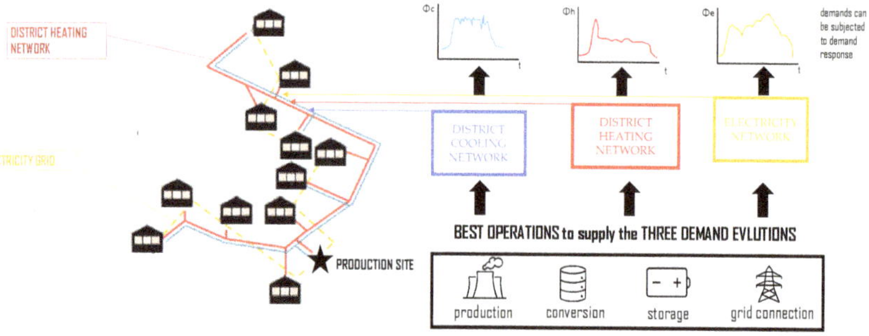

Figure 1. Schematic of the optimization of multi-energy districts.

1.3. Thermal Demand Response for Improving MED Operations

Consumers represent a major element of any smart energy system [16]. A very interesting approach to reducing the operating cost consists in modifying the plant load evolution in order to better fit optimal production. This result can be obtained through the application of demand response techniques, i.e., proper modifications of the building energy demand [17,18], which are still acceptable by the end-users. Demand response in multi-energy districts can provide significant results when applied to thermal networks [19] as thermal demand response has intended the possibility of relying on the thermal mass of buildings to modify the thermal demand profile of a building without significantly affecting the comfort conditions of the end-users (for more details, refer to the review paper [20]).

Such an approach can be implemented by acting on the thermal schedule (i.e., the times the thermal substations of the buildings connected with the district heating network are switched on and off) as well as the specific climatic settings on the building side. Several works in the literature report the advantages associated with thermal demand response: (1) thermal peak shaving, which allows one avoiding the use of less efficient technologies as show in the experimental works, where peak cut of 25–30% [21] and 15% [22] are obtained; (2) more constant daily demand; in [23] the load factor is shown to increase from 0.2 to 0.44; (3) primary energy savings (a 4% reduction is reported in [24]) and reduction of the energy cost (3–5% in [25], 9% in [22]).

The adoption of thermal demand response in the optimization of a multi-energy system provides additional benefits in terms of cost reduction due to the possibility of combining load variations with the use of the most advantageous technologies depending on the availability of resources and the efficiency of each plant. In the literature, only a few works addressing the optimization of the multi-energy district by means of thermal demand response can be found. These works are mainly focused on the search for the optimal operation of the available energy conversion systems, often combined with thermal storage. The combination of supply-side optimization with demand response in the multi-energy district is addressed in a few studies, such as [25,26]. In [27], a technique for the optimization of the multi-energy district subject to demand response is proposed, also taking into account the thermal network dynamics. In works such as [28,29], it is shown that thermal network dynamics are usually neglected. Thermal network dynamics is the factor that makes the thermal loads at the plants different (in some cases to a large extent) from the summation of the thermal demand at the buildings due to the presence of the following:

(a) different time delay of the water streams exiting the various buildings to reach the plant, depending on the building distance to the plant and their schedule;
(b) thermal losses in pipelines during both operation and non-operation hours;
(c) the thermal inertia due to a large amount of water inside.

Therefore, the difference between the load (at the plant) and the demand (at the buildings) is due to the network thermal dynamic impact, along with factors related to the substation/heating circuit. Especially in the case of large networks and discontinuous operations, it is actually important to consider this factor. When considering the network dynamic simulation, the computational cost significantly increases. An interesting approach is proposed in networks [28], where the computational costs start from 30 min, in the case of small-scale networks.

1.4. Work Contribution and Research Gaps

In this work, the authors present a methodology for the fast optimization of multi-energy districts where thermal demand response is applied; the contribution of the thermal network dynamics is performed by means of a simplified approach which allows one to obtain dramatic reductions in the computational time, without large penalties on the optimization results. The developed methodology relies on the solution of a programming optimization, estimating the best set of technologies for each time step, considering their proper efficiencies and costs and the influence of the storage availability.

The main strengths and novelties of the work are:

(1) district heating simulation is included in the multi-energy district optimization along with demand response, which is conducted just in a few other works in the literature. Except for very few works, thermal demand response is applied to a predefined set of the operational condition of the production side (i.e., once the production is available, demand response activities with the aim of minimizing costs). In this new perspective, demand response is included in the plant operational management;
(2) a specific approach is developed to reduce the computational costs to make it suitable for real-time network management and for the analysis of multiple realistic scenarios (e.g., in stochastic optimization), also in the case of large networks;

(3) a realistic case study in terms of the number and type of generators, energy demand profiles, energy prices, storage size, is considered.

Results show that the adoption of the optimized strategy with demand response for the management of the technologies and the thermal storage allows a cost reduction of about 3% and that the efficacy of the fast optimizer is close to the full optimizer (−3% vs. −3.6%). Computational time dramatically reduces from 2700 s to 1 s.

1.5. Paper Structure

The paper structure is reported here. The optimization approach used to simulate a multi-energy district with demand response is described in Section 2.1 (specifically, the part related to demand response is described in Section 2.1.2). The Thermo-fluid dynamic model of the network is described in Section 2.2. The innovative approach proposed in this work and the reason for the choices done are described in Section 2.3. In Section 3 the case study is presented; information on the energy system considered (Section 3.1) and specific details on the demand response approach adopted (Section 3.2) are reported. The results are discussed in Section 4, in terms of production patterns (Section 4.1), operating costs (Section 4.2), and computational costs (Section 4.3). A discussion section is added (Section 5) to discuss the extension of the results to other cases, the model strength/limitation, and the applicability of the model proposed in relation to other available approaches. Conclusions are reported in Section 6.

2. Methodology

2.1. Problem Definition

2.1.1. Best Operations

The best operation for the energy system, i.e., the one that minimizes the cost of the energy supplied, is obtained by applying an optimization. The objective function is the total cost of the energy systems (Equation (1)), namely the sum of the operating energy costs of the various energy vectors purchased by the system minus the cost of the energy vector sold to the extern:

$$\min(C_{tot}) \tag{1}$$

Costs are computed by multiplying the specific costs of gas and electricity times the corresponding energy flows. The specific costs are, for each time step, known; the quantities of energy purchased/sold are a consequence of the operating conditions. Therefore, power fluxes (electrical, heat, and cold) produced by each technology installed or stored/released by the thermal/electricity storages are the independent variables of the optimization. In the cases considered in this work, the conversion/storage technologies are five: gas Combined heat and production (CHP) units, gas heat only boiler (HOB), electric heat pump (EHP), and cold and thermal storage (see Section 3). All the time steps must be optimized in a unique optimization since the hot and cold storage operations make the problem time-dependent (i.e., the load at a certain time, t_1, is dependent on the load at whatever other time, t_2). The overall number of variables is thus the product of the number of technologies (n_{tech}) times the time step considered in the analysis (here 96). This constitutes group A of independent variables; the remaining part, group B, is related to the application of the demand response.

Considering, therefore, the presence of various technologies absorbing and producing/releasing energy in the system, the total cost can be written as Equation (2), where the cost for each timestep and each technology can be estimated as in Equation (3).

$$C_{tot} = \sum_{i=1}^{n\_ts} \sum_{j=1}^{n\_tech} C_{i\,j} \tag{2}$$

$$C_{ij} = c_i \; P_{ij} \tag{3}$$

where P is the power absorbed/released by each technology and c is the cost of the energy vector in input for each technology. In case of excess production, the specific cost is the sale cost of the energy vector times the amount of energy sold. The relation among sources and products of the technologies (therefore between chemical/thermal/electrical energy vectors) is a function of their efficiency. When the efficiency of the production and conversion technologies is independent of the load, the correlation between the source and product is linear. Otherwise, the results are non-linear. In this application, efficiencies are considered constant, according to the data reported in Table 1. Another important factor is, especially for heat pumps, the dependency of efficiency on the production temperature; in this case, since the water supply temperature is constant, the dependency has not been considered. However, the approach proposed is also valid for the non-linear problem, as clarified in Section 2.3.

Table 1. Characteristics of the Technologies adopted in the multi-energy district.

Technology	Acronym	Efficiency	Max Inlet Power (MW)	Energy Storable (MWh)
Natural gas combined heat and power plant	CHP	0.36 0.53	7	
Natural gas heat only boiler	HOB	0.92	6	
Electric heat pump	EHP	4.4	0.75	
Thermal storage for heating purposes	Hot storage	1	0.6	1.6
Thermal storage for cooling purposes	Cold storage	1	1	5

The constraints of the optimization concern:

a. The energy balances. The produced/purchased electricity, hot and cold energy, must be, at each time step, sufficient to supply the loads. The equation, for each energy vector v, and each time step j, has the form of Equation (4):

$$\begin{aligned}\Phi_{load\_v,j} = & \left(\sum_{i=1}^{nPrTech\_v} \Phi_{pr\ i}\right)_{v,j} + \left(\sum_{i=1}^{nStOut\_v} \Phi_{st\_out\ i}\right)_{v,j} - \left(\sum_{i=1}^{nStIn_v} \Phi_{st_{in} i}\right)_{v,j} \\ & + \left(\sum_{i=1}^{nAbsTech_v} \Phi_{abs\ i}\right)_{v,j} + \left(\Phi_{purch}\right)_{v,j} - \left(\Phi_{sold}\right)_{v,j}\end{aligned} \quad (4)$$

where $nPrTech\_v$, $nStOut\_v$, $nStIn\_v$, and $nConvTech\_v$ are, respectively, the technologies producing v, the storages releasing v, the storages are absorbing v and the conversion technologies absorbing v. Φ_{purch} and Φ_{sold} are the fluxes of the vector v that are respectively purchased and sold by the multi-energy district. Equation (4) is applied to the thermal, the electricity, and the cooling load, for each time step j.

In the case of thermal load (right-hand-side term), the production is performed with combined heat and power plant and heat only boiler (1st left-hand-side term), thermal storage is available (2nd and 3rd right-hand-side term), there are no technologies absorbing heat (4th right-hand-side term) and no other ways exist for purchasing/selling heat (5th and 6th right-hand-side term).

In the case of cooling load (right-hand-side term), the production is performed with an electric heat pump (1st left-hand-side term), and no other technologies exist that: act as cooling storage (2nd and 3rd right-hand-side term), absorb cooling (4th right-hand-side term) allows purchasing/selling cold (5th and 6th right-hand-side term).

In the case of electricity load (right-hand-side term), the production is performed with CHP (1st left-hand-side term), there is battery storage (2nd and 3rd right-hand-side term), there is the electric heat pump absorbing electricity (4th right-hand-side term), and it is possible to purchase/sell electricity from/to the grid (5th and 6th right-hand-side term).

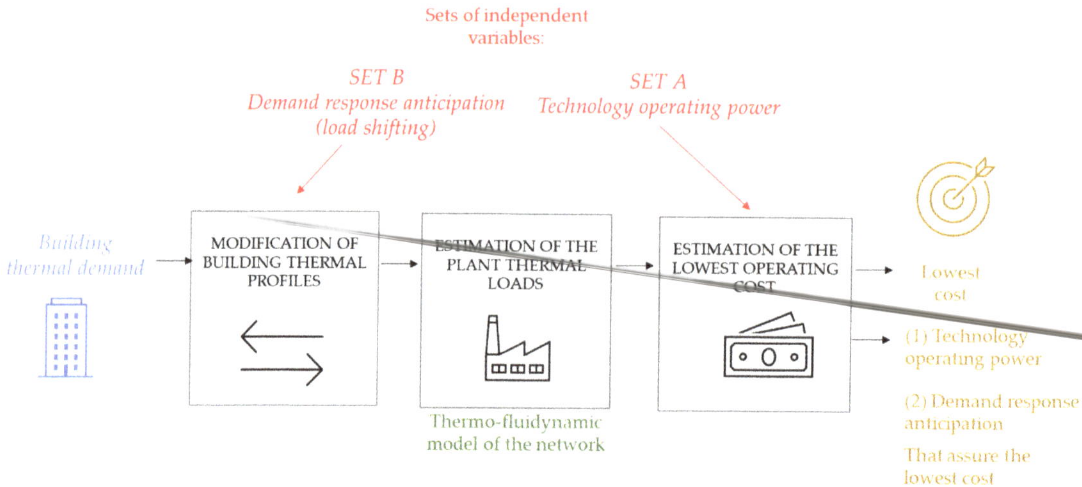

Figure 2. Schematic of the general logic for the solution of demand response-multi-energy system problems.

2.4. Fast Approach Proposed

In this work, a novel approach is proposed for fast simulation. This is composed of different steps, as schematized in Figure 3:

- Hydraulic optimization. The first step consists of the selection of a set of optimal load shifting, which minimizes the peak mass-flow rate. This allows to minimization of the mass-flow peak required by the users. The first step allows thus to estimate the heating demand anticipation of each building a_i such that

$$\min\left(\max\left(G_{plant}(t)\right)\right) = 0 \tag{12}$$

  The optimization is solved with a mixed integer linear programming using the solver Gurobi.

- Plant heating load estimation. The second step consists of the estimation of the plant heating load given the building thermal demand, which is a consequence of the best set of anticipations (output of step 1). The estimation of the plant thermal load is conducted through physical simulation of the district heating network. This is performed to take into account mixing, delays, and thermal transients, occurring within the pipelines. The thermo-fluid dynamic model of the district heating network is described in Section 2.2.
- Multi-energy district optimization. In the last step, the optimization of the operation of the multi-energy system, given the thermal demand estimated in step 2, is done. The aim is to find the best set of thermal power for each production/conversion/storage technology (P in Equation (3)) with the aim of minimization of the daily operating cost (Equation (1)). The problem solution is conducted with a Mixed Integer Linear Programming (MILP) approach. This means that the approach is also suitable in the case of non-linear efficiencies since these can be easily piecewise linearized. The details of the optimization problem are given in Section 2.1.

This approach is conceived with the aim of separately estimating the best set of thermal power for the multi-energy district and the best demand response to apply to the buildings. This allows for significantly reducing the computational time required for the solution of the problem, as shown in the Result section.

Here the idea behind the model just presented is discussed. The peak creation is a consequence of two aspects. The first is related to the temperature drop in the network during thermal transient, which occurs when a part of the system is not operating (e.g., during the night). In this case, the mass flow is zero or very low. The water cools down quite rapidly because of the larger thermal losses. During the system switching on, the demand is much higher to recover the normal operating temperature of the water. This has an impact on peak creation. The second is related to the larger mass flow rate. The substation valve, after the switching off/attenuation, opens completely to ensure a reheating of the secondary circuit/substation heat exchanger/heating devices, making dramatically increasing the mass flow rate within the pipelines [33]. The impact of the first contribution is mainly dependent on the quantity of water within the network and, therefore, on the network dimension and on the number of temperature fluctuations. In the case of small networks or small temperature fluctuations, the contribution of the first factor is not significant as the contribution due to the mass flow increase. This is proved by previous results of an author published in [29]: if in large networks, the dynamic of the system should be considered, in small networks, or networks with no large thermal transients, this can be neglected. The approach presented here is focused on networks where this contribution is not large.

The idea is that in these cases, it is possible to decouple the simulation of the thermal transient from the optimization of demand response without losing most of the optimization result accuracy. The best demand response anticipations are therefore estimated, considering only the second contribution to the peak shaving due to the mass flow rate increase; for this reason, the mass flow rate is minimized.

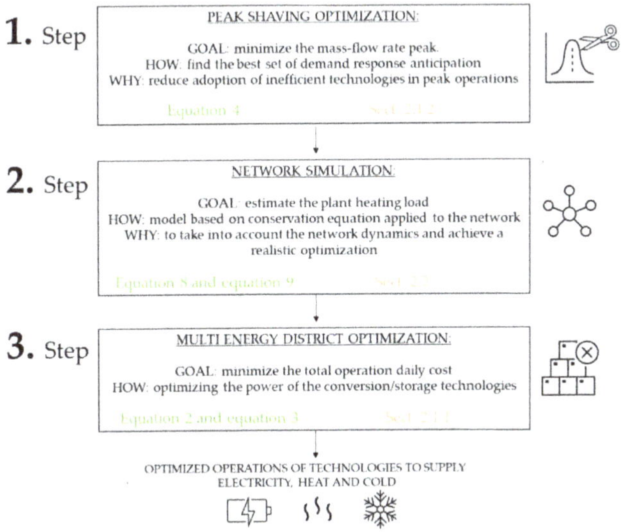

Figure 3. Schematics of the approach adopted for fast simulations.

3. Case Study

3.1. District Heating Details

In the present work, a small network is considered a case study since the method is specifically conceived for a system with moderate thermal transients, as detailed in Section 2.3. This includes industrial, office, and residential users. The total number of connected substations is 60. The district heating network is characterized by 202 nodes and 201 branches for a total length of 4.5 km. The multi-energy system is characterized by a set of technologies for the production/conversion and storage of energy.

- Heat production: heat can be supplied with a combined heat and power (CHP) unit and with a natural gas boiler (HOB).
- Electricity Production: electricity can be produced using the CHP and by grid purchase.
- Cooling production: cooling is produced using the Electric heat pump.

Furthermore, electrical storage and thermal storage are considered. Details on the installed technologies are provided in Table 1. A schematic of the connection of the conversion and storage units is provided in Figure 4. As concern the costs, the electricity costs range from 0.02 and 0.08 €/kWh for selling and from 0.15 to 0.2 €/kWh for electricity purchase. Gas cost is 0.021 €/kWh.

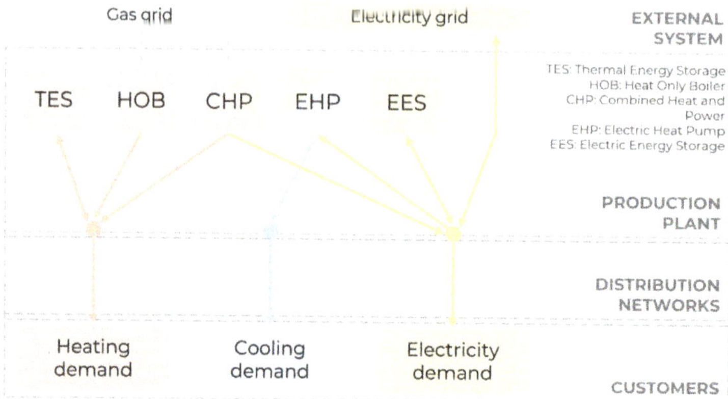

Figure 4. Schematic of the multi-energy considered in the analysis, with energy vectors and energy infrastructure (storage and conversion units).

3.2. Thermal Demand Response

The analysis is conducted on a typical mild winter day. The heating, cooling, and electricity loads are reported in Figure 2. Concerning the electricity demand, this never goes to zero. A certain demand is always present, and in particular, this is higher in the evening. The cooling load is much lower than the other loads since this is only required for industrial applications.

As concern the heating load, this has a significant thermal peak in the early morning. In areas with milder winters, such as the Mediterranean, the building occupants often have the habit of shutting down or remarkably attenuating the heating devices at night. This habit creates, as a consequence, a typical thermal demand profile, as the one reported in Figure 5. In the morning, the systems are switched on, and a significant thermal peak occurs because of the two main reasons largely discussed in Section 2.3. Peaks are usually supplied with low-efficiency plants; therefore, it is worth shaving them. This is mainly conducted by using demand response or thermal storage. Storages allow for flexible management but, on the other hand, require space and investment. In the present analysis, thermal storage demand side, management is adopted as a further technique to shave the peak and increase the overall efficiency.

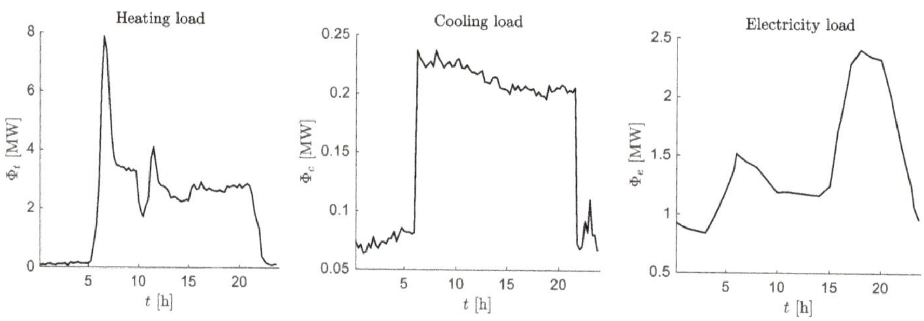

Figure 5. Load profiles.

With thermal demand response, the thermal demand of the users can be changed in the optimization tool according to the requirements of the system. In the present case study, demand shifting is considered, which allows the achievement of significant peak reduction without modifications in the system control strategy [34,35]. In order to prevent affecting the users' thermal comfort, only anticipations up to 30 min are allowed. No delays are considered to avoid unsatisfactory comfort conditions in the morning. In case anticipation is larger than 30 min, suitable tools for the simulation of the indoor conditions or direct smart sensors for measuring temperatures should be used to maintain satisfactory internal comfort. The demand shifting, as often happens in a real network, can be applied only on a discrete basis, every 5 min, due to reasons related to the controller adopted. Therefore, the possibilities for demand response are 5, 10, 15, 20, 25, and 30 min.

4. Results

4.1. Load Profiles

In this section, the optimal operation of the multi-energy district technologies that allow supplying of the three total load profiles at the minimum overall cost is reported. Figure 6 shows the results of the optimization performed with the fast approach proposed. In black, the load evolutions are reported. The black curve is filled in different colors, depending on the technology adopted to supply the energy. In striped colors, the energy that is produced in excess is reported. The excess energy is used (a) to be sold (e.g., electricity sold to the grid), (b) to be stored (e.g., thermal energy in the thermal storage), (c) to supply other conversion technologies (e.g., electricity must be produced in excess to supply the EHP). Cooling is entirely produced by using the EHP, which is the only available technology. The heating base load is produced using the CHP. The HOB is used to supply the morning peak, along with thermal storage. Concerning the electricity, the base load is produced by CHP. Electricity is stored when the load is lower to the CHP production and released for valley filling.

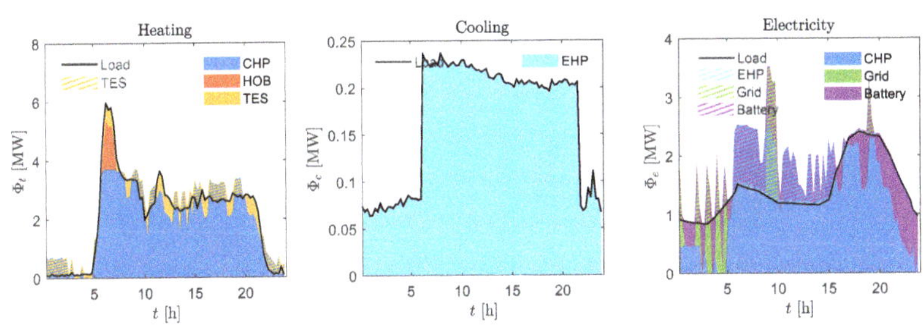

Figure 6. Fast optimization results: operations of the technology for the supply of heating, cooling, and electrical load.

Results obtained with the fast approach are compared with the results achieved with the conventional approach [19]. The same comments made for the fast optimization results apply to these results. Only small differences are present between the evolutions of Figures 6 and 7, mainly due to slightly different management of the storage. The same technologies are adopted to supply the multi-energy district, which has very similar load evolutions.

4.2. Total Cost

Figure 8 presents a comparison of the total costs obtained with the two optimizations depicted in Figures 6 and 7. The comparison also includes the case of operation optimization with demand response. Therefore, the three bars and the line represents respectively, are the overall daily operating costs and the daily cost related to the boiler technology in three cases: (a) optimization without demand response, (b) fast optimization with demand response, (c) with the full optimization with demand response. At first, it is possible to state that the adoption of demand response allows for a significant reduction of the operating cost. The cost without demand response is 2385 €/day, while the minimum cost of the fast and full otimizer are respectively 2314 €/day and 2298 €/day. This means that a reduction of 3% and 3.6% is achieved in the two cases. Both these differences are significant for the operation of an energy system. The difference in the total cost obtained using the fast optimization (2314 €/day) and the full otimizer (2298 €/day) is 16 €/day. The difference is not negligible but limited. In particular, it should be considered that in several cases, the uncertainty related to the thermal demand could make this difference smaller or even inverted; in these cases, the fast approach could represent a favorable option since it can be adopted and included in stochastic optimization. On the other hand, the cost reduction related to the use of heat-only boiler is significant between the fast and full optimization approach (36 €/day for the fast vs 25 €/day for the full). The difference is that the thermal transients have been neglected. This result suggests that, in large networks characterized by significant thermal transient, this could provide large cost increases. Nevertheless, the large computational time required by the full approach might force one to accept a sub-optimal result obtained with the fast approach (see next section).

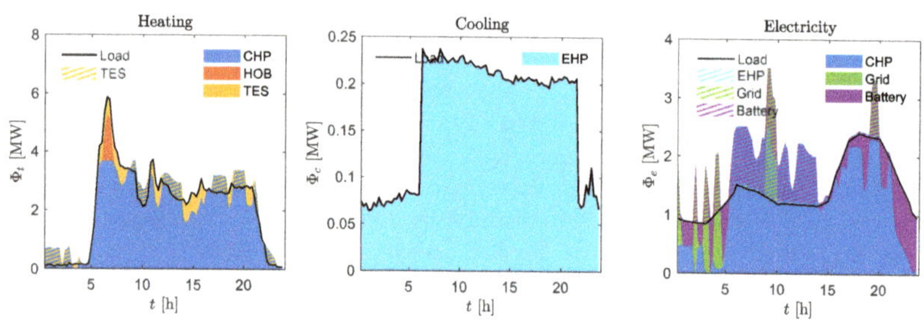

Figure 7. Full optimization results (Approach B): operations of the technologies for the supply of heating, cooling, and electrical load.

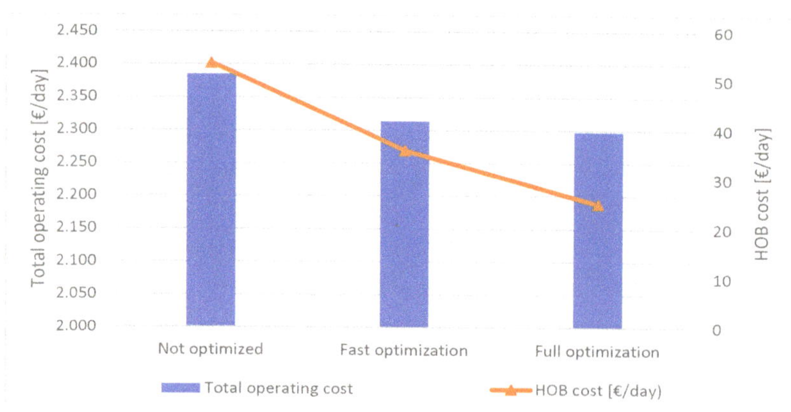

Figure 8. Comparison between total cost and HOB cost obtained without optimization, fast and full optimization.

4.3. Computational Resources

The computational costs for running the fast and the full optimizations are completely different. These are shown in Figure 9 for a PC laptop with a total of 16 GB of memory and an Intel i7-8565U CPU @ 1.80 GHz. The fast optimization only takes 1 sec to provide results. The full optimization approach requires about 45 min. This is not an unsustainable time, in general. It depends on the application. For obtaining a one-day ahead optimization, i.e., to plan the daily operation for the following day, this is a reasonable time. For having a real-time optimization, with the aim of updating the results of the previous simulation, given updated profile data, this is not suitable. In this case, fast optimization is preferable. The same applies in the case of multi-scenario simulation (e.g., for stochastic analysis). These aspects are discussed in Section 4.

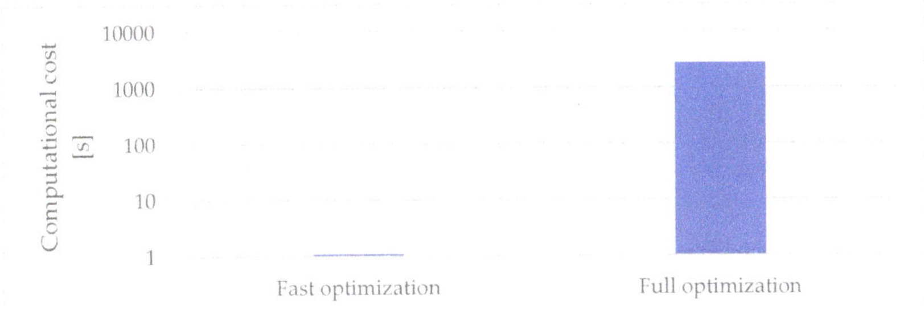

Figure 9. Comparison computational cost required for fast and full optimization.

5. Discussion

The idea of the approach developed in this work was born while observing (a) the high computational costs required to solve multi-energy district operation optimization with thermal demand response and (b) the different impacts of the thermal transient in different kinds of networks/loads. The authors decided to develop and test a fast approach, specifically for cases characterized by (a) a small network (i.e., with small pipe lengths and diameters) and (b) thermal loads that do not cause significant temperature variations within the network.

Comparison with/without demand response: The comparison shows that the demand response provides non-negligible overall cost savings. The saving range between 3 to 3.6%. As concern the extension of the result on another multi-energy district, the saving is expected to strongly depend on various aspects. The first is the entity of the peaks. The second is related to the technologies adopted to supply the network; when technologies with completely different efficiency are available, the saving due to proper peak management significantly increases. The third is the thermal storage availability; when the storage capacity is large, the benefits related to demand response reduce. In the present case study, there is a large thermal peak, and the efficiency of thermal production technologies (CHP and HOB) are very different. Therefore, the thermal storage is designed with a large size. In such a case, the application of demand response to the multi-energy district offers limited advantages with respect to a case in which the storage is small. Although the obtained reduction in the total costs (3.6%) is a good result, it is not an upper limit for the combined optimization of demand response and multi-energy systems operation. Reductions over 5% can be easily achieved when the size of thermal storage is optimized, considering the possibility of combining it with the application of demand response techniques.

Comparison fast/full optimization. A comparison of results for the present case study shows that the full model guarantees moderately higher performance (+0.6% saving) with a time requirement considerably higher (+99.9% of the time). Therefore, it is important to correctly select the approach, depending on the application.

The *full optimizer*, as already reported in Section 4.3, can be used for:

- *one-day ahead optimization: this is useful for planning the daily operation of the following day.*
- *analyses for further development*: the model can be extremely useful to simulate future benefits achievable by installing new technologies, adopting different demand response techniques, applying innovative control strategies, or changing the thermal network operating conditions.
- *design purposes:* the model can be adopted to design a multi-energy district, taking into account the future operation, which allows more focused planning.

The *fast optimizer* is suitable for different applications:

- *real time optimization*: this is useful to achieve fast indications for the current system operation. This can be very useful also in combination with a one-day ahead optimizer, with the aim of updating the results of the day ahead simulation, given the updated building demand profiles of the current day.
- *multi-scenario simulation*: there are cases in which the uncertainties on the building thermal demand or the characteristics of the devices/network are significant. In these cases, a multi-scenario or a stochastic optimization can be useful to minimize the impact of uncertainties and unknowns on the results. Fast models are required to keep reasonable the computational time of the simulation.
- *Large scale network with moderate thermal transient*: this refers to the case of large networks when the computational time required to run the full model would be unsustainable. In particular, when the thermal peak is mainly due to mass flow increase, and not this is not related to the transient temperature difference, this approach could be adopted.

When implementing the approach into reality, a demand forecast is mandatory. Therefore, in order to minimize the impact of demand uncertainty, a good precision demand forecast should be adopted. At the same time, morning peak prediction can reach a good level of accuracy since (a) the switching on time is known in advance and (b) the thermal peaks are less dependent on the environmental temperature than the demand at other times (e.g., during late morning or afternoon) since the night cooling down of the DH infrastructure (e.g., the pipelines) is more independent from the outdoor temperature than the building *steady state load*.

Another important point concerns the optimization purpose. In this case, the purpose is the operation optimization of an already existing system. In case the goal was the contemporary design and optimization of the system, the significant change required would be the addition of new integer variables, i.e., the investment cost of the technologies installed. The solution can therefore be performed with the same cascade MILP approach proposed in the current work.

Future research directions should include the analysis of multi-energy districts that include large-scale district heating. This is a challenging topic because of two reasons. First, in a large network characterized by significant load variations, the thermal inertia, if neglected, could significantly impact the results. Secondly, large thermal networks are characterized by complex topologies and a large number of pipes; this significantly increases the number of variables of the optimization problem.

6. Conclusions

This paper proposes a fast approach to perform instantaneous optimization of multi-energy districts with thermal demand response. The approach decouples the demand response and the operation optimization, neglecting, in the demand response optimization, the impact of the thermal inertia. Between the two optimizations, the network model is applied to ensure to supply of the actual amount of heat required by the network. The approach is tailored for small-scale district heating system or medium/large if the temperature variations are limited since, in these cases, the contribution of the thermal inertia is low and the thermal transients in the network do not play a crucial role in the peak creation.

Results show that the adoption of optimization in operation is useful to reduce the operation costs of the multi-energy system by order of 3%. The fast optimizer provides satisfactory results as concerns the operating cost reduction: −3% vs. −3.6 of the full approach. At the same time, the computational time required is much lower (−99.9%).

In conclusion, the full approach presented here can be considered suitable for all the cases in which the computational cost must be low, such as real-time operations or multi-scenario. Future development will consist of the individuation of a possible approach to address the same problem in large-scale DH.

Author Contributions: Conceptualization, M.C. and E.G.; methodology, M.C.; investigation, M.C. writing—original draft preparation, M.C.; writing—review and editing, E.G.; supervision, E.G. All authors have read and agreed to the published version of the manuscript.

Funding: This research received no external funding.

Institutional Review Board Statement: Not applicable.

Informed Consent Statement: Not applicable.

Data Availability Statement: Not applicable.

Conflicts of Interest: The authors declare no conflict of interest.

References

1. Pelda, J.; Stelter, F.; Holler, S. Potential of integrating industrial waste heat and solar thermal energy into district heating networks in Germany. *Energy* **2020**, *203*, 117812. [CrossRef]
2. Guelpa, E.; Bischi, A.; Verda, V.; Chertkov, M.; Lund, H. Towards future infrastructures for sustainable multi-energy systems: A review. *Energy* **2019**, *184*, 2–21. [CrossRef]
3. Clegg, S.; Mancarella, P. Integrated Electrical and Gas Network Flexibility Assessment in Low-Carbon Multi-Energy Systems. *IEEE Trans. Sustain. Energy* **2016**, *7*, 718–731. [CrossRef]
4. Mancarella, P. MES (multi-energy systems): An overview of concepts and evaluation models. *Energy* **2014**, *65*, 1–17. [CrossRef]
5. Werner, M.; Muschik, S.; Ehrenwirth, M.; Trinkl, C.; Schrag, T. Sector Coupling Potential of a District Heating Network by Consideration of Residual Load and CO_2 Emissions. *Energies* **2022**, *15*, 6281. [CrossRef]
6. Allegrini, J.; Orehounig, K.; Mavromatidis, G.; Ruesch, F.; Dorer, V.; Evins, R. A review of modelling approaches and tools for the simulation of district-scale energy systems. *Renew. Sustain. Energy Rev.* **2015**, *52*, 1391–1404. [CrossRef]
7. Ren, H.; Zhou, W.; Nakagami, K.; Gao, W.; Wu, Q. Multi-objective optimization for the operation of distributed energy systems considering economic and environmental aspects. *Appl. Energy* **2010**, *87*, 3642–3651. [CrossRef]
8. Bischi, A.; Taccari, L.; Martelli, E.; Amaldi, E.; Manzolini, G.; Silva, P.; Campanari, S.; Macchi, E. A detailed MILP optimization model for combined cooling, heat and power system operation planning. *Energy* **2014**, *74*, 12–26. [CrossRef]
9. Bischi, A.; Taccari, L.; Martelli, E.; Amaldi, E.; Manzolini, G.; Silva, P.; Campanari, S.; Macchi, E. A rolling-horizon optimization algorithm for the long term operational scheduling of cogeneration systems. *Energy* **2019**, *184*, 73–90. [CrossRef]
10. Buoro, D.; Casisi, M.; De Nardi, A.; Pinamonti, P.; Reini, M. Multicriteria optimization of a distributed energy supply system for an industrial area. *Energy* **2013**, *58*, 128–137. [CrossRef]
11. Arcuri, P.; Beraldi, P.; Florio, G.; Fragiacomo, P. Optimal design of a small size trigeneration plant in civil users: A MINLP (Mixed Integer Non Linear Programming Model). *Energy* **2015**, *80*, 628–641. [CrossRef]
12. Mitra, S.; Sun, L.; Grossmann, I.E. Optimal scheduling of industrial combined heat and power plants under time-sensitive electricity prices. *Energy* **2013**, *54*, 194–211. [CrossRef]
13. Yang, Y.; Zhang, S.; Xiao, Y. Optimal design of distributed energy resource systems coupled withenergy distribution networks. *Energy* **2015**, *85*, 433–448. [CrossRef]
14. Ren, H.; Gao, W. A MILP model for integrated plan and evaluation of distributed energy systems. *Appl. Energy* **2010**, *87*, 1001–1014. [CrossRef]
15. Marquant, J.F.; Evins, R.; Bollinger, L.A.; Carmeliet, J. A holarchic approach for multi-scale distributed energy system optimisation. *Appl. Energy* **2017**, *208*, 935–953. [CrossRef]
16. Schweiger, G.; Eckerstorfer, L.V.; Hafner, I.; Fleischhacker, A.; Radl, J.; Glock, B.; Wastian, M.; Rößler, M.; Lettner, G.; Popper, N.; et al. Active consumer participation in smart energy systems. *Energy Build.* **2020**, *227*, 110359. [CrossRef]
17. Gelazanskas, L.; Gamage, K.A.A. Demand side management in smart grid: A review and proposals for future direction. *Sustain. Cities Soc.* **2014**, *11*, 22–30. [CrossRef]
18. Siano, P. Demand response and smart grids—A survey. *Renew. Sustain. Energy Rev.* **2014**, *30*, 461–478. [CrossRef]
19. Capone, M.; Guelpa, E.; Mancò, G.; Verda, V. Integration of storage and thermal demand response to unlock flexibility in district multi-energy systems. *Energy* **2021**, *237*, 121601. [CrossRef]
20. Guelpa, E.; Verda, V. Demand response and other demand side management techniques for district heating: A review. *Energy* **2021**, *219*, 119440. [CrossRef]
21. Wu, Y.; Mäki, A.; Jokisalo, J.; Kosonen, R.; Kilpeläinen, S.; Salo, S.; Liu, H.; Li, B. Demand response of district heating using model predictive control to prevent the draught risk of cold window in an office building. *J. Build. Eng.* **2021**, *33*, 101855. [CrossRef]
22. Ala-Kotila, P.; Vainio, T.; Heinonen, J. Demand response in district heating market—Results of the field tests in student apartment buildings. *Smart Cities* **2020**, *3*, 157–171. [CrossRef]
23. Sweetnam, T.; Spataru, C.; Barrett, M.; Carter, E. Domestic demand-side response on district heating networks. *Build. Res. Inf.* **2019**, *47*, 330–343. [CrossRef]
24. Wernstedt, F.; Davidsson, P.; Johansson, C. Demand Side Management in District Heating Systems. In Proceedings of the 6th International Joint Conference on Autonomous Agents and Multiagent Systems (AAMAS 2007), Honolulu, HI, USA, 14–18 May 2007.

25. Hedegaard, R.E.; Kristensen, M.H.; Pedersen, T.H.; Brun, A.; Petersen, S. Bottom-up modelling methodology for urban-scale analysis of residential space heating demand response. *Appl. Energy* **2019**, *242*, 181–204. [CrossRef]
26. Thang, V.V.; Ha, T.; Li, Q.; Zhang, Y. Stochastic optimization in multi-energy hub system operation considering solar energy resource and demand response. *Int. J. Electr. Power Energy Syst.* **2022**, *141*, 108132. [CrossRef]
27. Reynolds, J.; Ahmad, M.W.; Rezgui, Y.; Hippolyte, J.L. Operational supply and demand optimisation of a multi-vector district energy system using artificial neural networks and a genetic algorithm. *Appl. Energy* **2019**, *235*, 699–713. [CrossRef]
28. Capone, M.; Guelpa, E.; Verda, V. Multi-objective optimization of district energy systems with demand response. *Energy* **2021**, *227*, 120472. [CrossRef]
29. Guelpa, E. Impact of network modelling in the analysis of district heating systems. *Energy* **2020**, *213*, 118393. [CrossRef]
30. Capone, M.; Guelpa, E.; Verda, V. Accounting for pipeline thermal capacity in district heating simulations. *Energy* **2021**, *219*, 119663. [CrossRef]
31. Versteeg, H.K.; Malalasekera, W. *An Introduction to Computational Fluid Dynamics: The Finite Volume Method*; Pearson Education Limited: London, UK, 2017.
32. Guelpa, E.; Sciacovelli, A.; Verda, V. Thermo-fluid dynamic model of large district heating networks for the analysis of primary energy savings. *Energy* **2019**, *184*, 34–44. [CrossRef]
33. Guelpa, E. Impact of thermal masses on the peak load in district heating systems. *Energy* **2021**, *214*, 118849. [CrossRef]
34. Aoun, N.; Bavière, R.; Vallée, M.; Brun, A.; Sandou, G. Dynamic Simulation of Residential Buildings Supporting the Development of Flexible Control in District Heating Systems. In Proceedings of the 13th International Modelica Conference, Regensburg, Germany, 4–6 March 2019; Volume 157, pp. 129–138. [CrossRef]
35. Aoun, N.; Aurousseau, A.; Sandou, G. Load shifting of space-heating demand in DHSs based on a building model identifiable at substation level Context Space-heating demand management. In Proceedings of the 4th Smart Energy Systems and 4th Generation District Heating Conference, Aalborg, Denmark, 13–14 November 2018.

designs

MDPI

Article

Techno-Economic Feasibility Study of a 1.5 MW Grid-Connected Solar Power Plant in Bangladesh

Md. Feroz Ali [1,*], Nitai Kumar Sarker [1], Md. Alamgir Hossain [2], Md. Shafiul Alam [3], Ashraf Hossain Sanvi [1] and Syed Ibn Syam Sifat [1]

1 Department of Electrical and Electronic Engineering, Pabna University of Science and Technology, Pabna 6600, Bangladesh; nitaikumarsarker@gmail.com (N.K.S.); sanvi91597@gmail.com (A.H.S.); sjsifat71@gmail.com (S.I.S.S.)

2 Queensland Micro- and Nanotechnology Centre (QMNC), Griffith University, Nathan, QLD 4111, Australia; mdalamgir.hossain@griffith.edu.au

3 Applied Research Center for Environment and Marine Studies, King Fahd University of Petroleum & Minerals (KFUPM), Dhahran 31261, Saudi Arabia; mdshafiul.alam@kfupm.edu.sa

* Correspondence: feroz071021@gmail.com

Abstract: This study addresses the pressing energy constraints in nations like Bangladesh by proposing the implementation of photovoltaic (PV) microgrids. Given concerns about environmental degradation, limited fossil fuel reserves, and volatile product costs, renewable energy sources are gaining momentum globally. Our research focuses on a grid-connected solar PV system model at Char Jazira, Lalpur, Natore, Rajshahi, Bangladesh. Through PVsyst 7.1 simulation software, we assess the performance ratio (PR) and system losses, revealing an annual solar energy potential of 3375 MWh at standard test condition (STC) efficiency. After considering losses, the system generates 2815.2 MWh annually, with 2774 MWh exported to the grid. We analyze an average PR of 78.63% and calculate a levelized cost of energy (LCOE) of 2.82 BDT/kWh [1 USD = 110 BDT]. The financial assessment indicates a cost-effective LCOE for the grid-connected PV system, with an annual gross income of 27,744 kBDT from selling energy to the grid and operating costs of 64,060.60 BDT/year. Remarkably, this initiative can prevent 37,647.82 tCO_2 emissions over the project's 25-year lifespan.

Keywords: carbon footprint; performance ratio; photovoltaic array system; PVsyst software; panel orientation; levelized cost of electricity/energy (LCOE); payback period; return on investment (ROI); cost-effectiveness; project lifespan

1. Introduction

In Bangladesh, fossil fuels presently dominate energy production, accounting for 99% of the total [1]. However, the country has established challenging renewable energy targets for 2030 and 2040, mandating major time and financial commitments [1]. To accomplish these goals, solar energy is poised to be extremely important [1]. As the world actively seeks sustainable energy solutions, solar power has emerged as a prominent player [2]. Grid-connected solar power plants are a focus of efforts worldwide to cut carbon emissions and switch to cleaner electricity sources since they can capture a lot of sunshine with little harm to the environment [3,4].

This study's main goal is to evaluate the feasibility of building a 1.5 MW solar power plant in Lalpur, Natore, Bangladesh, while taking into account its integration with the current grid system. The evaluation utilizes the robust simulation capabilities of the PVsyst program. Notably, Bangladesh experiences an average annual increase in energy demand of 4.68%, attributed to a burgeoning population and an improving economy [1]. Like many nations, Bangladesh faces the dual challenge of meeting escalating energy demands while mitigating greenhouse gas emissions [5]. Interest in the development of solar power has increased as a result of the country's adherence to the Paris Agreement and desire to increase the proportion of renewable energy in its energy mix [6]. Given Bangladesh's abundant solar irradiance, there is a significant opportunity to harness solar energy, bolstering the nation's energy security and aligning with its environmental objectives [7]. Aligned with the Paris Agreement and a commitment to increasing renewable energy in its mix, Bangladesh sees a substantial opportunity to harness its abundant solar irradiance. The study emphasizes the importance of renewable energy, especially solar power, to enhance the nation's energy security and align with environmental objectives.

The goal of this study is perfectly aligned with the concept of a positive energy district (PED) whose goal is to attain a net positive energy balance. This means that it produces more energy from sustainable technologies, such as wind turbines and solar panels, than it consumes for infrastructure, buildings, and transportation. Greenhouse gas emissions are decreased and a district's dependency on fossil fuels is lessened by incorporating solar power into its energy mix. This makes the district more environmentally friendly since it is in line with sustainability and environmental aims [8–11]. Local energy production is the goal of positive energy districts. By generating electricity locally, solar power plants help to achieve this objective. This improves energy independence and lessens reliance on centralized power systems. There can be significant cost reductions using solar electricity. Positive energy districts can drastically lower their energy costs by producing their own electricity. It is also possible to resell extra energy to the grid to generate revenue or offset energy expenses. A district's energy supply can be made more resilient with the use of solar electricity. Solar panels can continue to produce electricity even when the grid is down, acting as a dependable source of power [12–15]. A PED should incorporate solar power with other sustainable technology, including energy-efficient buildings, sophisticated energy-storage systems, and smart grid management, to maximize these advantages. A district can therefore aim to achieve net positive energy and create a more robust, cost-effective, and sustainable energy ecology [16–19].

By establishing the 1.5 MW solar power plant, a district or city can become more self-sufficient in energy generation. In a broader context, the effect of such a renewable energy project could lead to the creation of a district producing more energy than it consumes, thereby aligning with the principles of a positive energy district. The feasibility study in the research work includes technical and economic assessments, such as evaluating the technology, costs, and benefits of implementing the solar power plant. A similar methodology is crucial in planning and executing projects within a positive energy district, as they require a comprehensive analysis of various energy sources, consumption patterns, and cost-effectiveness. A thorough techno-economic feasibility analysis is essential before moving forward with the development of a solar power plant. These analyses provide a complete evaluation of the viability of the proposed project, taking into account technical, economic, and environmental factors [20,21].

The technical factors are taken into account as follows. (i) Location and site suitability: The location of the solar power plant is crucial. Factors like solar irradiance, weather patterns, available land, topography, and proximity to the grid infrastructure are evaluated. (ii) Technology and equipment selection: Assessing the appropriate solar panel technology, inverters, racking systems, and other equipment is important, considering efficiency, durability, and compatibility with the site. (iii) System design and engineering: Designing the layout of the solar array, considering factors like orientation, tilt angle, and shading analysis is crucial to optimize energy production. (iv) Grid integration and interconnection: Evaluating the compatibility of the solar power plant with the existing grid infrastructure is essential to ensure proper interconnection and grid stability. (v) Maintenance and operations: Considering the maintenance requirements, operational procedures, and life-cycle costs is necessary to ensure smooth operations and optimal performance over the plant's lifetime.

The economic factors are described as follows. (i) Cost analysis: The initial capital costs for equipment, installation, land, and other associated expenses are estimated. Additionally, ongoing operational and maintenance costs are also considered. (ii) Return on investment and financial viability: Calculating the projected revenue from electricity sales, considering feed-in tariffs or power purchase agreements, and analyzing the payback period and overall return on investment are important for economic feedback. (iii) Incentives and financing: Available incentives, subsidies, tax credits, and financing options are evaluated to determine the financial feasibility of the project. (iv) Economic risks: Risks associated with market fluctuations, changes in regulations, and potential cost escalations are assessed that could impact the financial viability of the project.

The environmental factors are considered as follows. (i) Greenhouse gas emissions: The reduction in greenhouse gas emissions is achieved by generating solar power compared to conventional fossil fuel-based power generation. (ii) Land use and ecosystem impact: The impact on local ecosystems, land use, and potential mitigations are evaluated for minimizing disruption or enhancing biodiversity in the area. (iii) Resource consumption: The environmental impact of manufacturing solar panels is assessed, including energy and material inputs, and recycling or disposal measures are considered at the end of their life cycle. (iv) Social and community impact: The social aspects such as community acceptance, job creation, and local economic benefits derived from the solar power plant are considered.

By evaluating the technical, economic, and environmental aspects, decision-makers can make well-informed choices regarding project implementation. Additionally, techno-economic analyses shed light on the potential risks, challenges, and opportunities associated with the project. A study at Pakistan's Islamia University in Bahawalpur advocates for the installation of a grid-connected PV system, emphasizing its effectiveness and ecological benefits [22]. This solar-powered system produces 4908 MWh each year, which is comparable to 147,600 teak trees planted during a lifetime or 15,369.3 kg of coal saved per day. The study's objective is to assist with the creation, evaluation, implementation, and upkeep of new grid-connected systems across numerous places [22]. Afghanistan faces a challenging energy situation, relying on electricity imports and incurring significant costs [23]. To address this, the government plans to generate 5000 MW of renewable energy by 2032, with a focus on 1500 MW of solar projects. Afghanistan's Daikundi province offers the best sunlight for solar energy production, as shown by a 700 kWp grid-connected solar power plant that was tested using the PVsyst program [23]. The effectiveness of a standalone solar PV system concerning the load requirements at the mechanical department office at Bikaner Engineering College is the subject of another study [24]. The study reveals an average yearly energy need of 1086.24 kWh, with the solar panel providing 1143.6 kWh. Multiple losses contributed to the system's reduced power capacity, with variations in performance ratio throughout the year [24].

In another research study [25], the viability of a 200 KWp on-grid monocrystalline silicon solar project in Dubai was assessed using the PVsyst program. The system exhibited an impressive yearly performance ratio of 81.67%. From its 32 parallel strings and 22 series modules, the solar facility generated 352.6 MWh annually, achieving an energy output of 1757 kWh/KWp per year. This study further offered detailed loss estimates for the entire year, providing valuable insights into the system's performance [25]. Another article focused on the viability of a solar PV plant in Pune, India, utilizing a 250 KWp Si-poly photovoltaic facility as the subject [26]. To model the plant, which consisted of 310 Wp modules coupled to 65 strings and 42 string inverters, the study used PVsyst 7.2 software and Meteonorm 8.0 data. Several tests were carried out to establish the best angle for power generation, taking into account things like incidence radiation, performance ratio, and grid input energy [26]. Additionally, a study used PVsyst software to analyze a 20 MW grid-connected solar facility in Devdurga, Karnataka, India [27]. The analysis provided insights into performance trajectories and geographical positioning, predicting a 110-acre area with a remarkable 76.28% performance ratio. The study also examined the system's behavior concerning tilt and orientation, further enriching our understanding of solar plant dynamics [27].

In a notable study [28], the grid connected photovoltaic array system, which is intended to produce 15 kWp at STC, was the main focus. The study carefully looked into several factors, including system output, output power losses, performance ratio, output energy, PV module output, field type adjustment, power distribution curves, temperature distribution, and module connection in string design [28]. Furthermore, another research effort involved the design and simulation of a 60 kWp solar power plant for a rural area in Uttar Pradesh, India, utilizing PVsyst software version 7.2.2. The analysis considered seasonal tilt angles and conducted a comprehensive performance study [29]. A separate research paper [30] delved into a study of a 1 MW photovoltaic plant located in Morocco's northern zone. The study utilized PVsyst software to evaluate the plant's performance, making comparisons among solar energy production at different sites. This study underscored the critical need for accurate predictions of system efficiency in renewable energy systems [30]. PVsyst simulation software was employed in yet another study [31], where the main goal was to build and simulate a hybrid photovoltaic system for a system of energy-efficient street lighting. The system comprised 16 Narada batteries, 13 series-connected modules, and 4 parallel strings. The modeling showed a yearly energy output of 26.68 MWh overall and an annual energy production of 1283 kWh/KWp specifically. Notably, the study highlighted performance ratios and battery cycle conditions, providing valuable insights into system behavior. In a different research endeavor [32], a 5 MW solar power project in Afghanistan's Ghor region was planned and simulated. The plant incorporated a 5300 kW Growatt converter type and 6399 KWp STP-320 W PV module. Various modeling tools, including PVsyst, PVGIS, and HOMER, were utilized to evaluate solar renewable energy sources. The findings showcased annual energy generation variations among the modeling tools, emphasizing the importance of accurate simulation for efficient planning.

Despite the existing studies utilizing PVsyst and exploring solar power plant designs and financial aspects, there is a research gap in conducting a comprehensive analysis encompassing detailed losses, financial analysis, cost-effectiveness, and carbon balance collectively. Additionally, previous works have not sufficiently addressed the potential challenges, risks, and mitigation strategies associated with the implementation of solar power plants. Our study bridges these research gaps by introducing a novel approach to solar energy project assessment, focusing on the specific site of Lalpur, Natore, Bangladesh. The geographical specificity of Char Jazira, a riverine island in the floodplains of the Jamuna River, is highlighted, emphasizing its ideal characteristics for a solar project. The inclusion of detailed meteorological data, covering parameters such as global horizontal irradiance, temperature, wind velocity, Linke turbidity, and relative humidity, adds depth to the study. The comprehensive techno-economic feasibility assessment, utilizing PVsyst software for modeling, brings a nuanced understanding of the solar power plant's viability. Performance analysis reveals seasonal variations and the environmental impact is quantified, emphasizing a significant reduction in carbon emissions. The study's economic viability metrics, including a quick payback period, underscore the practicality of the proposed solar project. In essence, the paper offers a holistic exploration of solar energy potential, integrating geographical, meteorological, economic, and environmental considerations, making notable contributions to sustainable energy solutions.

In this study, we have used the potent PVsyst program to model the operation of a 1.5 MW grid-connected solar power plant in Lalpur, Natore. PVsyst is highly esteemed for its capacity to model PV systems under varying conditions encompassing solar irradiance, temperature, shading, and system configuration [33]. The software enables accurate predictions of energy generation, performance ratios, and financial metrics, rendering it an invaluable tool in the feasibility assessment of solar projects [34]. However, a collective and detailed analysis, including losses, financial aspects, cost-effectiveness with a simple payback period, and carbon balance, has not been previously conducted in the mentioned research and studies.

We intend to develop the solar power plant layout after assessing the project site's solar resource potential and climate. Important aspects to take into account are the panel orientation, tilt angle, and shading analysis [35,36]. With the aid of the PVsyst program, we intend to simulate the energy generation profile of the solar power plant under various conditions. This simulation will help to estimate key financial metrics such as the LCOE, payback period, and ROI, enabling a comprehensive assessment of the project's economic viability [24]. Additionally, we want to pinpoint any obstacles, dangers, and suitable countermeasures related to the construction of the solar power plant [37–39].

This study is divided into several sections, each of which deals with a different component of the techno-economic feasibility assessment. These sections include an evaluation of the solar resource, plant design and simulation, financial analysis, and a discussion of findings. Through this structured approach, we aim to attain a comprehensive understanding of the project's viability and potential benefits, empowering informed decision-making in the pursuit of sustainable energy solutions.

2. Location and Data

2.1. Project Site Selection

The chosen project site is located in Lalpur, Natore, Rajshahi, Bangladesh, specifically known as Char Jazira. The geographical coordinates for this location are 24.0900° N latitude, 88.5800° E longitude, and an altitude of 22 m above sea level. The time zone in this region is UTC+6. In Bangladesh, a 'char' typically denotes a riverine island or sandbar that forms within the braided river systems, shaped by the dynamics of sedimentation and erosion in the river. The specific location for this project, Char Jazira, fits this description. Char Jazira is positioned in the floodplains of the Jamuna River, a major river in Bangladesh. The land is state-owned, and due to its nature as a riverine island, it is devoid of tree shadows, making it ideal for a solar project. The incident solar irradiation in this area is notably high. The location of Char Jazira, Lalpur, in Bangladesh has been shown in Figures 1 and 2 using a pinned location and red color marked area, respectively, from Google Maps.

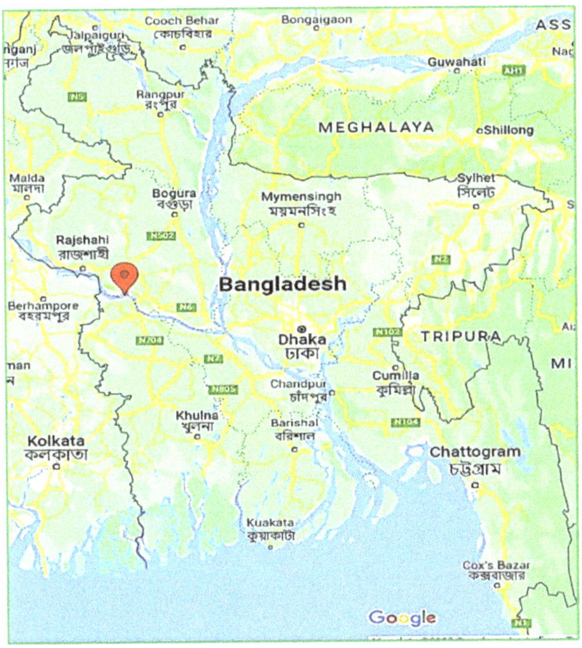

Figure 1. Pinned location of Char Jazira, Lalpur, in Bangladesh from Google Maps.

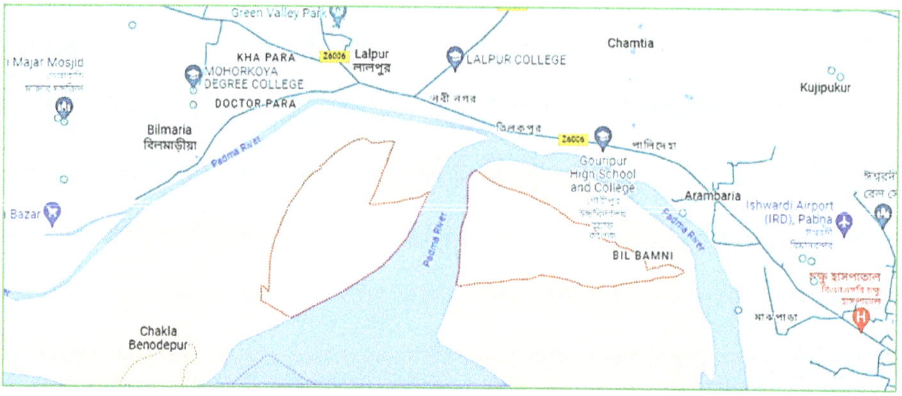

Figure 2. Red color marked area—location of Char Jazira, Lalpur, from Google Maps.

The average annual rainfall in Lalpur, Natore, Bangladesh, is 2200 mm (86.6 inches). The wettest months are June, July, and August, receiving an average rainfall of 300 mm (11.8 inches) each. On the other hand, the driest months, namely December, January, and February, witness an average rainfall of 50 mm (2.0 inches) each. Lalpur in Natore, Bangladesh, observes the lowest rainfall during these months.

Considering the above information regarding the project site, there is huge potential for one power plant in the location chosen for the solar power plant, and it must contribute to the country's crucial grid electricity. Thus, the analysis of the project on that site can be considered notable and novel.

2.2. Meteorological Data

PVsyst software relies primarily on solar irradiance data for the accurate modeling of solar photovoltaic (PV) systems. In addition to solar irradiance data, meteorological data are also incorporated to enhance the precision of system modeling [40–42]. The following Table 1 shows the key meteorological parameters collected and needed for the software simulation of the solar power plant.

Table 1. Collected and applied meteorological data.

Month	Global Horizontal Irradiance (kWh/m²/mth)	Diffuse Horizontal Irradiance (kWh/m²/mth)	Temperature (Degree Celcius)	Wind Velocity (m/s)	Linke Turbidity	Relative Humidity (%)
January	122.3	54.1	17.4	0.79	6.377	77.6
February	132.6	60.3	21.5	0.91	5.946	70.6
March	174.6	78.3	26.5	1.2	6.316	64.6
April	182.5	88.7	29.1	1.4	7	71.3
May	190	100.7	30.3	1.1	7	71.4
June	155.3	99.9	29.5	0.98	7	79.9
July	145.1	99.2	29.2	0.88	5.772	80.9
August	145.8	90.6	29	1.3	5.18	83.1
September	139.4	74.2	27.9	1.2	5.29	86.5
October	136	72.5	27	0.8	5.7	81.7
November	127.8	54	23.1	0.6	6.589	78.4
December	122.7	50.2	18.7	0.6	6.948	79.1

2.2.1. Solar Irradiance

Solar irradiance data is essential for PVsyst software to compute a PV system's energy production accurately. This data includes:

- Global horizontal irradiance (GHI): Total solar energy, including both direct and diffuse solar radiation, that is received on a horizontal portion of the Earth's surface. GHI, which is measured in watts per square meter (W/m^2), is an important aspect to consider when assessing the solar energy potential of a site [41,43].
- Direct normal irradiance (DNI): Solar radiation that strikes a surface perpendicular to the sun's rays. DNI is the amount of sunlight that enters the Earth straight from the sun, unaffected by air absorption or scattering. It is essential for applications and concentrated solar power (CSP) systems that need direct sunshine. Watts per square meter (W/m^2) are commonly used to measure DNI [44–46].
- Diffuse horizontal irradiance (DHI): Aside from direct sunlight, solar radiation is received from the sky. DHI is made up of solar energy that is reflected or dispersed and diffusely reaches the surface of the Earth. It is crucial for solar energy applications since it is measured on a horizontal surface, especially for PV systems that may take in both direct and diffuse sunlight. DHI can also be calculated in W/m^2 [44,47].

Understanding and utilizing these meteorological parameters in the PVsyst software is crucial for the precise modeling and assessment of solar energy systems. Applied GHI and DHI values from Table 1 are plotted in Figure 3.

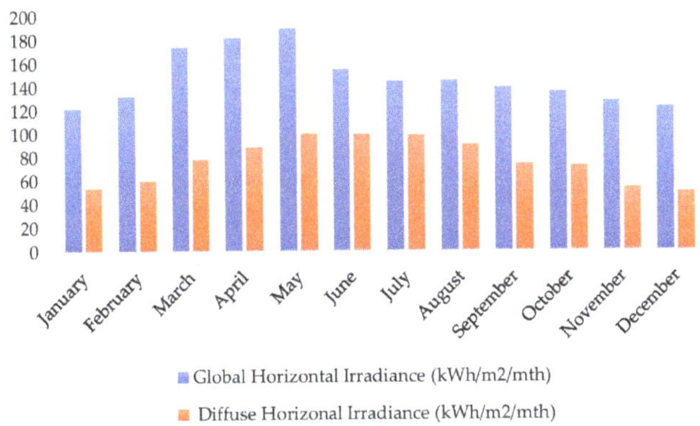

Figure 3. Solar irradiance.

2.2.2. Temperature

The performance of PV modules is greatly influenced by the ambient temperature. Temperature information is necessary for PVsyst software to accurately analyze the attributes of PV modules. The electrical characteristics of the PV modules can be determined with the use of these temperature data. In actuality, hot, bright days with little breeze tend to have higher operating temperatures for PV modules. A PV module's temperature is predicted by the nominal operating cell temperature (NOCT), a standardized rating system, under a variety of parameters, such as 20 °C ambient temperature, 800 W/m^2 of solar energy, and 1 m/s of wind speed. It is essential to comprehend a PV module's temperature because it has a significant impact on its electrical properties [48].

A detrimental effect of temperature arises on a PV module's efficiency. The energy band gap of the semiconductor material narrows with temperature, requiring less energy to produce a free electron-hole pair, which results in a decrease in efficiency. As a result, a PV module's power production decreases as temperature increases. The temperature coefficient of power (Pmax), which is often stated as a percentage per degree Celsius (°C), measures this decrease in power output caused by an increase in temperature [49].

Understanding how temperature changes affect a PV module's performance requires an understanding of the temperature coefficient of power. Applied temperature values in degrees Celsius (°C) from Table 1 for the software simulation are plotted in Figure 4.

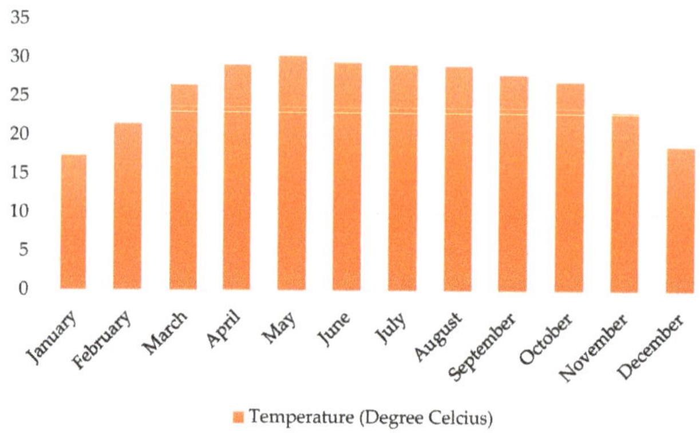

Figure 4. Temperature.

2.2.3. Wind Velocity

Wind velocity, or wind speed, significantly influences the performance of PV modules by affecting their operating temperature and cooling mechanisms. Accurate wind speed data are crucial for the PVsyst software to estimate the cooling effect on module performance and simulate its impact on energy production. Wind speed plays a beneficial role in solar power generation from panels. It aids in cooling the solar panels, consequently improving their efficiency. As solar panels become hotter, their efficiency tends to decrease. Wind helps dissipate this heat, maintaining or even enhancing the panel's performance. Wind also aids in removing dirt and debris from solar panels' surfaces, increasing their effectiveness. The electricity production of solar panels can increase by up to 20% when there is wind. The precise rise, however, relies on several variables, including the surrounding temperature, sun irradiation, and the type of solar panel being utilized. In general, higher wind speeds cause solar power generation to rise more quickly. The rise in power production does level off at a certain point, though. This is because the cooling effect of wind becomes less significant as the wind velocity continues to increase [50,51].

Understanding and incorporating wind speed data into simulations is critical for accurately assessing the impact of wind on solar panel efficiency and optimizing solar power generation. In Figure 5, the values of wind velocity from Table 1 collected for the location for every month of a certain year are plotted.

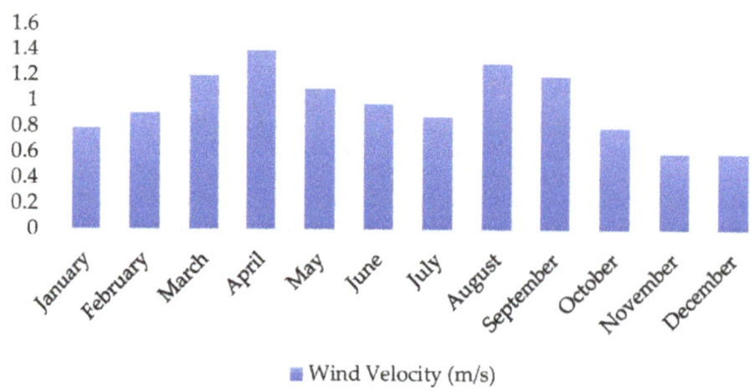

Figure 5. Wind velocity.

2.2.4. Linke Turbidity

In the context of solar energy modeling, the Linke turbidity coefficient serves as an atmospheric parameter that provides insights into water vapor concentrations and aerosols. While the clear sky model primarily uses location coordinates (latitude, longitude, and altitude), the Linke turbidity coefficient can introduce slight alterations. This coefficient typically ranges from 2.0, indicating a very dry and clear sky, to 5 or 6, indicating a humid and potentially polluted environment. A common default value used for many sites is 3.5 [52,53]. Turbidity, in a broader sense, characterizes the cloudiness or haziness of a liquid caused by numerous small suspended particles, including dust, sediment, organic materials, and plankton in water. Turbidity has a considerable impact on the amount and distribution of solar radiation that reaches solar PV panels or solar thermal collectors in the field of solar energy production.

Turbidity in the atmosphere can absorb and scatter light, reducing the amount of direct and diffuse solar irradiance that reaches solar panels. This decrease in solar irradiance ultimately diminishes the overall energy output of PV panels. Turbidity's impact is evident in the scattering of sunlight, potentially increasing the diffuse solar radiation component while decreasing direct solar radiation due to scattering. Certain solar energy systems, particularly those adept at capturing diffuse sunlight effectively, can benefit from this shift. Additionally, turbidity may appear as debris, dust, or dirt adhering to the PV panel's surface. The buildup of these particles reduces direct sunlight's ability to reach the solar cells, which has an impact on the PV system's effectiveness and power generation [54]. The performance of solar panels must be maximized, which means keeping them clean and removing as many of these impediments as possible. The values of Linke turbidity from Table 1 are plotted in Figure 6.

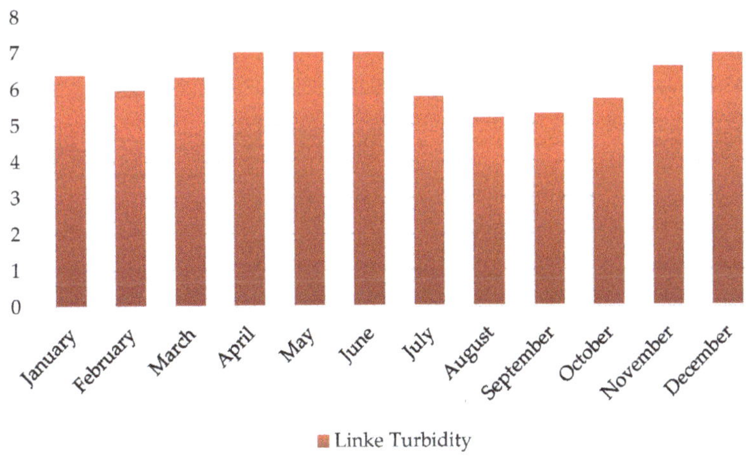

Figure 6. Linke turbidity.

2.2.5. Relative Humidity

A measurement known as relative humidity (RH) expresses the amount of moisture in the air in relation to the maximum amount of moisture the air may contain at a specific temperature. Its relevance for solar radiation and PV panel efficiency—two important factors impacting solar power generation—is stated as a percentage. The PV module temperature and efficiency are both significantly influenced by relative humidity. It has an impact on the module's electrical properties, which may therefore have an impact on energy output. The software's ability to correctly assess these effects depends on accurate input of relative humidity data.

In terms of solar radiation, relative humidity directly affects the amount of direct and diffuse solar radiation reaching solar panels. Cloud cover and a rise in atmospheric water vapor are frequently correlated with a higher relative humidity. Clouds can reflect and absorb sunlight, reducing the amount of direct solar irradiance that reaches solar panels and, as a result, reducing the energy output. Elevated relative humidity levels, often associated with increased cloud cover, lead to a higher proportion of diffuse solar radiation. Despite the reduction in direct sunlight due to cloud cover, diffuse radiation still contributes to the overall energy production of solar panels.

Relative humidity is closely intertwined with temperature. As temperature increases, the air can retain more moisture, resulting in a lower relative humidity for the same absolute moisture content. High temperatures, usually linked with low relative humidity, can enhance PV panel efficiency by reducing resistive losses within the solar cells. Regions with a high moisture content or coastal areas, often characterized by higher humidity levels, can experience dust and dirt accumulation on solar panels. Regular cleaning and maintenance of solar panels are crucial to ensure that accumulated dirt does not obstruct sunlight absorption and impact energy production [55,56]. Keeping the panels clean optimizes their performance and, in turn, the overall solar power generation efficiency. The collected and applied values of relative humidity from Table 1 are plotted in Figure 7.

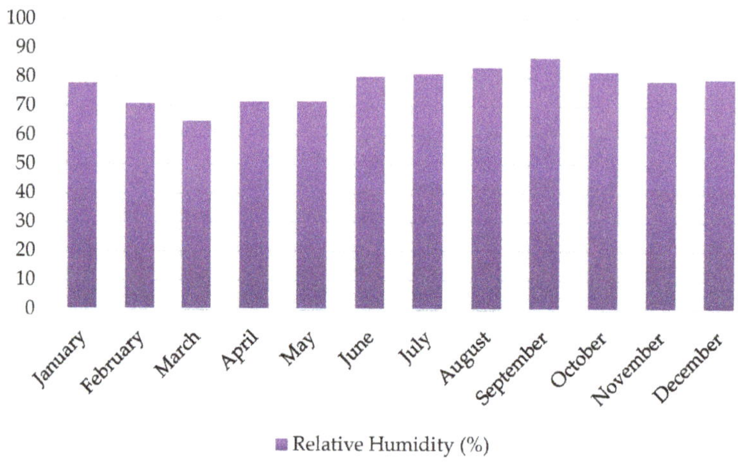

Figure 7. Relative humidity.

2.3. Sun Paths Diagram

Sun path analysis is crucial for determining the solar irradiation and shading conditions at a specific location over a given period. PVsyst makes use of solar geometry computations to determine the sun's position in relation to the site and time of interest. It precisely determines the sun's position throughout the year by taking into account variables like latitude, longitude, time zone, and date. The software can generate detailed graphical representations of the sun's path, including azimuth and sun height angles [57], as shown in Figure 8 for the selected location of Char Jazira, Lalpur. This graphic of solar paths shows the sun's daily position in the sky, as well as the seasons. PVsyst assists users in assessing the effects of shadowing, figuring out the best tilt and orientation for solar panels, and analyzing the overall solar energy potential at a particular site by visualizing the sun's path [58].

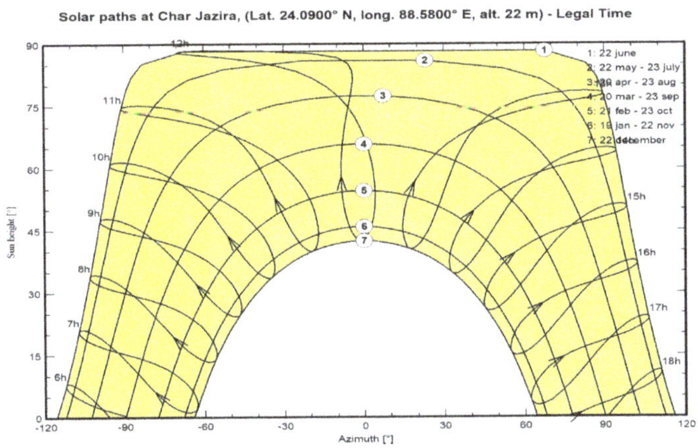

Figure 8. Sun paths diagram in legal time and rectangular coordinates.

3. Methodology

The steps of methodology for this research work are depicted in Figure 9 as a flowchart. The steps in the flowchart are described as follows. Site assessment: An important stage in guaranteeing the success of a solar project is to carry out an on-site assessment of the solar plant location in Lalpur, Natore, to obtain information on solar irradiance, shading analyses, weather patterns, and topographical factors. This information is essential for precisely estimating the region's solar potential and adjusting the layout and positioning of solar panels to increase their effectiveness and sustainability. Additionally, the evaluation will help in identifying potential challenges or limitations associated with the location that may need to be addressed during the planning and implementation phases of the solar project.

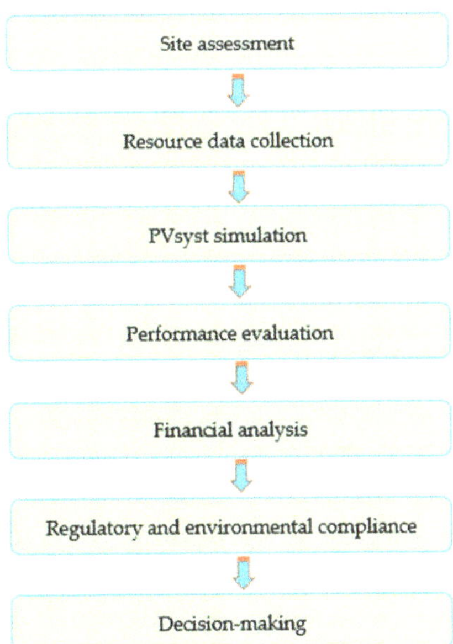

Figure 9. Flowchart of methodology.

Resource data collection: After the site assessment, collecting historical solar irradiance, meteorological data (temperature, wind velocity, Linke turbidity, relative humidity), and PV field orientation (tilt, azimuth) for the region to accurately model the plant's performance is essential for designing and operating a reliable and efficient solar power plant.

PVsyst simulation: Assessing the site and then giving input to PVsyst by using collected data to simulate the solar power plant's energy production, considering various factors like module type, array layout, inverter specifications, and system losses, allows for a precise and detailed projection of the plant's potential output. This simulation aids in optimizing the solar power system's configuration, ensuring it is tailored to the specific conditions of the Lalpur, Natore, location and can operate at peak efficiency, thus maximizing energy generation and overall performance.

Performance evaluation: After PVsyst simulation, the results obtained from PVsyst are crucial to assess the plant's annual energy yield, capacity factor, and performance ratio, considering local climatic conditions and system parameters. This evaluation provides valuable insights into the solar power plant's overall efficiency and productivity, allowing for informed decisions on potential optimizations or adjustments needed to enhance energy output and ensure the system operates at its highest capacity throughout the year, factoring in the specific weather patterns and environmental variables of the region.

Financial analysis: Evaluating the performance, and the estimation of project costs, including equipment procurement, installation, grid connection, operation, and maintenance, is essential. Calculation of the LCOE and payback period is crucial to evaluate economic viability, providing key financial indicators that help stakeholders to assess the long-term sustainability and profitability of the solar power project. These assessments ensure that the investment aligns with the financial goals and objectives of the project, enabling informed decision-making and potential adjustments to the project plan to achieve a feasible and economically sound solar power venture.

Regulatory and environmental compliance: Evaluation of legal and regulatory requirements for solar installations in Bangladesh is imperative, ensuring compliance with permits, grid codes, and environmental standards. Adhering to these guidelines not only ensures the legality and smooth operation of the solar power plant but also upholds environmental sustainability and safety standards set by the governing bodies. Comprehensive compliance guarantees that the solar project aligns with the country's policies, fostering a sustainable and responsible approach to solar energy deployment within the regulatory framework of Bangladesh.

Decision-making: Taking into account both technical and economic factors, based on the detailed analysis, an informed decision has to be made on the feasibility of building the 1.5 MW grid-connected solar power plant. Data on solar radiation, system design, energy output forecasts, costs, regulatory compliance, and environmental factors are all integrated into the study. This comprehensive assessment enables stakeholders to move forward with confidence in the project's success by allowing for a well-informed judgment of the practicality, sustainability, and economic viability of putting the solar power plant into operation.

During the PVsyst simulation, performance evaluation, and financial analysis of the above procedure, the auxiliary loads are considered as the loads of the lighting system, loads of the maintenance room and control room, etc. The auxiliary load of the plant is 10 kilowatts, i.e., the energy consumed by the auxiliaries can be estimated at 87.6 MWh/year (46.4 MWh/year from the grid and 41.2 MWh/year from solar power). The fixed feed-in tariff is 10 BDT/unit which is the injected electricity selling price to the grid. The consumption tariff is also considered to be 10 BDT/unit for the auxiliary load of the plant.

3.1. Proposed System Architecture

Solar irradiation is incident on the PV array, where sunlight is converted into electricity. The PV array output is connected to a DC–DC boost converter input. The inverter's input is connected to the output of the boost converter. In the inverter, DC voltage is converted into AC voltage. Power from the inverter's output is injected into the grid and also supplied to fulfill the auxiliary load demand. Figure 10 shows the schematic diagram of the proposed system where the architecture is illustrated into three sections, i.e., PV array, system, and user (load)/grid.

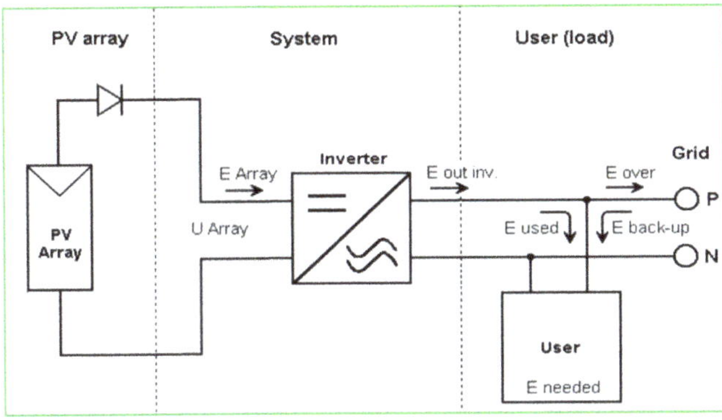

Figure 10. Schematic diagram of the system.

3.2. *System Configuration*

The proposed plant is a grid-connected system in which no 3D scene is defined and no shadings are considered. The system power is 1874 kWp. The system's power is 1874 kWp means the nominal power in STC is 1874 kWp, during the running of the system there are losses, and after the losses, the actual power remains 1500 kWp, that is, 1.5 MW.

3.2.1. PV Field Orientation

The field type of the PV module and array is a fixed plane. The plane tilt and azimuth of the field array are 20° and 0°, respectively, as shown in Figure 11, where a visual representation as well as quick optimization of the PV field orientation have been depicted. For near shading, no shadings are considered, and no 3D scene is defined in the PVsyst software for the rooftop of the house. The free horizon is considered for the simulation.

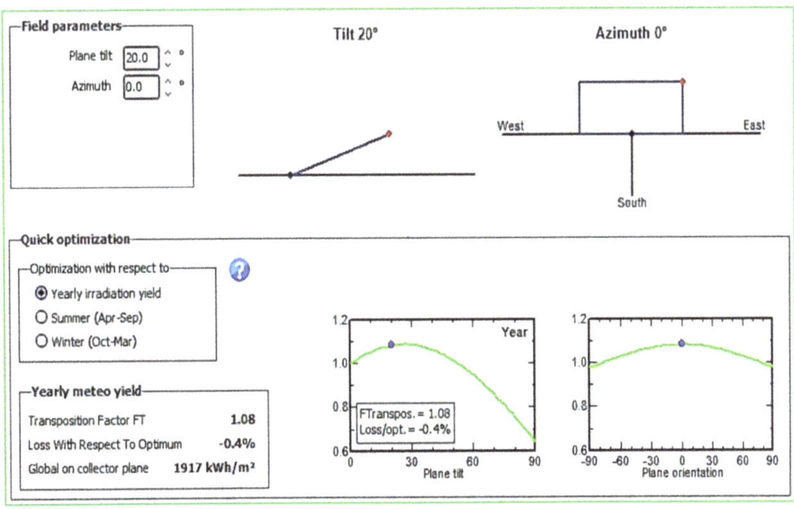

Figure 11. PV field orientation.

3.2.2. PV Array and Inverter Characteristics

The PV array was made by Eco Delta Power. Its model number, which was taken from the original PVsyst database, is ECO-300M-60. A single unit array has a nominal power (Pnom) of 300 Wp. There are 6248 active PV modules. These 6248 modules are set up in 284 strings, with each string made up of 22 modules connected in series. After installation, there are 10,227 m^2 of total PV modules and 9110 m^2 of cells. The nominal power in STC is 1874 kWp. The maximum power points of power (Pmpp), voltage (Vmpp), and current (Impp) for operating conditions at 50 °C are 1689 kWp, 643 V, and 2627 A, respectively. The PV array and inverter characteristics and their details in the design have been shown in Figure 12 which is collected from the PVsyst software.

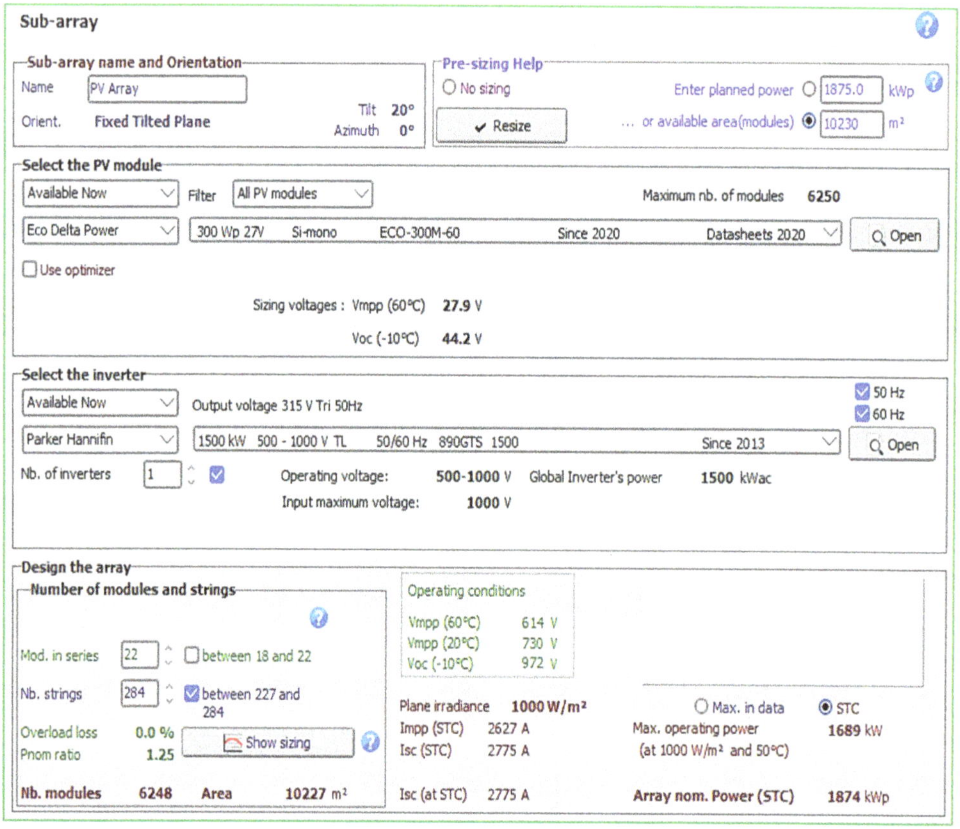

Figure 12. Information on the PV array and inverter.

The manufacturer of the inverter is Parker Hannifin. Its model number is 890GTS_1500 which is collected from the original PVsyst database. The Pnom of the inverter unit is 1500 KWac. The number of used inverters is 1. The operating voltage of the inverter is 500–1000 V. The Pnom ratio is 1.25.

3.3. *PV Field and Array Detailed Losses Parameter*

Each loss factor of the detailed losses parameters is carefully defined in the software input according to the PV system, as indicated in Table 2.

Table 2. Detailed losses parameter.

Thermal Loss Factor	
Module temperature according to irradiance	
Uc (const)	20.0 W/m^2K
Uv (wind)	0.0 $W/m^2K/m/s$
Ohmic Losses **Series Diode Loss**	
Voltage drop	0.7 V
Loss fraction	0.1% at STC
DC Wiring Losses	
Global array resistance	4.1 mΩ
Loss fraction	1.5% at STC

Figure 13 visualizes the wiring layout where 284 strings are created from the total of 6248 modules, each string consisting of 22 modules joined in a series sequence. The wiring diagram for groups of parallel strings, where the input of the inverter is connected to the combined outputs of the parallel strings, is displayed in Figure 14. The loss fraction of different types of losses parameters is defined in percentage for the software input according to the PV system, as indicated in Table 3.

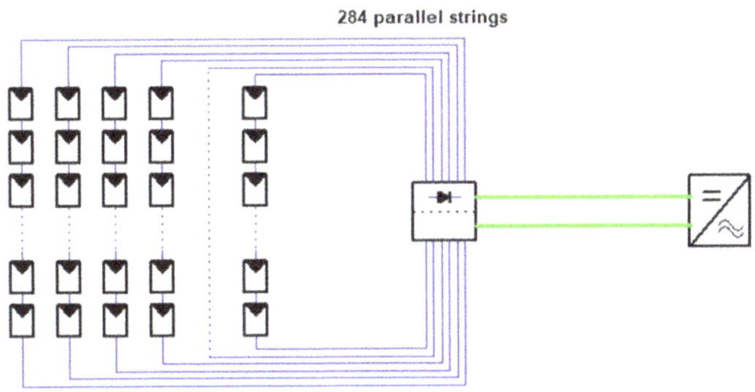

Figure 13. Wiring layout—parallel strings.

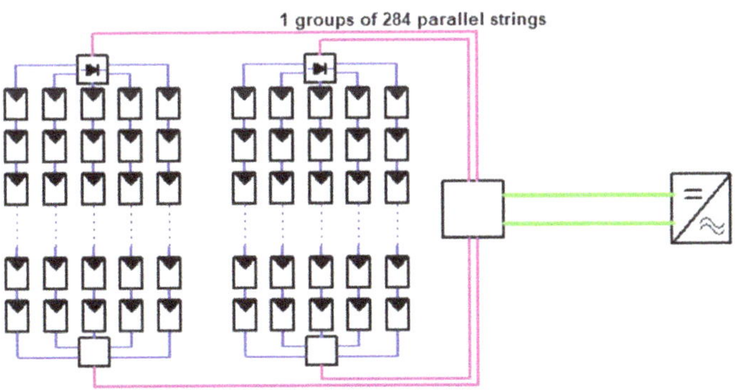

Figure 14. Wiring layout—groups of parallel strings.

Table 3. Different types of loss parameter.

Module Quality Loss	
Loss fraction	−0.8%
LID—Light Induced Degradation	
Loss fraction	2.0%
Module Mismatch Losses	
Loss fraction	2.0% at MPP
Strings Mismatch loss	
Loss fraction	0.1%
Array Soiling Losses	
Loss fraction	3.0%

IAM Loss Factor:

The decrease in irradiance that actually reaches the surface of PV cells in comparison to irradiance under normal incidence is known as the incidence effect (abbreviated IAM, for "Incidence Angle Modifier"). Reflections on the glass cover, which increase with the incidence angle, are the primary cause of this drop. The values considered for the IAM are given in the Table 4.

Table 4. Considered values for IAM.

0°	30°	50°	60°	70°	75°	80°	85°	90°
1.000	0.998	0.981	0.948	0.862	0.776	0.636	0.403	0.000

Incidence effect (IAM): Fresnel smooth glass, n = 1.526.

The graphing of the aforementioned information is displayed in Figure 15, also known as the IAM graph. PVsyst's IAM graph illustrates graphically how the PV modules' performance changes when the angle of incidence changes. Usually, it displays a curve that illustrates how solar energy decreases as an angle moves away from the optimal perpendicular position.

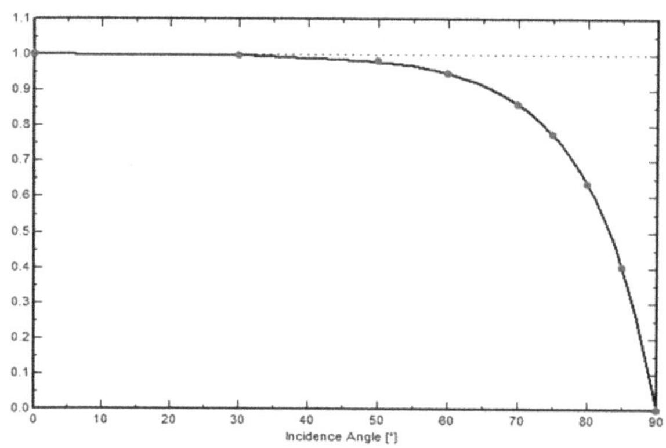

Figure 15. Incidence angle modifier.

An illustration of how several factors affect a PV array's overall performance can be found in PVsyst's losses graph. Plant designers may improve the design and operation of PV systems to maximize energy production by looking at the losses graph in PVsyst, which provides insights into the intricate relationships of several factors. This graphical representation makes it easier to comprehend in detail how each loss effect affects the PV array's overall performance. Plotting of the current vs. voltage curves for the loss types under consideration is shown in Figure 16, which displays the PV array behavior for each loss effect in various colors.

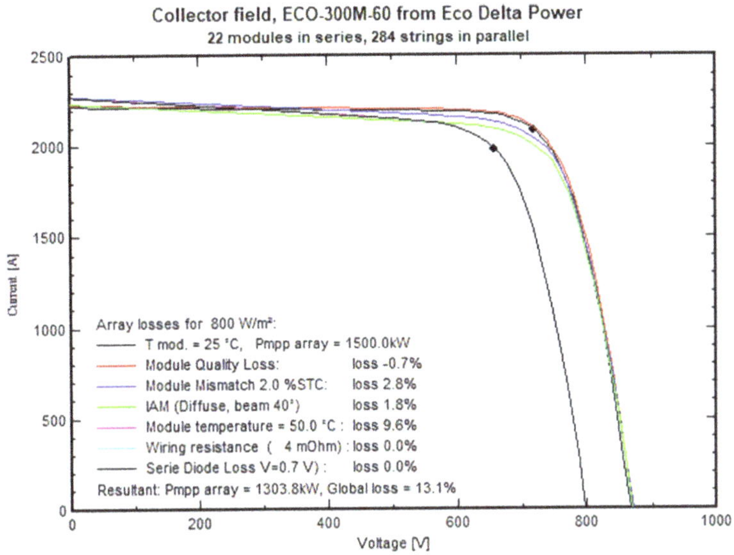

Figure 16. PV array behavior for each loss effect—losses graph.

3.4. Simulation Modeling

The annualized costs and discount factors from net present costs are calculated using the real discount rate. The variables "Nominal discount rate" and "Expected inflation rate" are used to generate the annual real discount rate, often known as the real interest rate or interest rate.

The following equation is used to obtain the real discount rate:

$$i = \frac{i' - f}{1 + f} \tag{1}$$

where i = real discount rate, i′ = nominal discount rate (the rate at which money could be borrowed), and f = expected inflation rate.

The value considered for the nominal discount rate (i′) is 8% and the expected inflation rate (f) is 2%. An annuity, which is a stream of equal yearly cash payments, has a present value that is determined using a ratio called the capital recovery factor. The following equation is used to calculate the capital recovery factor:

$$CRF(i, N) = \frac{i(1+i)^N}{(1+i)^N - 1} \tag{2}$$

where i = real discount rate, and N = number of years.

The number of years (N) is 25 years, which is the expected lifetime of the PV system.

The present value of all of a system's lifetime expenses minus the present value of all of its lifetime earnings equals the system's total net present cost (NPC). Costs include up-front expenses, expenses for repairs and replacements, expenses for operations and maintenance, expenses for fuel, fines for pollution, and the cost of grid electricity. The salvage value as well as grid sales revenue are included in revenues.

The entire annualized cost is used to calculate the levelized cost of energy. The entire annualized cost is equal to the annualized value of the total net present cost. The following equation is used to determine the total annualized cost:

$$C_{ann,tot} = CRF(i, N) \cdot C_{NPC,tot} \quad (3)$$

where $C_{NPC,tot}$ = the total net present cost [BDT], i = the annual real discount rate [%], N = the project lifetime [year], and CRF (i, N) = a function returning the capital recovery factor.

To calculate the LCOE, the program divides the annual cost of producing electricity (i.e., the total annual cost minus the cost of delivering the thermal load) by the total electric load serviced. The LCOE equation is:

$$LCOE = \frac{C_{ann,tot}}{E_{served}} \quad (4)$$

where $C_{ann,tot}$ = total annualized cost of the system [BDT/year], and E_{served} = total electrical load served [kWh/year].

The following equation is used to determine the total annualized cost when total present cost is considered:

$$C_{ann,tot} = CRF(i, N) \cdot C_{tot} \quad (5)$$

where C_{tot} = the total present cost [BDT], i = the annual real discount rate [%], N = the project lifetime [year], and CRF (i, N) = a function returning the capital recovery factor.

The software divides the annual cost of producing electricity (the total annual cost less the cost of serving the thermal demand) by the total amount of electrical energy produced to arrive at the LCOE. The equation for LCOE is:

$$LCOE = \frac{C_{ann,tot}}{E_g} \quad (6)$$

where $C_{ann,tot}$ = total annualized cost of the system [BDT/year], and E_g = total electrical energy generated [kWh/year].

4. Results and Discussion

The primary purposes of the PVsyst program are performance analysis and financial assessment of the solar facility. It supports pre-installation modeling by providing essential calculations, including space requirements, energy generation estimations based on allocated space, energy consumption calculations for auxiliary loads from both solar and the grid, injected energy estimations into the grid, and various loss calculations. In terms of financial analysis, the software furnishes data regarding the total installation cost, operational expenses, cost of energy production, payback period, and the potential reduction in CO_2 emissions.

4.1. Performance Analysis

The performance ratio measures how efficiently energy is produced or used in comparison to how much energy would be generated if the system were to run continuously at its nominal STC efficiency. The system's performance ratio in this instance is 0.786, or 78.63% efficiency as shown in Figure 17.

The normalized energy production in Figure 18 draws attention to three key figures: the PV array's collection loss of 1.03 kWh/kWp/day, the system loss of 0.08 kWh/kWp/day, and the useable energy generated by the inverter output of 4.11 kWh/kWp/day. A solar power plant with a greater normalized energy production will, in contrast to one with a lower normalized energy production, produce more energy per kilowatt of installed capacity. Consequently, a solar power plant demonstrating higher normalized energy production proves to be more productive and cost-effective in its operations.

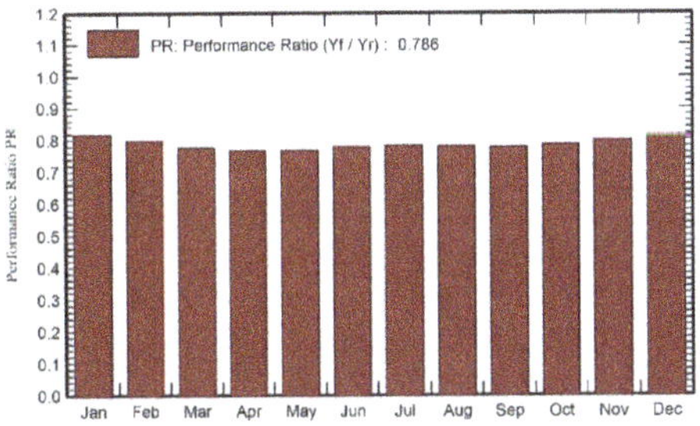

Figure 17. Performance ratio (PR).

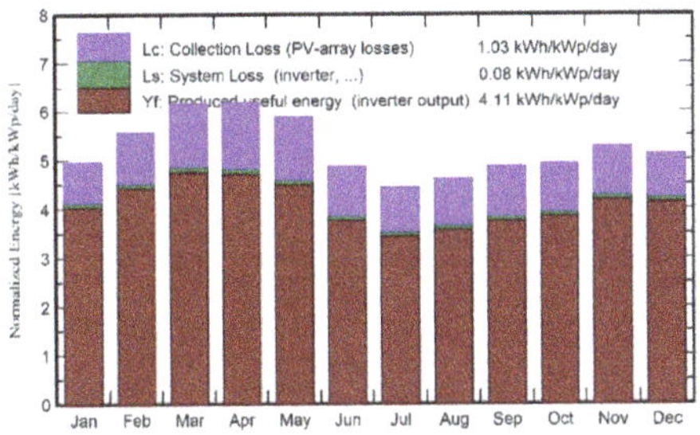

Figure 18. Normalized productions (per installed kWp).

Figure 19 depicts the relationship between the worldwide incident irradiance on the collector plane (measured in kWh/m^2/day) and the available solar energy (measured in kWh/day). Daily input/output diagrams are what this illustration is known as. The daily input/output diagram in the PVsyst software is a graphic representation of how the energy generated by a solar power plant varies in response to the daily incident irradiance. This tool is invaluable for comprehending the plant's performance and pinpointing areas with the potential for enhancement. Ideally, the daily input/output diagram should exhibit a nearly linear trend, albeit with a slight curvature at higher irradiance levels attributed to temperature effects. Any deviation from this pattern, particularly at elevated irradiances, may signify overload conditions on certain days. Monitoring this diagram over time provides insights into the solar power plant's performance. Changes in the diagram can help to identify issues like reduced energy production or increased losses, enabling timely intervention and system optimization.

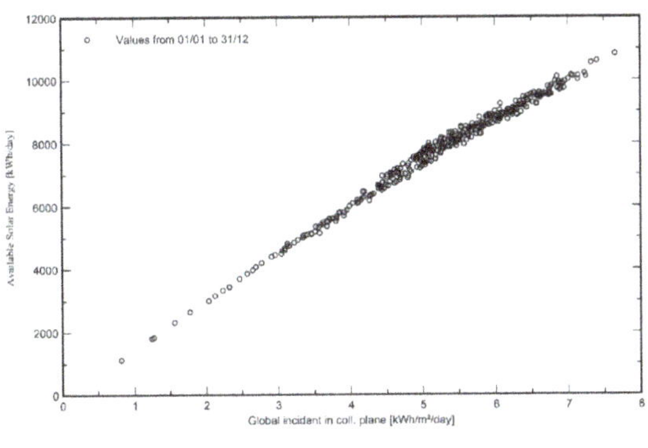

Figure 19. Daily input/output diagram.

Figure 20, illustrating system output power distribution, presents the relationship between the available solar energy in kW and the available solar energy in kWh per bin (kWh/Bin). In PVsyst software, the system output power distribution graph offers a visual representation of how the power output of a solar power plant is distributed over a specific time span.

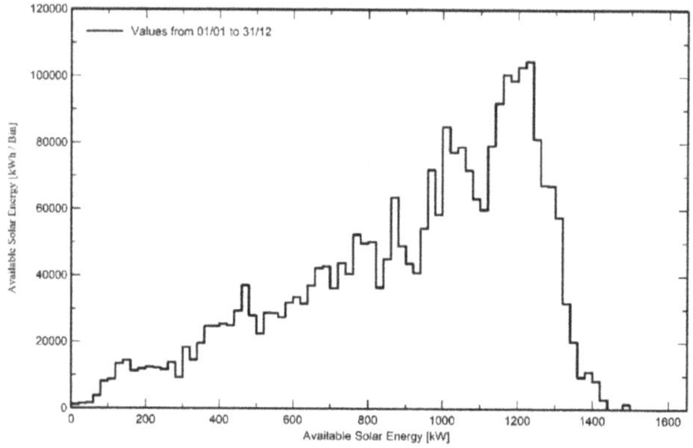

Figure 20. System output power distribution.

Table 5 provides detailed balances and main results for every month of a certain year to assess the performance and viability of photovoltaic (PV) systems. These balances and results offer a comprehensive overview of various parameters and outputs related to the solar energy system.

Table 5. Balances and main results.

Month	GlobHor kWh/m^2	DiffHor kWh/m^2	T_Amb °C	GlobInc kWh/m^2	GlobEff kWh/m^2	EArray MWh	E_User MWh	E_Solar MWh	E_Grid MWh	EFrGrid MWh
January	122.3	54.1	17.35	153.7	145.6	240.7	7.440	3.069	232.8	4.371
February	132.6	60.3	21.49	155.8	147.7	237.9	6.720	3.025	230.2	3.695
March	174.6	78.3	26.48	191.0	180.9	283.4	7.440	3.479	274.4	3.961
April	182.5	88.7	29.10	185.4	175.2	271.9	7.200	3.547	263.1	3.653
May	190.0	100.7	30.27	182.4	171.9	267.5	7.440	3.701	258.6	3.739
June	155.3	99.9	29.46	146.0	137.0	217.5	7.200	3.706	209.3	3.494
July	145.1	99.2	29.24	137.4	128.6	206.0	7.440	3.770	197.9	3.670
August	145.8	90.6	28.99	143.0	134.2	213.7	7.440	3.692	205.5	3.748
September	139.4	74.2	27.91	145.8	137.2	216.8	7.200	3.485	208.8	3.715
October	136.0	72.5	26.95	152.6	144.5	229.4	7.440	3.382	221.3	4.058
November	127.8	54.0	23.13	158.3	150.0	241.5	7.200	2.959	233.9	4.241
December	122.7	50.2	18.73	158.7	150.3	246.6	7.440	3.035	238.7	4.405
Year	1774.2	922.8	25.77	1910.0	1803.1	2872.9	87.600	40.850	2774.4	46.750

Legends: GlobHor—global horizontal irradiation; DiffHor—horizontal diffuse irradiation; T_Amb—ambient temperature; GlobInc—global incident in coll. plane; GlobEff—effective global, corr. for IAM and shadings; EArray—effective energy at the output of the array; E_User—energy supplied to the user; E_Solar—energy from the sun; E_Grid—energy injected into grid; EFrGrid—energy from the grid.

4.2. Loss Diagram

The loss diagram depicted in Figure 21 functions as a visual representation in the form of a system loss flow chart. It serves by breaking down and illustrating the diverse factors contributing to energy losses within a solar installation. Each component, including shading, temperature, module quality, and inverter efficiency, is visually depicted on this diagram, providing users with a clear view of the relative impact of each factor on energy generation [59]. By inputting specific data such as the geographical location, module specifications, and system design, users can generate a tailored loss diagram. This diagram proves to be an invaluable resource for solar professionals and engineers, aiding them in making well-informed decisions to optimize system design and operation. The ultimate goal is to maximize energy production and efficiency while minimizing losses [60]. As such, the loss diagram stands as a critical tool in the design and assessment of the viability of photovoltaic projects.

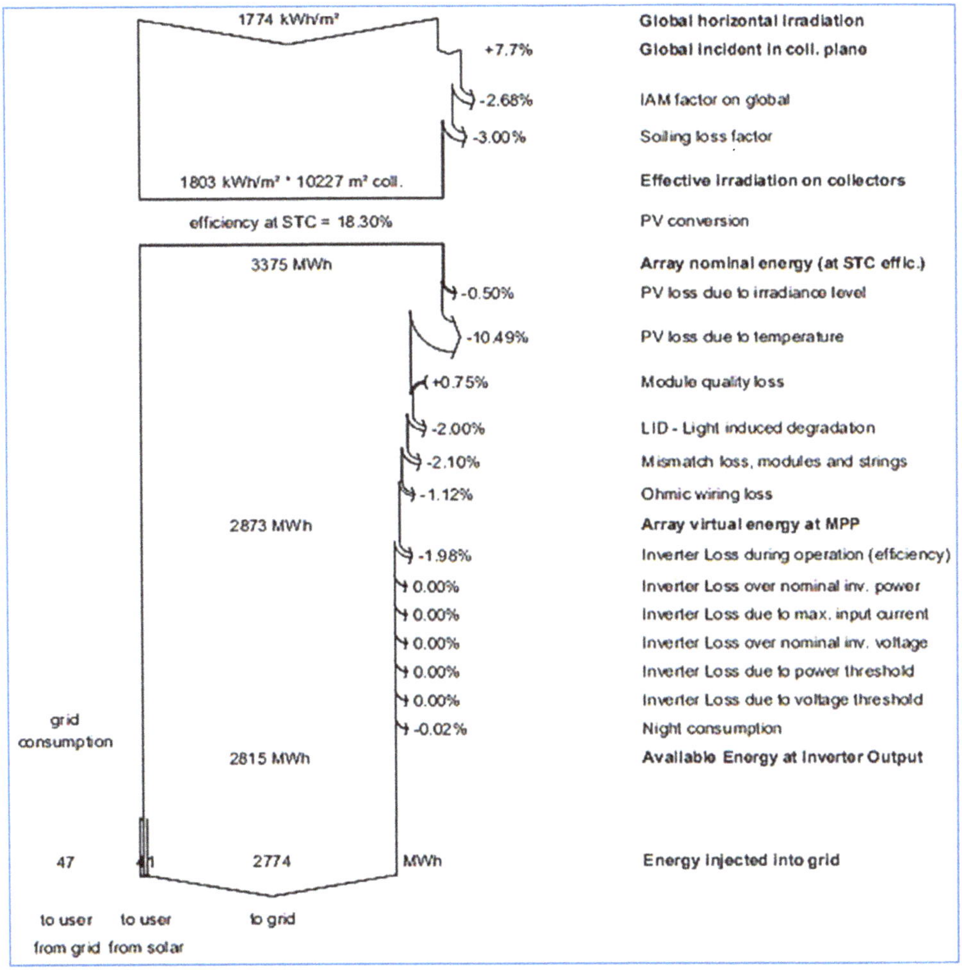

Figure 21. Loss diagram.

4.3. P50–P90 Evaluation

A probabilistic method for interpreting the simulation results across a number of years is the P50–P90 evaluation as shown in Table 6. A Gaussian probability distribution is displayed graphically by PVsyst for the number of years. The probability distribution for the plant, where probability is depicted in relation to E_Grid system production in GWh, is displayed in Figure 22.

Table 6. P50–P90 evaluation.

Meteo Data	
Meteo data source	Meteonorm 7.3, Sat = 100%
Kind	Not Defined
Year-to-year variability (Variance)	0.5%
Specified Deviation	
Global variability (meteo + system)	
Variability (Quadratic sum)	1.9%
Simulation and parameters uncertainties	
PV module modelling/parameters	1.0%
Inverter efficiency uncertainty	0.5%
Soiling and mismatch uncertainties	1.0%
Degradation uncertainty	1.0%
Annual production probability	
Variability	0.05 GWh
P50	2.77 GWh
P90	2.71 GWh
P95	2.69 GWh

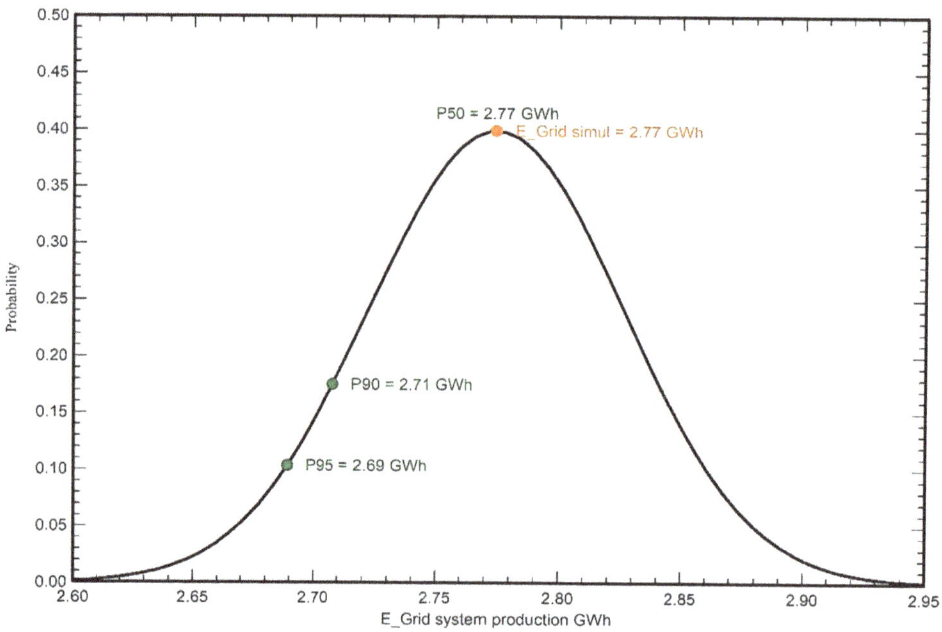

Figure 22. Probability distribution.

4.4. Financial Analysis

Financial analysis demonstrates the system's profitability. Additionally, it displays annual balances in detail between revenues generated by the pricing strategy specified in the tariffs section and costs described in the installation and operating costs section. Table 7 shows the installation costs, i.e., the costs of PV modules, inverters, and different settings of the system.

Table 7. Installation costs.

Item	Quantity Units	Cost BDT	Total BDT
PV modules ECO-300M-60	6248	12,003.84	75,000,000.00
Inverters 890GTS_1500	1	6,000,000.00	6,000,000.00
Installation settings	1	10,000,000.00	10,000,000.00
Total Depreciable asset			91,000,000.00 91,000,000.00

Table 8 shows the maintenance cost, i.e., the cleaning cost of the solar modules and the maintenance costs of other components of the system. The total installation cost amounts to 91,000,000 BDT, encompassing the initial expenses associated with setting up the solar power plant. This includes costs for solar panels, inverters, mounting systems, and other relevant components. The annual operating cost, considering a 2.00% inflation rate, is 64,060.60 BDT/year. The operating and maintenance cost over 25 years is 1,601,500 BDT. The overall cost over 25 years, inclusive of operation and maintenance, stands at 92,601,500 BDT. This incorporates the total installation cost and the estimated expenses of operating and maintaining the solar power plant over 25 years. This encompasses labor, materials, and other operational costs, with the inflation rate factored in to accommodate the expected rise in goods and services expenses. The energy consumed by auxiliary loads amounts to 87.6 MWh/year (46.4 MWh/year from the grid and 41.2 MWh/year from solar power), representing the energy utilized by auxiliary components within the solar power plant, such as the inverter and monitoring system. Annually, 2774 MWh of energy is supplied to the grid, symbolizing the quantity of energy the solar power plant generates and feeds into the grid. The LCOE stands at 2.82 BDT/kWh, signifying the average cost of generating electricity from the solar power plant throughout its operational life.

Table 8. Operating costs.

Item	Total BDT/Year
Maintenance Cleaning	50,000.00
Total (OPEX) Including inflation (2.00%)	50,000.00 64,060.60

Mathematical calculation of LCOE:
Here, $i' = 8\% = 0.08$, $f = 2\% = 0.02$.
Using Equation (1),

$$\begin{aligned} i &= \frac{i'-f}{1+f} \\ \Rightarrow i &= \frac{0.08-0.02}{1+0.02} \\ \Rightarrow i &= 0.0588. \end{aligned}$$

Now, $i = 0.0588$, $N = 25$ years.
Using Equation (2),

$$\begin{aligned} \text{CRF}(i, N) &= \frac{i(1+i)^N}{(1+i)^N-1} \\ \Rightarrow \text{CRF}(i, N) &= \frac{0.0588(1+0.0588)^{25}}{(1+0.0588)^{25}-1} \\ \Rightarrow \text{CRF}(i, N) &= 0.0773. \end{aligned}$$

Then,

$$\begin{aligned} C_{tot} &= 75,000,000 + 6,000,000 + 10,000,000 + 1,601,500 = 92,601,500 \text{ BDT} \\ i &= 0.0588 = 5.88\% \\ N &= 25 \text{ years} \\ CRF(i, N) &= 0.0773 \end{aligned}$$

Using Equation (3),

$$\begin{aligned} C_{ann,tot} &= CRF(i, N) \cdot C_{tot} \\ \Rightarrow C_{ann,tot} &= 0.0773 \times 92,601,500 \\ \Rightarrow C_{ann,tot} &= 7,158,095.95 \text{ BDT} \end{aligned}$$

Here,

$$\begin{aligned} C_{ann,tot} &= 7,158,095.95 \text{ BDT} \\ E_g &= [(2774 \times 20) + \{(2774 \div 2) \times 5\} + (41.2 \times 25)] \div 25 = 2537.8 \text{ MWh} = 2,537,800 \text{ kWh} \end{aligned}$$

Using Equation (4),

$$\begin{aligned} LCOE &= \frac{C_{ann,tot}}{E_g} \\ \Rightarrow LCOE &= \frac{7,158,095.95}{2,537,800} \\ \Rightarrow LCOE &= 2.82 \text{ BDT} \end{aligned}$$

which is similar to the PVsyst software simulated LCOE = 2.82 BDT/kWh.

The plant's net present value (NPV), or NPC, is calculated at 232,876,645 BDT by PVsyst software. NPV represents the present value of all future cash flows related to the solar power plant, obtained by discounting future cash flows using an appropriate discount rate. Additionally, the ROI stands at a significant 671.7%, showcasing the return on investment as a percentage. It is derived by dividing the solar power plant's net present value by the total project cost.

Detailed economic results for every year of the expected project lifetime have been calculated and demonstrated in Table 9.

Table 9. Detailed economic results (kBDT).

Year	Gross Income	Running Costs	Depreciable Allowance	Taxable Income	Taxes	After-Tax Profit	Self-Consumption Saving	Cumulative Profit	% Amorti.
2024	27,744	50	0	27,694	0	27,694	408	−62,898	30.9%
2025	27,744	51	0	27,693	0	27,693	408	−34,796	61.8%
2026	27,744	52	0	27,692	0	27,692	408	−6696	92.6%
2027	27,744	53	0	27,691	0	27,691	408	21,403	123.5%
2028	27,744	54	0	27,690	0	27,690	408	49,501	154.4%
2029	27,744	55	0	27,689	0	27,689	408	77,598	185.3%
2030	27,744	56	0	27,687	0	27,687	408	105,694	216.1%
2031	27,744	57	0	27,686	0	27,686	408	133,789	247.0%
2032	27,744	59	0	27,685	0	27,685	408	161,883	277.9%
2033	27,744	60	0	27,684	0	27,684	408	189,975	308.8%
2034	27,744	61	0	27,683	0	27,683	408	218,067	339.6%
2035	27,744	62	0	27,682	0	27,682	408	246,157	370.5%
2036	27,744	63	0	27,680	0	27,680	408	274,246	401.4%
2037	27,744	65	0	27,679	0	27,679	408	302,333	432.2%
2038	27,744	66	0	27,678	0	27,678	408	330,420	463.1%
2039	27,744	67	0	27,677	0	27,677	408	358,505	494.0%

Table 9. *Cont.*

Year	Gross Income	Running Costs	Depreciable Allowance	Taxable Income	Taxes	After-Tax Profit	Self-Consumption Saving	Cumulative Profit	% Amorti.
2040	27,744	69	0	27,675	0	27,675	408	386,588	524.8%
2041	27,744	70	0	27,674	0	27,674	408	414,671	555.7%
2042	27,744	71	0	27,672	0	27,672	408	442,752	586.5%
2043	27,744	73	0	27,671	0	27,671	408	470,831	617.4%
2044	27,744	74	0	27,670	0	27,670	408	498,909	648.3%
2045	27,744	76	0	27,668	0	27,668	408	526,986	679.1%
2046	27,744	77	0	27,667	0	27,667	408	555,061	710.0%
2047	27,744	79	0	27,665	0	27,665	408	583,134	740.8%
2048	27,744	80	0	27,663	0	27,663	408	611,206	771.7%
Total	693,595	1602	0	691,994	0	691,994	10,212	611,206	771.7%

This calculation is based on the energy sold to the grid, which amounts to 2774 MWh/year, and the selling price of energy to the grid, set at 10 BDT/kWh. The gross income is a crucial component of the solar power plant's cashflow, showcasing the revenue it generates. The operating cost of a 1.5 MW solar power plant over the course of one year is 64,060.60 BDT, while the gross income is 27,744 kBDT, i.e., 27,744,000 BDT. Table 9 and Figure 23 present a comprehensive cashflow analysis of the solar power plant. This cashflow analysis encapsulates all revenues and expenses associated with the solar power plant throughout its operational lifespan. It provides a detailed overview of the financial dynamics, incorporating the gross income generated from selling electricity to the grid. The representation of gross income as a positive cash flow emphasizes its role as a revenue stream for the solar power plant.

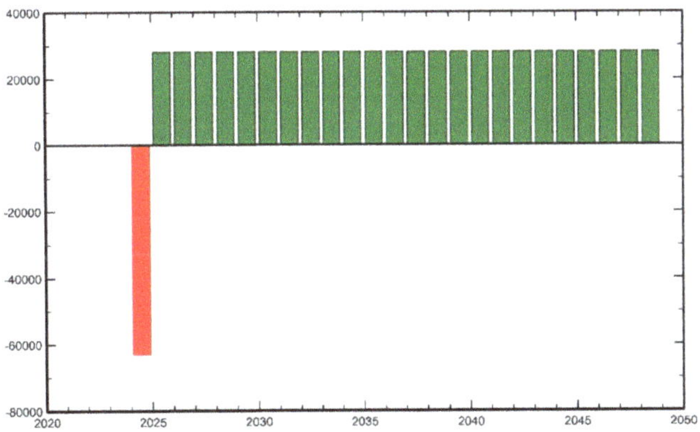

Figure 23. Yearly net profit (kBDT).

Figure 24 visually presents the cumulative cashflow of the solar power plant. This cumulative cash flow is derived by aggregating the annual cash flows throughout the predefined 25-year project lifespan. By summing up the annual cash flows, we obtain a clear depiction of how the solar power plant's overall financial performance is anticipated to evolve over the years. The cumulative cash flow serves as a critical metric for assessing the solar power plant's financial trajectory across its entire operational duration. It provides valuable insights into the financial dynamics and the cumulative impact of revenue generation and expenditure over time. The calculation of the cumulative cash flow is conducted using the specified equation, allowing for a comprehensive evaluation of the solar power plant's financial sustainability and profitability throughout its projected operational period.

$$\text{Cumulative cashflow} = \text{SUM (Annual cashflow)}$$

where annual cash flow is the cash flow associated with the solar power plant in each year of its lifetime. Figure 24 shows that the cumulative cashflow of the solar power plant is expected to be positive over its lifetime. This means that the solar power plant is expected to generate more revenue than it incurs in expenses. The cumulative cash flow graph is a useful tool for evaluating the financial performance of the solar power plant over its lifetime. Table 10 demonstrates the analysis of the NPV, payback period, and ROI based on inputting three different fixed feed-in tariff rates to the PVsyst software.

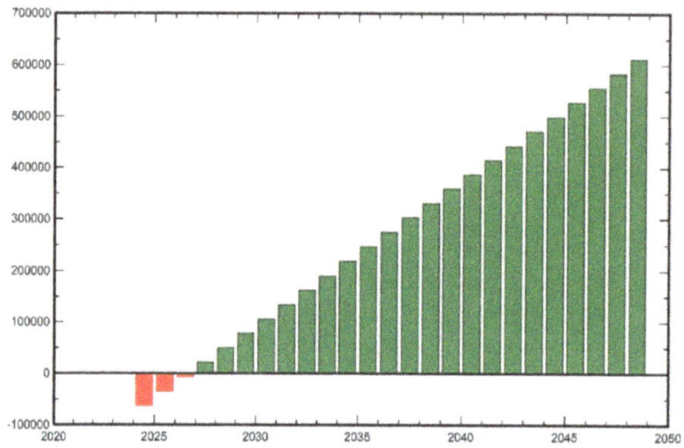

Figure 24. Cumulative cash flow (kBDT).

Table 10. Analysis of return on investment (ROI) based on a fixed feed-in tariff.

Electricity Sale Fixed Feed-in Tariff (BDT/kWh)	Net Present Value (NPV) (BDT)	Payback Period (Years)	Return on Investment (ROI) (%)
5.00	72,950,859	6.4	290.6
10.00	232,876,645	3.2	671.7
15.00	392,802,431	2.2	1052.8

4.5. Calculation of Payback Period

The payback period is the number of years needed to recover the net investment cost as stated in the installation and operational costs section. This period refers to the time it takes for users to recoup their investment through increased efficiency and accurate solar project assessments.

$$\begin{aligned}\text{Installation costs} &= \text{PV modules' cost} + \text{Inverters' cost} + \text{Settings' cost}\\ &= 75{,}000{,}000 + 6{,}000{,}000 + 10{,}000{,}000\\ &= 91{,}000{,}000 \text{ BDT}\end{aligned}$$

$$\text{Operation and maintenance costs} = 64{,}060.60 \text{ BDT/year}$$

$$\text{Operation and maintenance cost over 25 years} = 1{,}601{,}500 \text{ BDT}$$

$$\begin{aligned}\text{Total costs} &= \text{Installation costs} + \text{Operation and maintenance costs over 25 years}\\ &= 91{,}000{,}000 + 1{,}601{,}500\\ &= 92{,}601{,}500 \text{ BDT}\end{aligned}$$

Energy sold to grid = 2774 MWh

Energy consumed by auxiliary load = 41.2 MWh

Total energy generated = 2774 + 41.2 = 2815.2 MWh = 2,815,200 kWh

Total income = 2,815,200 $\times$ 10 = 28,152,000 BDT

Payback period = $\frac{Total\ costs}{Total\ income} = \frac{92,601,500}{28,152,000} = 3.2\ years$

After considering all installation and operational costs, and total income, the payback period is 3.2 years. PVsyst aids in minimizing payback periods by optimizing solar installations and maximizing energy production.

4.6. Carbon Balance

The carbon balancing tool is essential in streamlining the computation of the expected decrease in CO_2 emissions brought on by the installation of a PV system. The life cycle emissions (LCE) concept, which includes all carbon dioxide emissions connected to a particular amount or kind of energy, is the foundation of this computation. These emissions are assessed throughout the component's complete life cycle, including manufacturing, utilization, maintenance, and disposal, providing a comprehensive view [61].

By utilizing the carbon balance tool, we can determine that the PV system will effectively displace an equivalent amount of current grid power with electricity generated internally. This tool's justification is that carbon dioxide emissions will be reduced overall if the carbon footprint of the PV system per kWh is lower than that of obtaining energy from the grid [62]. This presents a meaningful way to estimate and promote the positive environmental impact of transitioning to solar energy. The saved CO_2 emission in tCO_2 is plotted against the project lifetime in years in Figure 25.

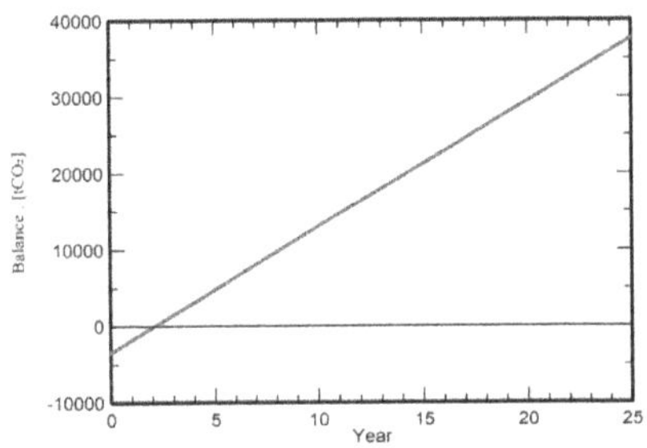

Figure 25. Saved CO_2 emission vs. time.

Therefore, the difference between generated and saved CO_2 emissions is the total carbon balance for a PV plant.

Table 11 shows the lifecycle emissions details for all the items of the system; it is given in $kgCO_2$ or tCO_2 and represents the total amount of CO_2 emissions caused by the construction and operation of the components of PV installation.

Table 11. System lifecycle emissions details.

Item	LCE	Quantity	Subtotal ($kgCO_2$)
Modules	1713 $kgCO_2$/kWp	1874 kWp	3,210,322
Supports	3.90 $kgCO_2$/kg	62,480 kg	243,377
Inverters	3.90 $kgCO_2$/units	1.00 unit	386
		Total	3454.085

The LCE grid is expressed in gCO_2/kWh and indicates the average CO_2 emissions per energy unit for grid-produced electricity.

The LCE system is expressed in tCO_2 and represents the total CO_2 emissions brought on by the structure's construction and operation of the PV installation.

Total energy generated = Energy sold to grid + Energy consumed by auxiliary load
= 2774 + 41.2
= 2815.2 MWh
= 2,815,200 kWh

Project lifetime = 25 years

LCE Grid = 584 gCO_2

LCE System = 3454.1 tCO_2 = 3,454,100,000 gCO_2

Now,

The carbon balance = (Total energy generated × Project lifetime × LCE Grid) − LCE System)
= (2,815,200 kWh × 25 years × 584 gCO_2) − 3,454,100,000 gCO_2
= 41,101,920,000 gCO_2 − 3,454,100,000 gCO_2
= 37,647,820,000 gCO_2 [as 1 tone = 1,000,000 g]
= 37,647.82 tCO_2

Table 12 shows a comparison between the proposed Char Jazira solar power plant and various completed and operational solar park projects in Bangladesh. Location, generation capacity and status of some notable solar farms and future solar initiatives in Bangladesh have been mentioned in Table 13.

Table 12. Comparison of different completed and running solar park projects and this proposed Char Jazira solar power plant in Bangladesh [63,64].

SL.	Project Name	Capacity (MWp)	Location	Latitude, Longitude	Agency	Expected Energy Generation and CO_2 Emission Reduction During System Life	Expected Energy Generation and CO_2 Emission Reduction until the Data Collection Day
1	200 MW (AC) Solar Park by Beximco Power Co. Ltd.	200	Sundarganj, Gaibandha	25.328795° N, 89.541671° E	BPDB	4 TWh, 2 M tCO_2	156 GWh, 74 k tCO_2
2	30MW (AC) Solar Park by Intraco CNG Ltd. & Juli New Energy Co. Ltd.	30	Gangachara, Rangpur	25.855312° N, 89.222482° E	BPDB	654 GWh, 309 k tCO_2	36 GWh, 17 k tCO_2
3	100 MW (AC) Solar Park by Energon Technologies FZE & China Sunergy Co.Ltd (ESUN)	100	Mongla, Bagerhat	22.650135° N, 89.761117° E	BPDB	2 TWh, 1 M tCO_2	199 GWh, 94 k tCO_2
4	Sirajganj 6.13 MW (AC) Grid-connected Solar Photovoltaic Power Plant	7.6	Sirajganj, Sirajgonj	24.386177° N, 89.748409° E	NWPGCL	166 GWh, 78 k tCO_2	22 GWh, 10 k tCO_2
5	35 MW AC Solar Park by Consortium of Spectra Engineers Limited & Shunfeng Investment Limited	35	Shibalaya, Manikganj	23.848491° N, 89.913733° E	BPDB	763 GWh, 361 k tCO_2	102 GWh, 48 k tCO_2
6	50 MW (AC) Solar Park by HETAT-DITROLIC-IFDC Solar Consortium	50	Gauripur, Mymensingh	24.75894° N, 90.59746° E	BPDB	1 TWh, 516 k tCO_2	166 GWh, 79 k tCO_2

Table 12. *Cont.*

SL.	Project Name	Capacity (MWp)	Location	Latitude, Longitude	Agency	Expected Energy Generation and CO_2 Emission Reduction During System Life	Expected Energy Generation and CO_2 Emission Reduction until the Data Collection Day
7	Kaptai 7.4 MWp (6.63 MW AC) Grid-connected Solar PV Power Plant	7.4	Kaptai, Rangamati	22.493286° N, 92.218809° E	BPDB	161 GWh, 76 k tCO_2	37 GWh, 17 k tCO_2
8	8 MW Solar Park by Parasol Energy Ltd.	8	Panchagarh, Panchagarh	26.376098° N, 88.591665° E	BPDB	174 GWh, 82 k tCO_2	40 GWh, 19 k tCO_2
9	20MW (AC) Solar Park by Joules Power Limited (JPL)	20	Teknaf, Cox's Bazar	20.980463° N, 92.252503° E	BPDB	436 GWh, 206 k tCO_2	116 GWh, 55 k tCO_2
10	1.5 MW Grid-connected Solar Power Plant in Lalpur, Natore [Proposed]	1.874	Char Jazira, Lalpur, Natore	24.0900° N, 88.5800° E	Proposed	70.375 GWh, 33074.061 tCO_2	Proposed

Table 13. Noteworthy solar farms and upcoming solar projects in Bangladesh [1,64].

Location	Generation Capacity (MW)	Project Status
Gangachara, Rangpura	30	Ongoing
Dharmapasha, Sunamganja	32	Ongoing
Gauripur, Mymensingh	50	Ongoing
Chuadangaa	50	Future
Netrokonaa	50	Future
Mongla, Bagerhata	100	Ongoing
Fenia	100	Future
Narsingdia	120	Future
Sundarganj, Gaibandhaa	200	Ongoing

5. Conclusions

The escalating consumption of fossil fuels and surging consumer demand have led to frequent occurrences of load shedding. To combat this issue and ensure a continuous supply of electricity, solar energy stands as a promising solution. In this study, we explored the potential of solar energy generation, focusing on addressing the power deficit and reducing reliance on conventional energy sources. The annual solar energy availability of 3375 MWh emphasizes the substantial capacity of solar power. The inverter's annual output of 2815.2 MWh signifies its ability to generate a significant amount of electricity annually. However, the overall system's reduced power capacity of 559.8 MWh/year is attributed to diverse losses encountered in the system. Despite this, a notable portion of solar energy, 2774 MWh/year, is successfully integrated into the grid.

An analysis of performance ratios revealed seasonal variations, with the highest PR of 82% in January and the lowest of 77% in April. The yearly average PR of 78.63% indicates the system's consistent performance and efficiency. Additionally, the solar fraction (SF) of 46.63% underlines the substantial contribution of solar energy in meeting electricity demands. The LCOE at 2.82 BDT per unit demonstrates the economic viability and cost-effectiveness of this solar PV system. A reasonable specific cost per Wp of 48.5 BDT during the investment phase further supports the financial feasibility of the solar project.

Over the plant's projected 25-year lifespan, the expected energy generation of 70.375 GWh showcases the long-term benefits of solar energy. Remarkably, the calculated payback period of 3 years, 2 months, and 13 days emphasizes the rapid return on investment, highlighting the economic advantages of this sustainable energy solution.

The significant environmental impact is a crucial aspect, with an impressive reduction of 37,647.82 tCO_2 emissions over the 25-year lifespan, reflecting the positive contribution of solar energy in reducing carbon footprints and promoting environmental sustainability. In conclusion, this solar technology proves to be a green and clean energy source, making substantial strides toward a more sustainable and eco-friendly future.

Solar power plants have revolutionized clean energy production but face several shortcomings. One primary limitation is their intermittent nature due to dependency on sunlight. Energy-storage systems like batteries help to mitigate this issue but are expensive and have limited storage capacities. Efficiency remains another challenge; while solar panels have improved, they still cannot convert all sunlight into usable energy.

Future research aims to address these limitations. Improving energy-storage systems is crucial for a consistent power supply. Advancements in battery technology or alternative storage methods such as molten salt, compressed air, or hydrogen are under exploration. Enhancing solar panel efficiency via new materials, designs, and integrated technologies could significantly boost energy output. Innovations in flexible and transparent panels might enable diverse applications. Research on reducing environmental impact involves designing eco-friendly installation methods, recycling solar components, and utilizing less land-intensive solar technologies like floating solar farms or integrating them into infrastructures like highways or buildings.

Author Contributions: Conceptualization, M.F.A. and N.K.S.; methodology, N.K.S.; software, M.F.A., N.K.S., A.H.S. and S.I.S.S.; validation, M.F.A., M.A.H. and M.S.A.; formal analysis, M.F.A.; investigation, N.K.S., A.H.S. and S.I.S.S.; resources, N.K.S.; data curation, N.K.S. and A.H.S.; writing—original draft preparation, N.K.S., A.H.S. and M.F.A.; writing—review and editing, M.F.A., A.H.S. and M.A.H.; visualization, M.F.A. and N.K.S.; supervision, M.F.A., M.A.H. and M.S.A.; project administration, M.F.A.; funding acquisition, M.F.A., M.A.H. and M.S.A. All authors have read and agreed to the published version of the manuscript.

Funding: This work is supported by Pabna University of Science and Technology (PUST) under the research grant for the fiscal year 2023–2024.

Data Availability Statement: The authors confirm that the data supporting the findings of this study are available within the article.

Conflicts of Interest: The authors declare no conflict of interest.

Abbreviations

°	Degrees
°C	Degrees Celsius
A	Ampere
AC	Alternating current
Ah	Ampere hour
BDT	Bangladeshi Taka
BPDB	Bangladesh Power Development Board
CO_2	Carbon dioxide
COE	Cost of energy
CRF	Capital recovery factor
DC	Direct current
DHI	Diffuse horizontal irradiance
DNI	Direct normal irradiance
gCO_2	Gram carbon dioxide
GHI	Global horizontal irradiance
GIS	Geographical information system
GW	Gigawatt

GWh	Gigawatt hour
IAM	Incidence angle modifier
Impp	Current at maximum power point
IRR	Internal rate of return
kBDT	Kilo BDT
kW	Kilowatt
kWac	Kilowatt alternating current
kWdc	Kilowatt direct current
kWh	Kilowatt hour
kWp	Kilowatt peak
LCE	Life cycle emissions
LCOE	Levelized cost of energy
MWh	Megawatt hour
mΩ	Milliohm
NPC	Net present cost
NPV	Net present value
NWPGL	North-West Power Generation Company Ltd.
O&M	Operation and maintenance
OPEX	Operation expenditure
PED	Positive energy district
Pmpp	Power at maximum power point
Pnom	Nominal power
PR	Performance ratio
PV	Photovoltaic
RH	Relative humidity
ROI	Return on investment
SF	Solar fraction
STC	Standard test condition
tCO_2	Ton carbon dioxide
V	Volt
Vmpp	Voltage at maximum power point
W	Watt
Wh	Watt hour

References

1. Solar Energy in Bangladesh: Current Status and Future. Available online: https://energytracker.asia/solar-energy-in-bangladesh-current-status-and-future/?fbclid=IwAR2dpVoe0scUzqZ_VGKQm2M3jkzN0UawrjmlbM3h9i0F_wGdD874jBmZUzg (accessed on 4 September 2023).
2. Hossain, M.A.; Pota, H.R.; Hossain, M.J.; Blaabjerg, F. Evolution of Microgrids with Converter-Interfaced Generations: Challenges and Opportunities. *Int. J. Electr. Power Energy Syst.* **2019**, *109*, 160–186. [CrossRef]
3. Karmaker, A.K.; Islam, S.M.R.; Kamruzzaman, M.; Rashid, M.M.U.; Faruque, M.O.; Hossain, M.A. Smart City Transformation: An Analysis of Dhaka and Its Challenges and Opportunities. *Smart Cities* **2023**, *6*, 1087–1108. [CrossRef]
4. Alam, M.S.; Chowdhury, T.A.; Dhar, A.; Al-Ismail, F.S.; Choudhury, M.S.H.; Shafiullah, M.; Hossain, M.I.; Hossain, M.A.; Ullah, A.; Rahman, S.M. Solar and Wind Energy Integrated System Frequency Control: A Critical Review on Recent Developments. *Energies* **2023**, *16*, 812. [CrossRef]
5. Hasan, M.; Tanvir, A.A.; Siddiquee, S.M.S.; Zubair, A. Efficient Hybrid Renewable Energy System for Industrial Sector with On-Grid Time Management. In Proceedings of the 2015 3rd International Conference on Green Energy and Technology (ICGET), Dhaka, Bangladesh, 11 September 2015; pp. 4–9. [CrossRef]
6. Duman, A.C.; Güler, Ö. Economic Analysis of Grid-Connected Residential Rooftop PV Systems in Turkey. *Renew. Energy* **2020**, *148*, 697–711. [CrossRef]
7. Ihsan, M.A.; Eram, A.F.; Khadem, M.M. Viability Study of Net Metering Scheme in Residential Scale: A Bangladeshi Perspective. In Proceedings of the 2022 International Conference on Computational Intelligence and Sustainable Engineering Solutions (CISES), Greater Noida, India, 20–21 May 2022; pp. 361–365. [CrossRef]
8. JPI Urban Europe. Positive Energy Districts (PED). Available online: https://jpi-urbaneurope.eu/ped/ (accessed on 9 November 2023).

9. Chen, Y.; Wang, J.; Lund, P.D. Sustainability Evaluation and Sensitivity Analysis of District Heating Systems Coupled to Geothermal and Solar Resources. *Energy Convers. Manag.* **2020**, *220*, 113084. [CrossRef]
10. Bottecchia, L.; Gabaldón, A.; Castillo-Calzadilla, T.; Soutullo, S.; Ranjbar, S.; Eicker, U. Fundamentals of Energy Modelling for Positive Energy Districts. In *Smart Innovation, Systems and Technologies*; Springer: Singapore, 2022; Volume 263, pp. 435–445. [CrossRef]
11. Lindholm, O.; Rehman, H.U.; Reda, F. Positioning Positive Energy Districts in European Cities. *Buildings* **2021**, *11*, 19. [CrossRef]
12. Moreno, A.G.; Vélez, F.; Alpagut, B.; Hernández, P.; Montalvillo, C.S. How to Achieve Positive Energy Districts for Sustainable Cities: A Proposed Calculation Methodology. *Sustainability* **2021**, *13*, 710. [CrossRef]
13. Alpagut, B.; Akyürek, Ö.; Mitre, E.M. Positive Energy Districts Methodology and Its Replication Potential. *Multidiscip. Digit. Publ. Inst. Proc.* **2019**, *20*, 8. [CrossRef]
14. Bagheri, M.; Shirzadi, N.; Bazdar, E.; Kennedy, C.A. Optimal Planning of Hybrid Renewable Energy Infrastructure for Urban Sustainability: Green Vancouver. *Renew. Sustain. Energy Rev.* **2018**, *95*, 254–264. [CrossRef]
15. Castillo-Calzadilla, T.; Garay-Martinez, R.; Andonegui, C.M. Holistic Fuzzy Logic Methodology to Assess Positive Energy District (PathPED). *Sustain. Cities Soc.* **2023**, *89*, 104375. [CrossRef]
16. Zhang, X.; Penaka, S.R.; Giriraj, S.; Sánchez, M.N.; Civiero, P.; Vandevyvere, H. Characterizing Positive Energy District (PED) through a Preliminary Review of 60 Existing Projects in Europe. *Buildings* **2021**, *11*, 318. [CrossRef]
17. Marotta, I.; Guarino, F.; Longo, S.; Cellura, M. Environmental Sustainability Approaches and Positive Energy Districts: A Literature Review. *Sustainability* **2021**, *13*, 13063. [CrossRef]
18. Krangsås, S.G.; Steemers, K.; Konstantinou, T.; Soutullo, S.; Liu, M.; Giancola, E.; Prebreza, B.; Ashrafian, T.; Murauskaitė, L.; Maas, N. Positive Energy Districts: Identifying Challenges and Interdependencies. *Sustainability* **2021**, *13*, 10551. [CrossRef]
19. Neumann, H.M.; Hainoun, A.; Stollnberger, R.; Etminan, G.; Schaffler, V. Analysis and Evaluation of the Feasibility of Positive Energy Districts in Selected Urban Typologies in Vienna Using a Bottom-Up District Energy Modelling Approach. *Energies* **2021**, *14*, 4449. [CrossRef]
20. Poudyal, R.; Loskot, P.; Parajuli, R. Techno-Economic Feasibility Analysis of a 3-KW PV System Installation in Nepal. *Renew. Wind. Water Sol.* **2021**, *8*, 5. [CrossRef]
21. Imam, A.A.; Al-Turki, Y.A.; Sreerama Kumar, R. Techno-Economic Feasibility Assessment of Grid-Connected PV Systems for Residential Buildings in Saudi Arabia—A Case Study. *Sustainability* **2020**, *12*, 262. [CrossRef]
22. Faiz, F.U.H.; Shakoor, R.; Raheem, A.; Umer, F.; Rasheed, N.; Farhan, M. Modeling and Analysis of 3 MW Solar Photovoltaic Plant Using PVSyst at Islamia University of Bahawalpur, Pakistan. *Int. J. Photoenergy* **2021**, *2021*, 6673448. [CrossRef]
23. Baqir, M.; Channi, H.K. Analysis and Design of Solar PV System Using Pvsyst Software. *Mater. Today Proc.* **2022**, *48*, 1332–1338. [CrossRef]
24. Kumar, R.; Rajoria, C.S.; Sharma, A.; Suhag, S. Design and Simulation of Standalone Solar PV System Using PVsyst Software: A Case Study. *Mater. Today Proc.* **2020**, *46*, 5322–5328. [CrossRef]
25. Satish, M.; Santhosh, S.; Yadav, A. Simulation of a Dubai Based 200 KW Power Plant Using PVsyst Software. In Proceedings of the 2020 7th International Conference on Signal Processing and Integrated Networks (SPIN), Noida, India, 27–28 February 2020; pp. 824–827. [CrossRef]
26. Jagadale, P.; Choudhari, A.; Jadhav, S. Design and Simulation of Grid Connected Solar Si-Poly Photovoltaic Plant Using PVsyst For Pune, India Location. *Renew. Energy Res. Appl.* **2022**, *3*, 41–49.
27. Grover, A.; Khosla, A.; Joshi, D. Design and Simulation of 20MW Photovoltaic Power Plant Using PVSyst. *Indones. J. Electr. Eng. Comput. Sci.* **2020**, *19*, 58–65. [CrossRef]
28. Chauhan, A.; Sharma, M.; Baghel, S. Designing and Performance Analysis of 15KWP Grid Connection Photovoltaic System Using Pvsyst Software. In Proceedings of the 2020 Second International Conference on Inventive Research in Computing Applications (ICIRCA), Coimbatore, India, 15–17 July 2020; pp. 1003–1008. [CrossRef]
29. Kapoor, S.; Sharma, A.K.; Porwal, D. Design and Simulation of 60kWp Solar On-Grid System for Rural Area in Uttar-Pradesh by "PVsyst". *J. Phys. Conf. Ser.* **2021**, *2070*, 012147. [CrossRef]
30. Belmahdi, B.; El Bouardi, A. Solar Potential Assessment Using PVsyst Software in the Northern Zone of Morocco. *Procedia Manuf.* **2020**, *46*, 738–745. [CrossRef]
31. Tamoor, M.; Bhatti, A.R.; Farhan, M.; Miran, S.; Raza, F.; Zaka, M.A. Designing of a Hybrid Photovoltaic Structure for an Energy-Efficient Street Lightning System Using PVsyst Software. *Eng. Proc.* **2021**, *12*, 10–14. [CrossRef]
32. Mohammadi, S.A.D.; Gezegin, C. Design and Simulation of Grid-Connected Solar PV System Using PVSYST, PVGIS and HOMER Software. *Int. J. Pioneer. Technol. Eng.* **2022**, *1*, 36–41. [CrossRef]
33. PVsyst—Logiciel Photovoltaïque. Available online: https://www.pvsyst.com/ (accessed on 1 September 2023).
34. Dalal, S.; Jadhav, V.; Raut, R.; Narkhede, S. Analysis of 1kw solar rooftop system by using pvsyst. In Proceedings of the 2nd International Conference on Communication & Information Processing (ICCIP), Tokyo, Japan, 27–29 November 2020; pp. 1–8.
35. Solar Panel Orientation—Energy Education. Available online: https://energyeducation.ca/encyclopedia/Solar_panel_orientation (accessed on 1 September 2023).
36. Tilt and Orientation and Solar Energy. Available online: https://www.viridiansolar.co.uk/resources-1-3-tilt-and-orientation.html (accessed on 1 September 2023).

37. Bansal, N.; Pany, P.; Singh, G. Visual Degradation and Performance Evaluation of Utility Scale Solar Photovoltaic Power Plant in Hot and Dry Climate in Western India. *Case Stud. Therm. Eng.* **2021**, *26*, 101010. [CrossRef]
38. Talavera, D.L.; Muñoz-Cerón, E.; de la Casa, J.; Lozano-Arjona, D.; Theristis, M.; Pérez-Higueras, P.J. Complete Procedure for the Economic, Financial and Cost-Competitiveness of Photovoltaic Systems with Self-Consumption. *Energies* **2019**, *12*, 345. [CrossRef]
39. Cui, Y.; Zhu, J.; Meng, F.; Zoras, S.; McKechnie, J.; Chu, J. Energy Assessment and Economic Sensitivity Analysis of a Grid-Connected Photovoltaic System. *Renew. Energy* **2020**, *150*, 101–115. [CrossRef]
40. Meteo Database > Import Meteo Data > Meteonorm Data and Program. Available online: https://www.pvsyst.com/help/meteo_source_meteonorm.htm (accessed on 1 September 2023).
41. Intro—Meteonorm (En). Available online: https://meteonorm.meteotest.ch/en/ (accessed on 1 September 2023).
42. NASA Technical Reports Server (NTRS). Surface Meteorology and Solar Energy (SSE) Data Release 5.1. Available online: https://ntrs.nasa.gov/citations/20080012141 (accessed on 1 September 2023).
43. Solargis. Solar Resource Maps and GIS Data for 200+ Countries. Available online: https://solargis.com/maps-and-gis-data/download/bangladesh (accessed on 1 September 2023).
44. Gostein, M.; Hoffmann, A.; Farina, F.; Stueve, B. Measuring Global, Direct, and Diffuse Irradiance Using a Reference Cell Array. In Proceedings of the 2021 IEEE 48th Photovoltaic Specialists Conference (PVSC), Fort Lauderdale, FL, USA, 20–25 June 2021; pp. 923–927. [CrossRef]
45. PV Performance Modeling Collaborative (PVPMC). Direct Normal Irradiance. Available online: https://pvpmc.sandia.gov/modeling-guide/1-weather-design-inputs/irradiance-insolation/direct-normal-irradiance/ (accessed on 9 November 2023).
46. Zhang, C.; Yan, Z.; Ma, C.; Xu, X. Prediction of Direct Normal Irradiation Based on CNN-LSTM Model. In Proceedings of the Proceedings of the 2020 5th International Conference on Multimedia Systems and Signal Processing, Chengdu, China, 28–30 May 2020; pp. 74–80. [CrossRef]
47. PV Performance Modeling Collaborative (PVPMC). Diffuse Horizontal Irradiance. Available online: https://pvpmc.sandia.gov/modeling-guide/1-weather-design-inputs/irradiance-insolation/diffuse-horizontal-irradiance/ (accessed on 9 November 2023).
48. Dubey, S.; Sarvaiya, J.N.; Seshadri, B. Temperature Dependent Photovoltaic (PV) Efficiency and Its Effect on PV Production in the World—A Review. *Energy Procedia* **2013**, *33*, 311–321. [CrossRef]
49. Boston Solar. How Temperature & Shade Affect Solar Panel Efficiency. Available online: https://www.bostonsolar.us/solar-blog-resource-center/blog/how-do-temperature-and-shade-affect-solar-panel-efficiency/ (accessed on 9 November 2023).
50. Wind Effect on Solar Panels—AESOLAR. Available online: https://ae-solar.com/wind-effect-on-solar-panels/ (accessed on 9 November 2023).
51. Bhattacharya, T.; Chakraborty, A.K.; Pal, K. Effects of Ambient Temperature and Wind Speed on Performance of Monocrystalline Solar Photovoltaic Module in Tripura, India. *J. Sol. Energy* **2014**, *2014*, 817078. [CrossRef]
52. Eltbaakh, Y.A.; Ruslan, M.H.; Alghoul, M.A.; Othman, M.Y.; Sopian, K.; Razykov, T.M. Solar Attenuation by Aerosols: An Overview. *Renew. Sustain. Energy Rev.* **2012**, *16*, 4264–4276. [CrossRef]
53. Page, J.K. Radiation Data. In *Solar Energy Conversion II*; Elsevier: Amsterdam, The Netherlands, 1981; pp. 23–35. [CrossRef]
54. ScienceDirect Topics. Linke Turbidity Factor—An Overview. Available online: https://www.sciencedirect.com/topics/engineering/linke-turbidity-factor (accessed on 9 November 2023).
55. Shrestha, A.K.; Thapa, A.; Gautam, H. Solar Radiation, Air Temperature, Relative Humidity, and Dew Point Study: Damak, Jhapa, Nepal. *Int. J. Photoenergy* **2019**, *2019*, 8369231. [CrossRef]
56. Sohani, A.; Shahverdian, M.H.; Sayyaadi, H.; Garcia, D.A. Impact of Absolute and Relative Humidity on the Performance of Mono and Poly Crystalline Silicon Photovoltaics; Applying Artificial Neural Network. *J. Clean. Prod.* **2020**, *276*, 123016. [CrossRef]
57. Bowen, A. Fundamentals of solar architecture. In *Solar Energy Conversion*; Elsevier: Amsterdam, The Netherlands, 1979; pp. 481–553. [CrossRef]
58. Yousef, B.A.A.; Radwan, A.; Olabi, A.G.; Abdelkareem, M.A. Sun Composition, Solar Angles, and Estimation of Solar Radiation. In *Renewable Energy—Volume 1: Solar, Wind, and Hydropower: Definitions, Developments, Applications, Case Studies, and Modelling and Simulation*; Academic Press: Cambridge, MA, USA; Elsevier: Amsterdam, The Netherlands, 2023; Volume 1, pp. 3–22. [CrossRef]
59. Mermoud, A. *Modeling Systems Losses in PVsyst*; Institute of the Environmental Sciences Group of energy–PVsyst, Universitè de Genève: Geneva, Switzerland, 2013; pp. 1–15. Available online: https://energy.sandia.gov/wp-content/gallery/uploads/Mermoud_PVSyst_Thu-840-am.pdf (accessed on 4 September 2023).
60. Soualmia, A.; Chenni, R. Modeling and Simulation of 15MW Grid-Connected Photovoltaic System Using PVsyst Software. In Proceedings of the 2016 International Renewable and Sustainable Energy Conference (IRSEC), Marrakech, Morocco, 14–17 November 2016; pp. 702–705. [CrossRef]
61. Solar.Com. What Is the Carbon Footprint of Solar Panels? Available online: https://www.solar.com/learn/what-is-the-carbon-footprint-of-solar-panels/ (accessed on 2 September 2023).
62. Project Design > Carbon Balance Tool. Available online: https://www.pvsyst.com/help/carbon_balance_tool.htm (accessed on 2 September 2023).

63. National Database of Renewable Energy, SREDA. Solar Park. Available online: http://www.renewableenergy.gov.bd/index.php?id=1&i=1&pg=1 (accessed on 4 September 2023).
64. Ministry of Power, Energy & Mineral Resources. Sustainable and Renewable Energy Development Authority (SREDA)-Power Division. Available online: http://www.sreda.gov.bd/ (accessed on 4 September 2023).

designs

MDPI

Article

Mapping the Potential of Zero-Energy Building in Greece Using Roof Photovoltaics

Angeliki Kitsopoulou [1], Dimitris Pallantzas [1], Evangelos Bellos [2,*] and Christos Tzivanidis [1]

[1] Department of Thermal Engineering, School of Mechanical Engineering, National Technical University of Athens, Heroon Polytechniou 9, 15780 Zografou, Greece; akitsopoulou@mail.ntua.gr (A.K.); d_pallantzas@mail.ntua.gr (D.P.); ctzivan@central.ntua.gr (C.T.)

[2] Department of Mechanical Engineering, School of Engineering, University of West Attica, 250 Thivon & Petrou Ralli, 12244 Athens, Greece

* Correspondence: bellose@uniwa.gr

Abstract: The present study investigates the incorporation of renewable rooftop photovoltaic systems in fully electrified residential buildings and estimates the zero-energy demand building potential in relation to the climatic data of Greece. Specifically, the aim of the analysis is to calculate the maximum possible number of stories and therefore the total building height for a complete transformation to zero-net-energy building. The energy analysis, which is conducted using the DesignBuilder software, focuses on single-floor up to seven-story buildings. The importance of the present work lies in the acknowledgment of the diversity of the Greek residential sector, the adherence to national energy policies, and the European goal of fully electrified buildings. The examined case studies are equipped with electrically driven air-to-air heat pumps serving the space heating and cooling demands and with an air-to-water heat pump covering the domestic hot water requirements. The investigated locations are the four main cities of Greece, Athens, Thessaloniki, Chania, and Kastoria, which represent the country's four climatic categories. The conducted analysis allows for the mapping of the zero-energy building potential for the climatic data of Greece, demonstrating the possibility of striking a positive building energy balance through the integration of on-site renewable energy sources and the production of necessary electrical energy. The novelty of the present work lies in the identification of a key factor, namely, the building height, which determines the feasibility of transforming multifamily buildings into zero-energy buildings. According to the analysis results, the critical number of stories is calculated at six for Chania, five for Athens, four for Thessaloniki, and two for Kastoria. Regarding a three-story residential building, the incorporation of a renewable photovoltaic system can result in an annual surplus electricity production of 13,741 kWh (Chania), 10,424 kWh (Athens), and 6931 kWh (Thessaloniki), and a corresponding coverage of 100% (Chania), 69.0% (Athens), 38.9% (Thessaloniki) and 0% (Kastoria).

Keywords: zero net electricity; rooftop PV systems; electrification; residential building retrofit; on-site renewable energy

1. Introduction

The residential building sector constitutes 27.9% of the final energy consumption within the European Union, ranking as the third most significant sector in terms of energy demand [1]. For eight out of the twenty-seven European Union member states, the residential sector is characterized as the largest national energy consumer. For these member states, the energy consumption share of residencies varies from 29.3%, for the case of Germany, to 35.1% for the case of Croatia [2]. In Greece, residential energy consumption accounts for 28.8% of the nation's total final energy use [3]. Heating, cooling, and domestic hot water (DHW) preparation make up approximately 79% of households' energy usage in buildings [4].

The electrification of the building sector is strengthened by the adoption of electrically driven energy systems, for instance, heat pumps, to satisfy space heating and cooling and DHW needs [5]. Currently, only a quarter of Europe's residential sector total energy demand is electrically satisfied [1]. The contribution of renewables to electricity generation amounts to 44.2% for the European continent and 43.7% for Greece [6]. The increasing proportion of heat-pump-served buildings [7] combined with the implementation of deep retrofit initiatives aiming at the thermal performance upgrading of buildings act as key enablers for the decarbonization of Europe's energy mix [8]. Additionally, the incorporation of on-site renewable energy systems can aid in the transformation of the existing residential building stock into net-zero-energy and energy-self-sufficient buildings [9]. Zero-energy buildings are characterized by on-site renewable energy production that adequately covers its energy demand for space heating and cooling, DHW production, and electricity for appliances and lighting. Through the use of energy storage systems or the net-metering facility of the electrical grid, zero-energy buildings consume as much as or less than they produce.

Greece is characterized by a high solar potential and an available area for rooftop PV system installation of 128 km^2 [10]. The available PV rooftop installation area could potentially cover more than 30% of the country's final electricity consumption [11]. However, one of the primary obstacles hindering solar development, particularly in administrative processes such as rooftop solar installations in Greece, is the issue of grid availability [12]. Many regions in Greece are experiencing rejections of rooftop solar PV applications due to insufficient electricity grid capacity [12]. Additionally, the adoption of residential rooftop PV systems is interrupted by the relatively high investment cost [13], which according to Sagani et al. [14] is financially viable only over a minimum installed power of 5 kW. Nonetheless, the anticipated decrease in investment costs projected by the scientific community in the upcoming years, along with stable electricity prices, will gradually enhance the profitability of PV roof system installation [15]. According to Greece's energy regulation authority [16], for 2022, the installed photovoltaic power was equal to 5288 MW, while from 2011 to 2022, the PV installed capacity of Greece increased by 687%. The electricity production capability, especially at a residential level, is a driving force in the positive energy transformation of buildings and cities [17]. Regarding domestic electricity production by PVs, in 2023, Greece ranked first in Europe with a percentage (18.4%) more than double the European Union average (8.6%) and more than three times the global average (5.4%) [18]. Figure 1 illustrates the annual electricity production by PVs as well as the CO_2 emissions abated by PVs in Greece from 2013 to 2023. According to the greenhouse gas emission factors for electricity consumption given by the European Commission, for Greece, the most recent calculated emission factor for electricity consumption is 0.411 kg CO_2/kWh [19].

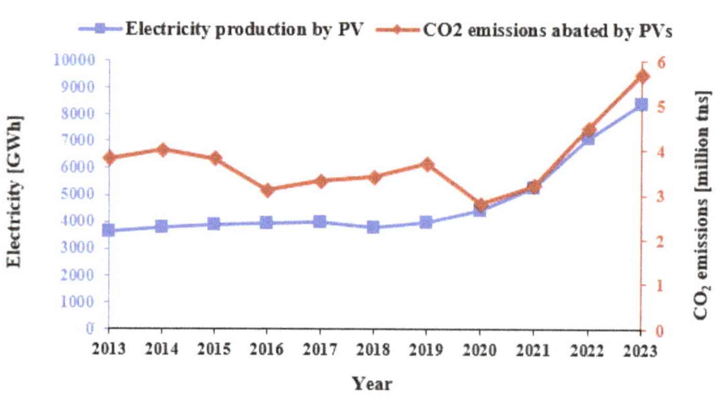

Figure 1. Electricity production by PVs and CO_2 emissions abated by PVs in Greece for 2013–2023 (data were retrieved from Refs. [18,20]).

Building height is a pivotal determinant of the potential of PV systems, impacting both energy consumption and the available area for photovoltaic deployment. The transformation of multi-story residencies into zero-net-energy buildings is a challenge that can be successfully addressed through a holistic energy efficiency improvement plan. For instance, according to the study of Bellos et al. [21], the transformation of a four-story multifamily building in Athens into a zero-energy building first requires a holistic energy renovation of the building thermal envelope and systems to restrict the overall energy demand. In another case, Dermentzis et al. [22] monitored for consecutive years two multifamily buildings in Austria equipped with electrically driven heat pumps and a rooftop PV system. They also underlined that a zero-net-electricity balance is feasible for high-efficiency building constructions, such as the Passive House standard.

Following that, Thebault and Galliard [23] investigated the integration of residential photovoltaic systems on low-, mid-, and high-rise buildings in France. They drew the conclusion that the installation of rooftop PV systems is the financially and energetically optimal strategy for low- and mid-rise buildings, whereas, for high-rise buildings, the incorporation of photovoltaics on the building façade is essential for building self-consumption and self-sufficiency. Solar potential for electrical production significantly determines the possibility of zero-energy classification in buildings. For this reason, Feng et al. [24] examined solar electricity production for the various climatic categories of Chinese cities and evaluated the integration of photovoltaic systems and the potential of achieving zero net electricity demand for multi-story family buildings. According to their analysis, the available solar irradiation and household electricity requirements are decisive factors affecting the PV rooftop performance and its sufficiency to fully serve mid-rise buildings up to seven stories in height.

Regarding the climatic conditions of Greece, researchers have investigated the incorporation of photovoltaic systems as a part of a building energy retrofit strategy. More specifically, Pallis et al. [18] examined various energy retrofit measures of a five-story office building and parametrically assessed their energy and economic performance for the four climatic categories of Greece. They concluded that the installation of photovoltaic systems is considered a cost-effective energy retrofit solution, regardless of the fuel type of building energy systems. The aforementioned conclusion is also drawn by a study by Gaglia et al. [19] regarding the national residential building sector. Other studies have focused on the energy renovation of public buildings through the integration of renewable energy systems coupled with energy storage systems [25], and the establishment of effective retrofit strategies based on the building typology [26]. Additionally, another research study by Sougkakis et al. [27] investigated the feasibility of near-zero or positive energy communities, through the examination of building-scale and community-scale retrofit action, for a case study in the city of Alexandroupolis.

The combinatorial use of photovoltaic systems and battery systems is considered a driving force toward a decarbonized and sustainable residential sector [28]. Additionally, the installation of residential capacity stands as a driving force, and subsequently the absorbed energy from the grid, which may allow for securing public grid stability. In 2019, the total installed residential capacity of battery energy storage systems was equal to nearly 2 GWh, while residential battery energy storage systems in combination with PVs accounted for approximately 7% of the total residential PV systems in operation in Europe [29]. Despite the increasing adaptation of battery energy storage systems, the market potential remains enormous, particularly considering that over 90% of European buildings still lack solar systems [29].

The most popular and commercially available battery systems are the lithium-ion and sodium–nickel chloride batteries [30]. Regarding lithium-ion batteries, according to Liu et al. [31], their superiority over other battery configuration choices lies in their long lifespan, their ability to perform effective charge and discharge cycles, and the fact that they are an appropriate option for off-grid systems. For instance, Forrousso et al. [32], parametrically investigated the sizing of a lithium-ion battery system for various climatic conditions in Morocco, to design a zero-net-energy building and maximize the load coverage factor and the energy autonomy. Additionally, Orth et al. [30] examined 26 lithium-ion battery configurations to determine the sizing, conversion, control, and standby losses in residential-scale use. They identified a significant potential for enhancement in residential PV battery systems, especially with regard to conversion efficiency and standby power consumption. Moreover, Chreim et al. [33] investigated the advantages of individual and shared renewable energy systems integrated with appropriate battery storage systems. They underlined that the sizing parameter is crucial for the optimum performance and economic efficiency of the system, which can be compromised due to the shortage of available solutions in the market.

The present analysis investigates the incorporation of PV roof systems in a typical building typology for a variable number of stories in four different locations. For this purpose, the DesignBuilder software [34] is used for yearly building energy simulation, in combination with accurate weather data extracted from the PVGIS tool, namely, the incident solar irradiation on the PV modules, for the calculation of electricity production by the PV system [35]. The aim of this study is to define the maximum number of stories that allow the transformation of a fully electrified building into a zero-energy building, an issue that has not been resolved for the residential building sector of Greece. The examined building locations are the cities of Chania, Athens, Thessaloniki, and Kastoria, which represent the four different Mediterranean climatic categories of Greece. The examined building is equipped with electrically driven air-to-air heat pumps that serve the building's heating and cooling demands and an air-to-water heat pump for the satisfaction of DHW requirements. The conducted analysis allows for mapping the zero-energy building potential based on climatic data for Greece and introduces the possibility of positive electricity production through the integration of on-site renewable energy sources. The present study can serve as a valuable guideline for the identification of Greece's potential to increase the number of zero-energy buildings and achieve European goals for the decarbonization and electrification of the residential sector. Additionally, the present study takes into consideration the versatility of Greek buildings, the wide range in the number of story levels, as well as the national energy standards. Lastly, the outputs of the present study can serve as a useful guideline for cities with similar climatic conditions and contribute to the correlation between zero-energy potential and climate types.

2. Materials and Methods

2.1. Description of Building

We examine a variable-story building, with the number of stories varying from one to seven, designed and analyzed in DesignBuilder software [34]. The simulation analysis is conducted for an entire year with a six-minute simulation step. Air temperature is set as the control temperature for the operation of the building space heating and cooling energy systems, while the selected solution algorithm is the conduction transfer function. According to the national building inventory [36], the main categories in which buildings are distributed in Greece are single-story buildings (50.5%), two story-buildings (33.3%), low-rise buildings with three to five stories (13.2%), and mid-rise buildings with over six stories (2%). Figure 2a–d depict the case of a three-story building and its geometry in the four directions of the horizon. Every story consists of two apartments with the same area of 75 m^2 and a common space area of 10 m^2, as depicted in Figure 2e, serving the communication between floors. The building is characterized by a rectangular shape of 16 m in length and 10 m in width, a flat roof, and no basement. The windows' total area represents 20% of the building's outer surface, with the greatest window area being located on the south side in the form of doors with floor-length windows. The entire south side has a 16 m length balcony of 1 m width that serves as an overhang for shading purposes. Additionally, on the east and west sides, the windows are also shaded either with a small balcony or an overhang of 0.6 m width.

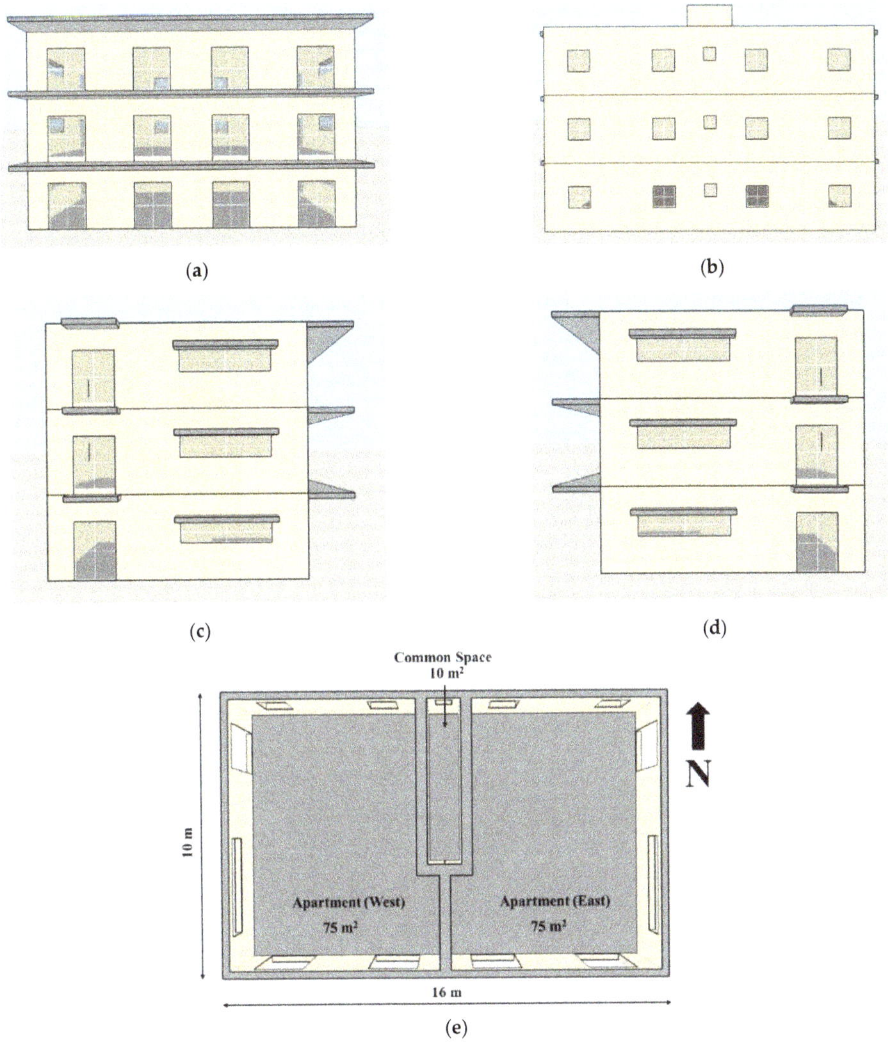

Figure 2. Depiction of the examined topology for the case of the three-story building: (**a**) south side, (**b**) north side, (**c**) west side, (**d**) east side, (**e**) plan view.

The building construction is composed of reinforced concrete, with double brick layered walls. Also, the insulation material used is expanded polystyrene with a thermal conductivity value of 0.045 W/mK, while the windows are air-filled double-glazed. The thermal properties of the building constructions are determined by national regulations according to the climatic category of the examined location. Specifically, the national technical guidelines determine the maximum allowed thermal transmittance values of the building's external walls, roof ground slab, and windows, which are given in Table 1 [37]. In Table 2, the mean thermal transmittance value of the entire building thermal envelope is presented according to the number of floors and city [37]. Additionally, the ratio of the total outer surface to volume and the percentage of the window area are given. The total solar transmittance value of the windows is considered to be 0.6 for every location. Additionally, the solar absorptance and emittance values are considered to be 0.6 and 0.8, respectively, for the external surface of the entire building envelope for every city [37].

Table 1. Total thermal transmittance value (U-value) of the building construction per city.

Building Construction	Chania (Zone A)	Athens (Zone B)	Thessaloniki (Zone C)	Kastoria (Zone D)
External walls	0.55	0.45	0.40	0.35
Roof slab	0.45	0.40	0.35	0.30
Ground slab	1.10	0.80	0.65	0.60
Windows	2.80	2.60	2.40	2.20

Table 2. Mean thermal transmittance value (U-value) of the building per number of floors and city, total outer surface-to-volume ratio, and window area ratio.

Floors	Mean U-Value [W/m^{2}K]				Outer Surface/Volume [m^2/m^3]	Window Area [%]
	Chania	Athens	Thessaloniki	Kastoria		
1	0.67	0.56	0.50	0.44	0.95	27.4
2	0.76	0.65	0.57	0.51	0.64	40.8
3	0.80	0.69	0.62	0.56	0.53	48.8
4	0.84	0.72	0.65	0.58	0.48	54.0
5	0.86	0.75	0.67	0.60	0.45	57.8
6	0.88	0.76	0.69	0.62	0.43	60.6
7	0.89	0.78	0.70	0.63	0.41	62.8

Every apartment is occupied by a three-member family with an occupant load of 80 W [37] and a daily occupancy factor of 75% [38]. The lighting and appliance equipment are characterized by specific loads of 4 W/m^2 and 2 W/m^2 and a mean operating daily factor of 50% [38]. These loads are calculated for the total gross area of the building, which is 160 m^2 per story. Every apartment is served by a reversible unitary air-to-air heat pump and an air source heat pump for the domestic hot water demand. The temperature setpoint for the heating season is set to 20 °C and for the cooling season to 26 °C. The common spaces are not equipped with a heating and cooling system and therefore the gross treated floor area per story corresponds to 150 m^2. The daily demand for domestic hot water (DHW) is set to 50 L per person at a serving temperature of 45 °C. Table 3 summarizes the building's operating data.

Table 3. Building operating parameters.

Parameters	Values
Gross treated floor area per story [m^2]	150
Occupancy [persons/apartment]	3
Occupant load [W/person]	80
Lighting specific load [W/m^2]	4
Lighting operation fraction [%]	50
Appliances specific load [W/m^2]	2
Appliances operation fraction [%]	50
DHW daily demand [liters/person]	50
DHW temperature set-point [°C]	45
Heating temperature set-point [°C]	20
Heating temperature set-point [°C]	26
Nominal COP/EER for heating/cooling mode	4.0

The modeling of the building energy systems is based on the DesignBuilder performance curves [34]. Regarding the heating and cooling energy system, the selected heating, ventilation, and air conditioning (HVAC) template is a unitary air-to-air heat pump with a nominal coefficient of performance (COP) and a nominal energy efficiency ratio of 4.0 for the heating and cooling modes, respectively. The nominal heating and cooling capacity of the heat pump is auto-sized to effectively serve the heating and cooling needs for every examined location. The DHW analysis is executed with the HVAC template of the water heater heat pump, with a selected nominal COP of 4.0 [39].

The building rooftop is covered by a solar photovoltaic system for electricity production. The examined photovoltaic technology comprises monocrystalline cells with a module surface area of 1.94 m^2 and a nominal electricity generation of 360 W [40]. In Table 4, the technical characteristics of the examined photovoltaic modules are given. The photovoltaic panels are installed with an azimuth angle of 0°, direction to the south, and the optimal per-city slope, as given in Table 5. The maximum possible installed nominal electrical power is calculated at 16.56 kW, as depicted in Figure 3. The electrical losses of the conversion of the variable direct current output of the PV solar panels into a utility frequency alternating current are set to 5%.

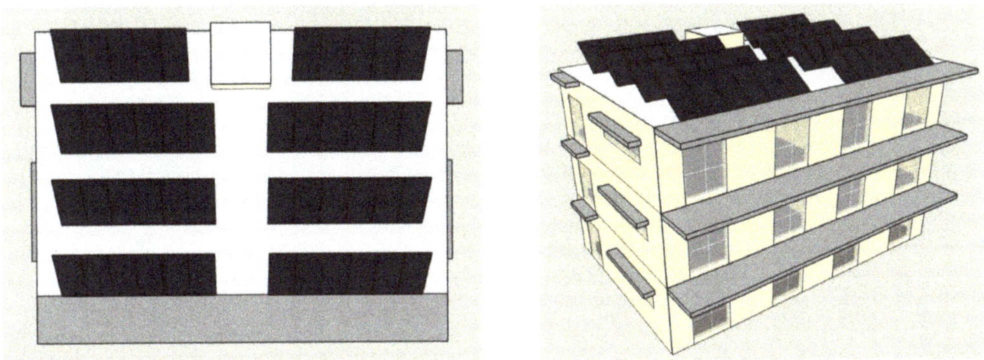

(**a**) (**b**)

Figure 3. (**a**) Plan view and (**b**) axonometric view of the roof photovoltaic system at the maximum installed power in the case of a three-story building.

Table 4. Technical characteristics of examined photovoltaic panel system.

Parameters	Values
Maximum power, P_{max} [W]	360
Open-circuit voltage, V_{oc} [V]	47.2
Short-circuit current, I_{sc} [A]	9.79
Voltage at point of maximum power, V_{mpp} [V]	38.9
Current at point of maximum power, I_{mpp} [A]	9.26
Module reference efficiency, η [%]	18.5
Module length [m]	1.956
Module width [m]	0.922
Nominal operating cell temperature [°C]	44
Temperature coefficient of P_{max} [%/°C]	−0.39
Temperature coefficient of V_{oc} [%/°C]	−0.29
Temperature coefficient of I_{sc} [%/°C]	0.05

Table 5. Basic climatic parameters of each examined location.

City	HDDs	CDDs	Mean Yearly Temperature [°C]	Mean Yearly Water Temperature [°C]	Horizontal Irradiation [kWh/m^2]	Optimum PV Panel Slope [°]	Maximum PV Panel Irradiation [kWh/m^2]
Chania	648	308	18.3	18.6	1831	28	2009
Athens	1041	588	18.0	17.8	1833	33	2081
Thessaloniki	1793	314	15.0	15.6	1648	35	1904
Kastoria	2726	134	11.6	13.7	1580	34	1808

2.2. Climatic Conditions

The present study examines a multifamily building of typical reinforced concrete-brick construction in the environmental conditions of four main Greek cities, which represent the four climatic categories of Greece. Table 5 gives the basic climatic parameters of the four examined cities extracted from the PVGIS tool [35], including the heating degree days (HDDs) and cooling degree days (CDDs), the yearly mean air temperature, the yearly mean water temperature, and the yearly specific horizontal irradiation. Regarding the installation of photovoltaic panels, the optimum PV panel slope is also given, as well as the respective yearly specific in-plane irradiation. Additionally, in Appendix A, the average daily ambient air temperature and the total solar daily irradiation are illustrated in Figure A1 for each of the four examined cities.

2.3. Simulation Strategy

The present analysis investigates the installation of a roof photovoltaic system for a typical building construction with one to seven stories in height. The building is located in one of four different Greek cities, which represent the four different climatic categories of Greece. The energy systems serving the building space heating and cooling and the DHW preparation are electrically driven and therefore the building produces no on-site carbon emissions. The aim of this study is to define the maximum number of stories that the photovoltaic system can fully serve or, in other words, the maximum number of stories for which the building presents zero net electricity demand. Validation evidence proving the credibility of the models used in the present study was given in Ref. [39]. Additionally, regarding the simulation of the photovoltaic models and the electricity production, the calculations of the DesignBuilder [34] software have been compared to the results provided by the PVGIS tool [35], demonstrating deviations of less than 1.0%.

2.4. Basic Mathematical Formulation

The thermal transmittance (U-value) of the building constructions is calculated as [37]

$$U = \frac{1}{h_{out} + \sum \frac{t_i}{k_i} + h_{in}} \tag{1}$$

where (k) stands for thermal conductivity [W/m·K] and (i) represents the material layers that each building construction comprises. The coefficients of convective heat transfer are selected from ISO 6946:2017 [41] and the national technical guidelines [37]. For instance, for the horizontal roof construction, the convective heat transfer coefficient is equal to h_{in} = 7.7 W/m^{2}K for the internal surface, and to h_{out} = 25 W/m^{2}K [41] for the external surface.

The coefficient of performance (COP) of the unitary reversible heat pump for the heating mode is calculated as the proportion of the heat pump's instantaneous heating thermal power input to electricity consumption:

$$COP = \frac{Q_{heat}}{P_{el_{heat}}} \tag{2}$$

Similarly, the energy efficiency ratio (EER) of the cooling mode is calculated by the proportion of the instantaneous cooling thermal power to electricity consumption:

$$\text{EER} = \frac{Q_{cool}}{P_{el_{cool}}} \tag{3}$$

The seasonal coefficient of performance (SCOP) and seasonal energy efficiency ratio (SEER) represent the yearly average COP and EER values of the unitary heat pump operation and are calculated based on the corresponding yearly energy loads.

$$\text{SCOP} = \frac{E_{heat}}{E_{el_{heat}}} \tag{4}$$

$$\text{SEER} = \frac{E_{cool}}{E_{el_{cool}}} \tag{5}$$

The building is fully electrified because its energy needs for space heating and cooling, DHW, lighting, and appliances are all electrically satisfied. The building adheres to the net metering billing mechanism and therefore all the electricity produced by its PV system is considered to be firstly exploited for the building's own energy needs. The building's total gross demand for electricity ($E_{el,gross}$) is calculated as the sum of its electricity demand for space heating and cooling, DHW preparation, lighting, and appliances needs.

$$E_{el_{gross}} = E_{el_{heat}} + E_{el_{cool}} + E_{el_{DHW}} + E_{el_{lighting}} + E_{el_{appliances}} \tag{6}$$

The net electricity demand ($E_{el,net}$) is calculated by subtracting the PV-generated electricity (E_{PV}) from the gross electricity demand:

$$E_{el_{net}} = E_{el_{gross}} - E_{PV} \tag{7}$$

3. Results

3.1. Energy Analysis of the Examined Buildings in Different Locations

This section presents the calculations regarding the thermal loads and energy behavior of the examined building typology for a variable number of floors (one to seven floors) and for the four Greek cities of Chania, Athens, Thessaloniki, and Kastoria. Specifically, Figures 4–7 illustrate the specific heating and cooling energy demand according to the number of floors and the specific electricity demand for heating and cooling according to the number of floors for the four examined locations. Also, in Appendix B, the thermal analysis calculations for each city are given in tables.

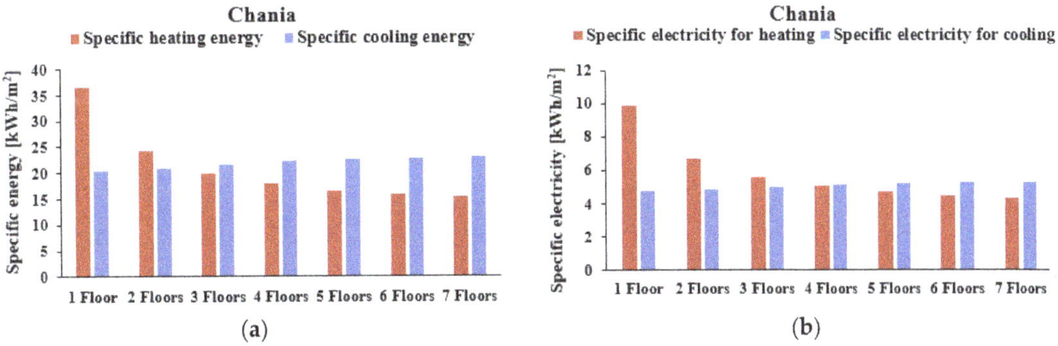

Figure 4. (**a**) Specific heating and cooling energy demand, and (**b**) specific electricity demand for heating and cooling, for the city of Chania.

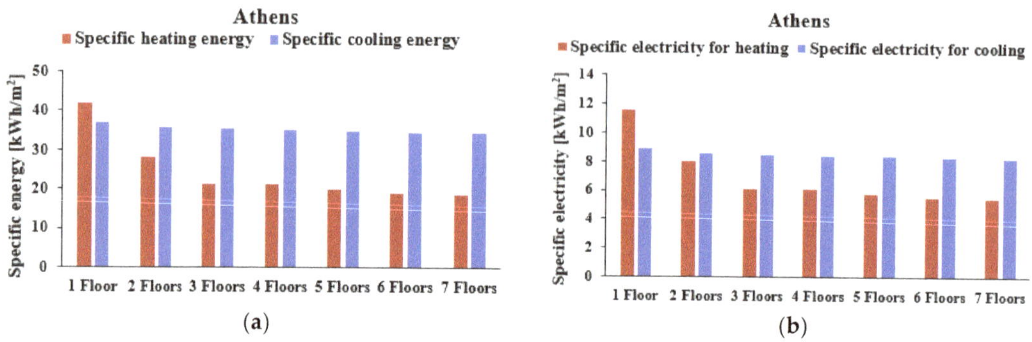

Figure 5. (**a**) Specific heating and cooling energy demand, and (**b**) specific electricity demand for heating and cooling, for the city of Athens.

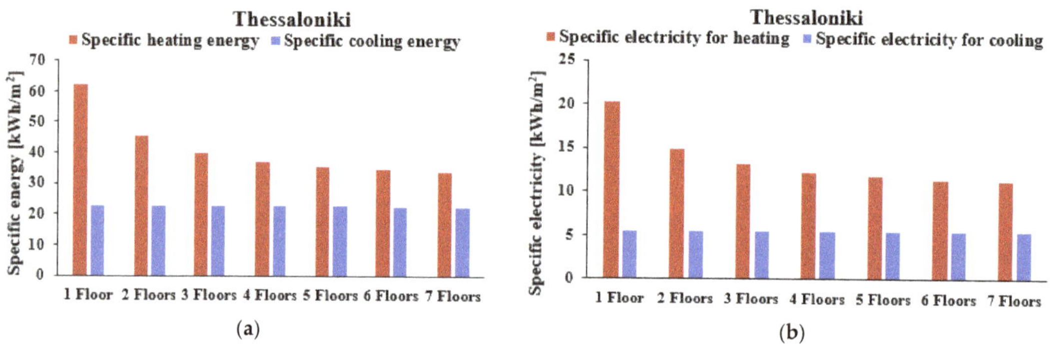

Figure 6. (**a**) Specific heating and cooling energy demand, and (**b**) specific electricity demand for heating and cooling, for the city of Thessaloniki.

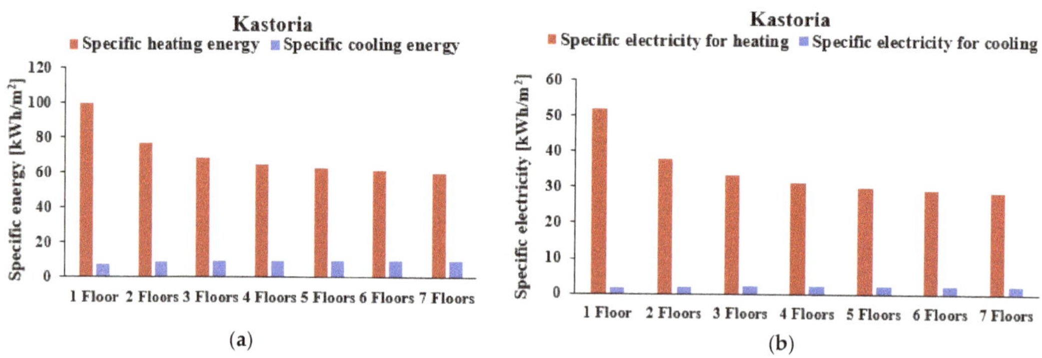

Figure 7. (**a**) Specific heating and cooling energy demand, and (**b**) specific electricity demand for heating and cooling, for the city of Kastoria.

Firstly, for the climatic conditions of Chania, the city with the highest yearly average ambient air temperature (18.3 °C) and lowest HDDs among all examined locations, the specific heating energy demand is given in Figure 4, which presents ranges between 15.32 kWh/m^2 for the case of the seven-story building and 36.39 kWh/m^2 for the case of the single-story building. The electricity demand for heating is calculated as between 4.33 kWh/m^2 and 9.86 kWh/m^2. For the case of Chania, the mean COP of the unitary air-to-air heat pumps is calculated as 3.59. The specific cooling energy demand varies between 20.44 kWh/m^2 for the case of the single-story building and 22.97 kWh/m^2 for the case of the seven-story building. It is observed that the specific cooling energy demand presents a reverse behavior in contrast to the specific energy demand for heating, namely, that the increase in the number of floors leads to an increase in the specific energy demand for cooling and to a decrease in the specific energy demand for heating. This is explained by the fact that the ratio of the window area to the building's total outer surface increases with the addition of stories. Specifically, for the case of a single-story building, the respective ratio is 27.4%, while for the seven-story building, it reaches 62.8%. This fact, in combination with the high total solar irradiation observed in Chania, has an adverse effect on the building's cooling energy demand but a positive result regarding the heating energy demand. The specific electricity demand for cooling varies between 4.75 kW/m^2 for the case of the single-story building and 5.29 kWh/m^2 for the case of the seven-story building. The mean EER of the unitary air-to-air heat pumps is calculated at 4.33.

For the case of Athens, both the heating and cooling energy demand are increased, as depicted in Figure 5, in comparison with the thermal loads of Chania. Specifically, the specific heating energy demand varies between 18.69 kWh/m^2 for the seven-story building and 41.73 kWh/m^2 for the single-story building. The electricity demand for heating is calculated as 5.42 kWh/m^2 for the seven-story building and peaks at 11.58 kWh/m^2 for the case of a single floor. The mean COP value for the heat pump heating mode in Athens is 3.51. This value is slightly decreased in contrast to Chania, which is a result we expected. Athens is characterized by the highest CDDs among all examined locations and therefore demonstrates the highest energy demand for space cooling. Precisely, the specific cooling energy demand for the city of Athens is found to vary between 34.41 kWh/m^2 for the seven-story building and 37.09 kWh/m^2 for a single floor. The electricity demand for cooling is calculated to vary between 8.92 kWh/m^2 and 8.30 kWh/m^2. The mean EER value for the heat pump's cooling mode is 4.15.

For the city of Thessaloniki, as shown in Figure 6, the annual average ambient air temperature is 3.0 K and 3.3 K lower than the respective values of Athens and Chania. The specific heating energy demand varies between 18.69 kWh/m^2 for the seven-story building and 41.73 kWh/m^2 for the single-story building, while the corresponding electricity demand for heating is calculated to be in the range of between 11.58 kWh/m^2 and 5.42 kWh/m^2. For the city of Thessaloniki, the mean COP value of the unitary heat pump heating mode is 3.04, which is distinctively lower in comparison to the cities of Chania and Athens. The specific cooling energy demand for the city of Thessaloniki is between 22.49 kWh/m^2 for a seven-story building and 22.77 kWh/m^2 for a single-story building. These results are comparable with the corresponding values for Chania, which was anticipated because both cities demonstrate similar values of CDDs. The electricity demand for cooling for Thessaloniki is calculated as between 5.41 kWh/m^2 and 5.45 kWh/m^2, presenting almost no fluctuation. The mean EER value regarding the cooling mode of the air-to-air heat pumps is 4.16.

Following that, Figure 7 illustrates the thermal energy load calculations for the city of Kastoria, which presents the coldest climatic conditions among the examined locations. Firstly, the specific heating energy demand is calculated as 59.94 kWh/m^2 for the seven-story building and 99.28 kWh/m^2 for the most energy-demanding case of the single-story building. The respective specific electricity demand for heating is in the range of between 28.60 kWh/m^2 and 51.83 kWh/m^2, and the mean COP value is 2.05. On the other hand, the specific cooling energy is significantly lower, since Kastoria demonstrates the mildest summer conditions among all examined locations Specifically, for a single-story building, the specific cooling energy demand is 7.87 kWh/m^2, while for the most energy-demanding case of seven stories, it is 10.07 kWh/m^2. The corresponding specific electricity demands are 1.93 kWh/m^2 and 2.44 kWh/m^2, while the mean EER value is 4.11. Conclusively, the case of the three-story building and the city of Athens, for which the electricity demand for heating/cooling is equal to 6.10/8.51 kWh/m^2, is taken as a reference to facilitate a comparison among all examined locations. The variations in electricity demand for heating/cooling are equal to −8.52%/−41.13% (Chania), 115.41%/−36.08% (Thessaloniki), and 448.52%/−72.74% (Kastoria).

Figure 8 compares and contrasts the electricity demands for heating and cooling, as well as the aggregated electricity demand for both heating and cooling. The electricity demand values are presented in relation to a building's geometrical parameter, which concerns the ratio of its total outer surface, namely the surfaces of external walls, the roof, and the ground slab, divided by its volume. The greatest value of 0.95 m^2/m^3 corresponds to the single-story building, while the lowest value of 0.41 m^2/m^3 corresponds to the seven-story building. According to Figure 8a, the building with the highest energy demand during the heating season is located in Kastoria, with electricity consumption ranging from 7774 kWh to 30,034 kWh. The second most energy-demanding city is Thessaloniki, with an electricity demand for heating in the range of 3037 kWh to 11,759 kWh. Athens ranks third, with its electricity usage for heating falling between 1737 kWh and 5686 kWh. Chania follows closely with a demand ranging from 2014 kWh to 4545 kWh.

On the other hand, according to Figure 8b, Athens is characterized by the highest electricity demand for cooling, ranging from 1338 kWh to 8711 kWh. Following this, the electricity usage for cooling in the cases of Thessaloniki and Chania is similar, ranging from 817 kWh to 5680 kWh for Thessaloniki and from 712 kWh to 5559 kWh for Chania. Lastly, for the city of Kastoria, the electricity demand for cooling varies from 289 kWh to 10,104 kWh. Figure 8c illustrates the aggregation of electricity demand for both heating and cooling, highlighting Kastoria as the most energy-demanding case study. This figure allows a direct comparison among the four examined locations and the number of stories. Regarding the case of the single-story building, the total electricity demand for heating and cooling ranges from 2191 kWh (Chania) to 8063 kWh (Kastoria), a difference that is equivalent to a 368% increase. Conversely, as the number of stories increases, the disparity between the most and least energy-demanding locations diminishes. Specifically, for Chania, the total electricity demand is calculated at 10,104 kWh, compared to 32,599 kWh for Kastoria, resulting in a more moderate 323% increase. This result highlights the fact that single-story buildings are the most vulnerable to ambient conditions. However, as the number of stories increases, the impact of the ambient conditions is slightly mitigated.

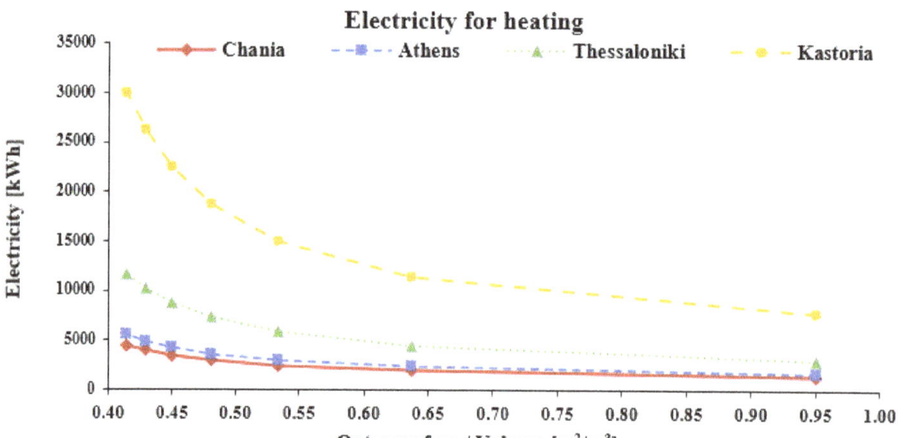

(a)

Figure 8. *Cont.*

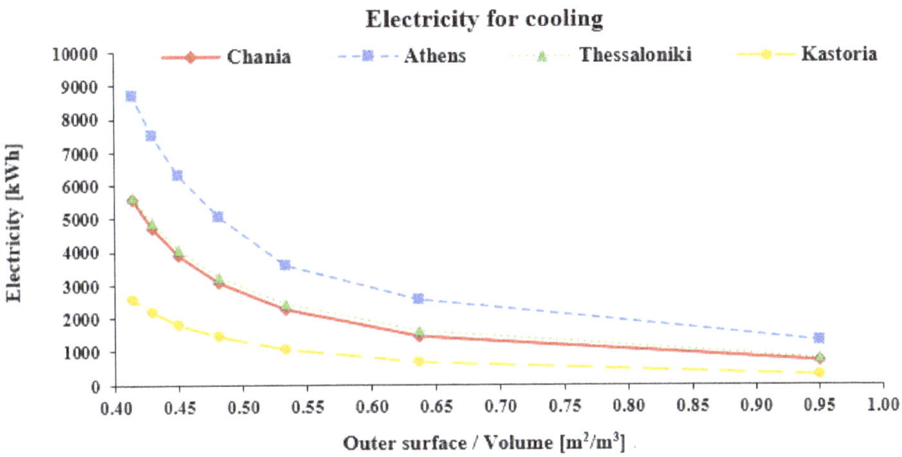

(b)

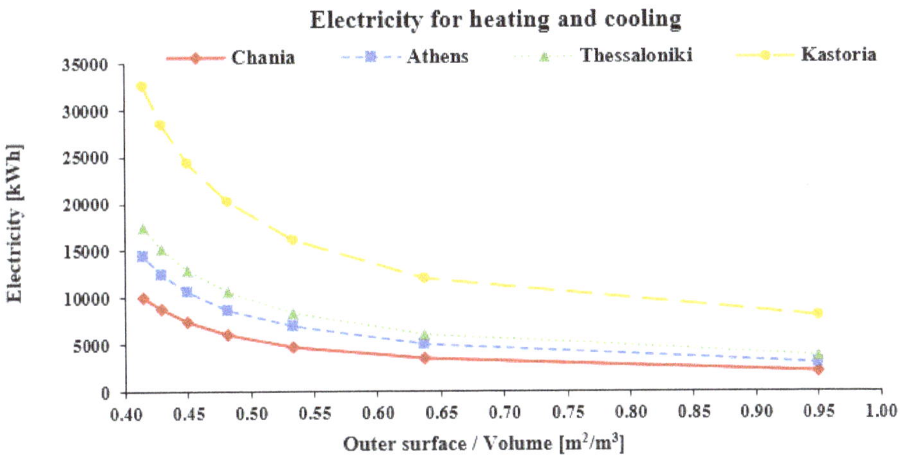

(c)

Figure 8. (**a**) Electricity demand for heating, (**b**) electricity demand for cooling, and (**c**) electricity demand for both heating and cooling, for the four examined locations.

Regarding the energy demand for the preparation of domestic hot water, Figure 9 depicts the energy input required per floor per city, as well as the necessary electricity demand. These values are presented according to the number of floors because it is considered that every floor is occupied by a specific number of people. The deviations between the DHW energy demand per city are due to the difference in the yearly mean water temperature of water. Precisely, according to Table 5, the mean yearly water temperature is equal to 18.6 °C (Chania), 17.8 °C (Athens), 15.6 °C (Thessaloniki), and 13.7 °C (Kastoria). According to Figure 9, for the city of Chania, the DHW energy demand is calculated as 3248 kWh per floor, for Athens as 3338 kWh per floor, for Thessaloniki as 3603 kWh per floor, and for Kastoria as 3847 kWh per floor. The corresponding electricity demand for DHW preparation is equal to 902 kWh per floor for Chania, 951 kWh per floor for Athens, 1185 kWh per floor for Thessaloniki, and 1895 kWh per floor for Kastoria.

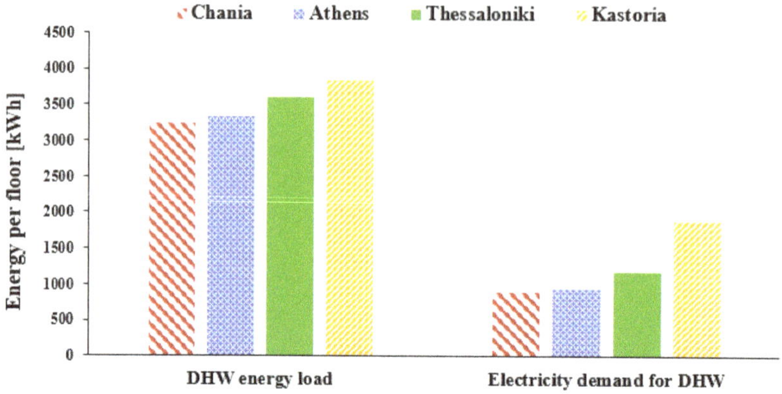

Figure 9. DHW energy load and electricity demand per floor for the four examined locations.

3.2. Analysis of PV Electrical Production Potential

In this section, the potential of the electricity production of the roof photovoltaic system is analyzed for each of the four examined locations. The installation of rooftop PVs is investigated with the aim of transforming low- and mid-rise multifamily buildings into zero-net-energy buildings. The dense urbanization of Greek cities, especially Athens, highlights the roof area as the only available building area for the installation of on-site renewable energy systems. The dense urban development in combination with the complex architecture of the Greek building sector, characterized by a lack of uniformity and the abundancy of balconies and semi-closed spaces [42], mean building-integrated—for instance, façade-integrated—photovoltaic systems are not appropriate solutions.

The yearly specific horizontal irradiation, as well as the optimum titled angle of the PV panels, are equal to 1831 kWh/m^2 and 28° (Chania), 1833 kWh/m^2 and 33° (Athens), 1648 kWh/m^2 and 35° (Thessaloniki), and 1580 kWh/m^2 and 34° (Kastoria), according to Table 5. The entire roof is equipped with PV panels of nominal power equal to 16.56 kW. Figure 10 presents the maximum electricity production in contrast to the gross electricity demand for a number of total floors and locations. This figure allows us to identify the maximum number of floors that can be fully served by the PV system and lead to a zero net electricity demand or positive electricity production. Specifically, for the city of Chania, the electricity production by the PV system is 27,098 kWh, for Athens, it is 25,521 kWh, for Thessaloniki, it is 24,730 kWh, and for Kastoria, it is 24,638 kWh. The calculated electricity production can fully satisfy the electrical requirements of a six-story building in Chania, a five-story building in Athens, a four-story building in Thessaloniki, and a two-story building in Kastoria. Following on from this, Figure 11a illustrates the surplus electricity production per number of floors per city, and Figure 11b the specific positive electricity production per one floor with the considered floor area of 150 m^2. More specifically, according to Figure 11, for the critical number of floors, the positive electricity production is calculated at 1155 kWh or 7.7 kWh/m^2 (Chania, six floors), 354 kWh or 3.36 kWh/m^2 (Athens, five floors), 1517 kWh or 10.11 kWh/m^2 (Thessaloniki, four floors), and 4885 kWh or 32.57 kWh/m^2 (Kastoria, two floors). Additionally, Table 6 summarizes the gross electricity demand, including the electricity demand for space heating and cooling, DHW preparation, lighting, and appliances needs, as well as the surplus electricity production per number of floors and location. It should be noted that the integration of the photovoltaic system into the building rooftop and its shading effect on the rooftop are taken into consideration during the thermal analysis to effectively calculate the building energy demand for heating and cooling. PV modules operate as shading components that decrease the amount of solar irradiation intercepted by the building rooftop and therefore alleviate solar heat gains for the building interior.

Table 6. Gross electricity demand and positive electricity production per city and number of floors.

Parameters	Number of Floors						
	1	2	3	4	5	6	7
				Chania			
Gross electricity demand [kWh]	5054	9188	13,358	17,544	21,741	25,943	30,147
Electrical positivity [kWh]	22,044	17,910	13,741	9554	5357	1155	0
				Athens			
Gross electricity demand [kWh]	5987	10,808	15,097	20,380	25,167	29,955	34,781
Electrical positivity [kWh]	19,534	14,713	10,424	5141	354	0	0
				Thessaloniki			
Gross electricity demand [kWh]	7000	12,395	17,799	23,213	28,624	34,043	39,462
Electrical positivity [kWh]	17,730	12,334	6931	1517	0	0	0
				Kastoria			
Gross electricity demand [kWh]	11,919	19,753	27,667	35,594	43,615	51,595	59,592
Electrical positivity [kWh]	12,719	4885	0	0	0	0	0

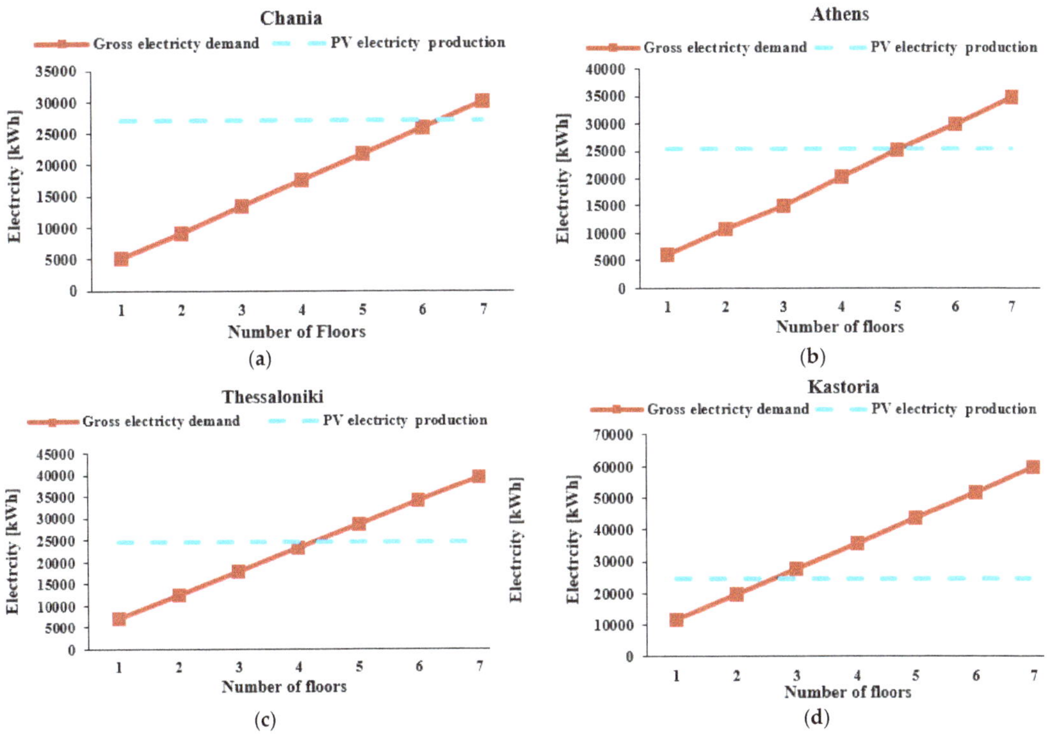

Figure 10. Gross electricity demand per number of floors and maximum PV electricity production for the cities of (**a**) Athens, (**b**) Athens, (**c**) Thessaloniki, and (**d**) Kastoria.

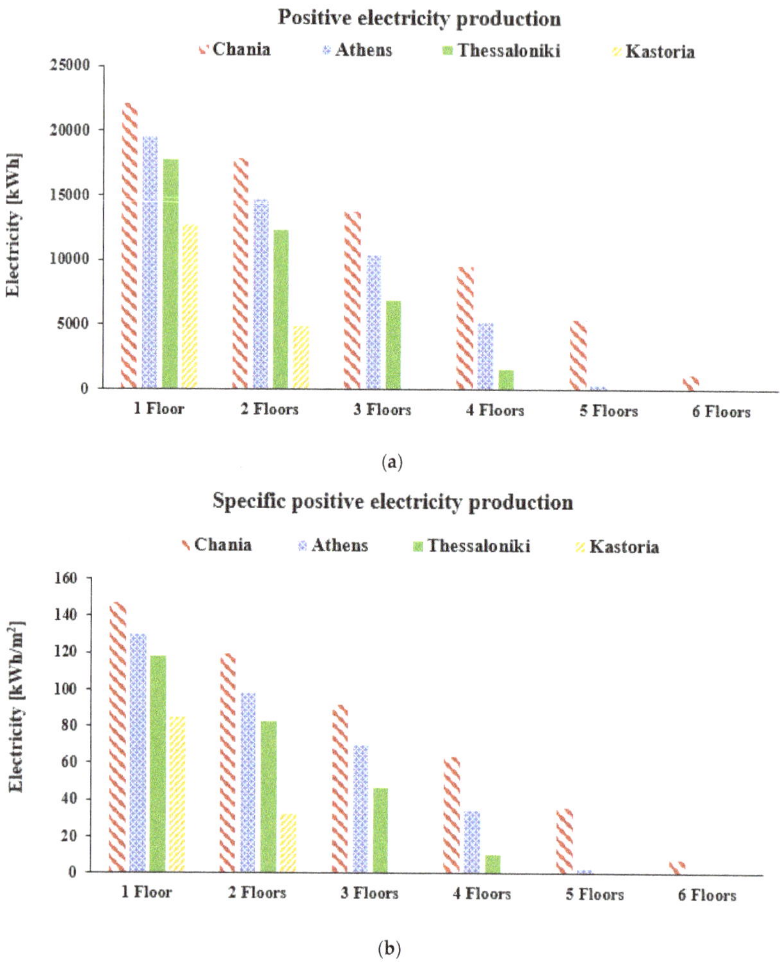

Figure 11. (**a**) Positive electricity production and (**b**) specific positive electricity production per one-floor area.

4. Discussion

Positive energy districts (PEDs) are urban areas of positive energy equilibrium, comprising highly efficient buildings and integrated renewable energy systems for energy production [43]. The concept of positive energy buildings lies in the need to create sustainable cities, with a low carbon footprint, able to positively contribute to the energy supply of their surroundings. Addressing the challenges of converting a city to an energy-self-sufficient and sustainable urban environment requires the adoption of multiple, custom-made solutions that apply not only to city-scale restrictions but also to the national energy mix and energy strategy for security and resilience [44]. Despite the indisputable advantages of positive energy districts, their development is restricted to a research level, with only a few case studies within projects with European funding being applied [45].

Regarding the case of Greece, currently, three cities are engaged in international projects of developing exemplar energy districts utilizing concepts of sustainable energy deployment: Heraklion [46], Trikala [47], and Thessaloniki [48]. The located efforts contribute to the country's successful implementation of its national energy and climate plan goal, setting a cornerstone for future urban development goals. As in the majority of similar positive energy district projects [45], the transformation of a city's building stock into ZEBs is a pivotal element for the feasibility of a PED [17]. It has been proven that the combinational adoption of passive or building envelope retrofitting techniques, as well as the installation of highly efficient active energy systems, supports PEDs as an achievable and economically viable target [49]. In that direction, innovative building materials with advanced thermal and optical properties, highly efficient energy systems for heating, cooling, and domestic hot water production, renewable energy systems for energy production, as well as energy storage systems need to be appropriately utilized to enhance buildings and subsequently a city's energy performance and footprint.

In this study, the parameter of a building's height or number of stories was investigated concerning the transformation potential into zero-net-energy buildings. Greece is characterized by an old, energy-consuming building stock with immense potential for energy savings [50]. According to the thermal analysis results, the mean annual specific energy demand of a residential building was calculated for heating at 20.92 kWh/m^2 (Chania), 24.29 kWh/m^2 (Athens), 41.21 kWh/m^2 (Thessaloniki), and 70.35 kWh/m^2 (Kastoria), and for cooling at 21.95 kWh/m^2 (Chania), 35.32 kWh/m^2 (Athens), 22.62 kWh/m^2 (Thessaloniki), and 9.48 kWh/m^2 (Kastoria). These results are corroborated by various studies that examine the thermal performance of residential buildings in the Mediterranean climate. For instance, in a study by Ascione et al. [49], two three-floor villas equipped with a reversible air-source electric heat pump with a nominal COP/EER equal to 2.2 and 2.0 are examined for the climatic conditions of Greece and South Italy. For the newly constructed villa in Greece, the annual specific electricity demand for heating/cooling is 20.0 kWh/m^2/ 34.0 kWh/m^2, while for the ancient villa in South Italy, the corresponding values are 60.0 kWh/m^2 and 17.0 kWh/m^2. Furthermore, in a previous study concerning a flat roof single-family property in Athens, the specific energy demand for heating was found to range between 18.09 kWh/m^2 and 31.02 kWh/m^2, while for cooling, it was between 31.65 kWh/m^2 and 37.21 kWh/m^2 [39]. Regarding the colder Mediterranean climate of Barcelona, the yearly specific heating demand for a mid-rise multifamily building ranges between 28.0 kWh/m^2 and 68.0 kWh/m^2, while the specific cooling energy demand is between 1.0 kWh/m^2 and 6.0 kWh/m^2 [51]. Finally, the specific heating/cooling energy demands for a deeply renovated four-story multifamily building in Athens are found to be 11.0 kWh/m^2 and 24.4 kWh/m^2, respectively [21]. In the latter study, a photovoltaic system is installed on an available rooftop area of 160 m^2, converting the multifamily building into an energy-positive building with a yearly electricity production of 24.4 MWh or 152.4 kWh per m^2 of available rooftop area. In the present study and for the case of Athens, the respective figures were found to be 25.5 MWh and 159.5 kWh per m^2 of available rooftop area.

Despite the fact that low-rise buildings are the most energy-consuming type of residential buildings [52], they are characterized by an analogically larger available rooftop area for the installation of photovoltaic systems for energy production in comparison to mid- or high-rise buildings. In other words, the analogy of available rooftop area for electricity-production exploitation in comparison to the total treated floor area of a building is a key factor for its successful transformation into a zero-energy building. This aspect poses a question and a serious challenge of whether cities of high population and building density can be successfully converted into PEDs. Comprehensive research regarding the role of the number of building stories in impacting the feasibility of creating positive energy districts needs to be conducted.

Developing the equipment of the building sector with electrically driven heat pumps is proven to be a necessary step towards the decarbonization of the sector and the increase in its energy efficiency. In contrast to carbon-fueled energy systems, heat pumps, characterized by a high coefficient of performance, when combined with the enhancement of a building's thermal behavior, result in the restricted consumption of electrical energy. The necessary electricity can be derived from on-site electricity production through the installation of a renewable photovoltaic system on a building rooftop, securing a building's energy autonomy and low energy footprint.

The present work focused on four main cities of Greece, which represent the country's four distinct climatic categories. According to the present study's results, the coverage of the available rooftop area with a photovoltaic system can satisfy the yearly electricity demand of a variable number of stories of a multifamily building. The critical number of stories is directly connected to the meteorological conditions of a building's location, and the findings of the present study can be useful inputs for a future study in which a correlation between the zero-energy potential of multi-story residencies and climate types could be identified. Regarding the outputs of the present study, the maximum number of stories is six for the moderate winter and summer conditions of the city of Chania, characterized by a mean yearly ambient temperature of 18.3 °C and an annual PV in-plane solar irradiation of 1999.3 kWh. For cooling-dominant Athens, where the mean yearly ambient temperature and annual PV in-plane solar irradiation are equal to 18.0 °C and 2070.3 kWh, respectively, the maximum height of a zero-energy building is five stories. For the colder climates of Thessaloniki and Kastoria, where the mean ambient temperature and annual PV-incident irradiation are 15.0 °C and 11.6 °C and 1905.9 kWh and 1789.9 kWh, respectively, the maximum multifamily building height is calculated at four or two stories. These results can be used as a guideline for cities that demonstrate similar climatic characteristics, namely, similar cooling and heating degree days, yearly solar irradiation, and ambient temperature, as well as the PV electricity production potential.

The widespread and successful integration of photovoltaic renewable systems in Greece's national energy mix concerns both smaller- as well as larger-scale installations. In this direction, the Greek Ministry of Environment and Energy issued a public call with its EUR 238 million subsidy program (from 20% to 100% of the investment cost) for rooftop photovoltaics that either support household applications combined with a storage system, or professional farmers with or without the combination of a storage system, for self-consumption with the application of net metering [53]. Regarding large-scale electricity production and storage projects, Greece has committed to completing by mid-2025 two important constructions of photovoltaic units with an individual capacity of 252 MW, in combination with integrated molten-salt thermal storage units, an extra-high-voltage substation, and a 309 MW photovoltaic unit with an integrated lithium-ion battery energy storage system [54]. These projects are predicted to increase the annual net renewable electrical energy in Greece by 1.2 TWh, increasing the country's capacity in energy storage and thereby enhancing grid stability and availability [54]. Additionally, two important European Union-funded projects of grid interconnections between Greece's mainland and islands (Cyclades and Crete) are being realized to combat grid unreliability and high electricity costs [55]. Lastly, two important cornerstones in the substantial improvement of Greece's electrical grid should inarguably involve upgrade works to the current infrastructure as well as the liberation of the national electricity market [56].

5. Conclusions

The present analysis has investigated the integration of rooftop photovoltaic systems for on-site electricity production and the transformation of residential buildings into zero-net-energy buildings. For this purpose, low- and mid-rise residential buildings of one to seven stories were simulated for the four climatic categories of Greece. The aim of the study was to identify the zero-net-energy potential of residencies in Greece by taking into consideration the variability of the residential sector and the national energy efficiency guidelines. The conclusions drawn in this study are summarized below:

- Full electrification can be achieved through the installation of electrically driven air-to-air heat pumps for serving space heating and cooling and of an air source heat pump for domestic hot water preparation. The energy performance of the energy systems, as well as the heating, cooling, and domestic hot water energy demand, are mainly determined by the climatic conditions of a location.
- For Athens and the case of the three-story building, the gross electricity demand is calculated at 33.55 kWh/m^2, representing an 11.5% increase compared to Chania.

Conversely, for Thessaloniki, the total electricity demand is 17.9% higher than in Athens, and for Kastoria, it is 83.3% higher than in Athens.
- Zero-net-energy building transformation is feasible for fully electrified low-rise, up-to-two-story, multifamily buildings in each of the four climatic categories of Greece.
- The installation of photovoltaic systems on the available rooftop space of multifamily buildings can successfully satisfy the overall electricity demand of up to six stories for Chania, up to five stories for Athens, up to four for Thessaloniki, and up to two for Kastoria.
- Positive electricity production is restricted with the increase in stories. Greater electricity production is achieved for the milder climatic conditions of Chania, with Athens, Thessaloniki, and, lastly, Kastoria following.

The installation of rooftop PV is an effective solution for the achievement of energy self-sufficiency in buildings. Additionally, for the densely developed main cities of Greece, rooftops are the only option for successfully integrating photovoltaic systems into the building envelope without diminishing their energy performance due to uncontrolled shading from closely placed neighboring buildings. In a future study, the zero-net-energy transformation of high-rise residential buildings needs to be addressed. For the respective case, alternative photovoltaic solutions in combination with stricter and energy-efficient standards need to be examined. Addressing these issues is of pivotal importance to the development of positive energy building stocks, which is a prerequisite for positive energy cities.

Author Contributions: Conceptualization, A.K., D.P. and E.B.; methodology, A.K., D.P. and E.B.; software, A.K. and D.P.; investigation, A.K. and D.P.; writing—original draft preparation, A.K., D.P., E.B. and C.T.; writing—review and editing, A.K., D.P., E.B. and C.T.; supervision, E.B. and C.T. All authors have read and agreed to the published version of the manuscript.

Funding: This research received no external funding.

Data Availability Statement: Data available after request.

Conflicts of Interest: The authors declare no conflict of interest.

Nomenclature

h_{in}	Inside convection coefficient, W/m^2K
h_{out}	Outside convection coefficient, W/m^2K
I	Current, A
k	Thermal conductivity, W/mK
P_{el}	Electrical power, W
Q	Load, W
t	Thickness of construction layers, m
U-value	Thermal transmittance value, W/m^2K
V	Voltage, V
Subscripts and superscripts	
max	Maximum
mpp	Maximum power point
heat	Heating
cool	Cooling
oc	Open circuit
sc	Short-circuit
Abbreviations	
CDDs	Cooling degree days
COP	Coefficient of performance for heating mode
DHW	Domestic hot water
EER	Energy efficiency ratio for cooling mode
HDDs	Heating degree days

HVAC	Heating, ventilation, and air conditioning
PED	Positive energy district
SCOP	Seasonal coefficient of performance for heating mode
SEER	Seasonal energy efficiency ratio for cooling mode

Appendix A. Weather Data

This appendix includes the basic weather data of the examined locations. Figure A1 depicts the ambient temperature and the total solar irradiation on the horizontal surface, while Figure A2 illustrates the monthly PV in-plane solar irradiation for the four studied locations.

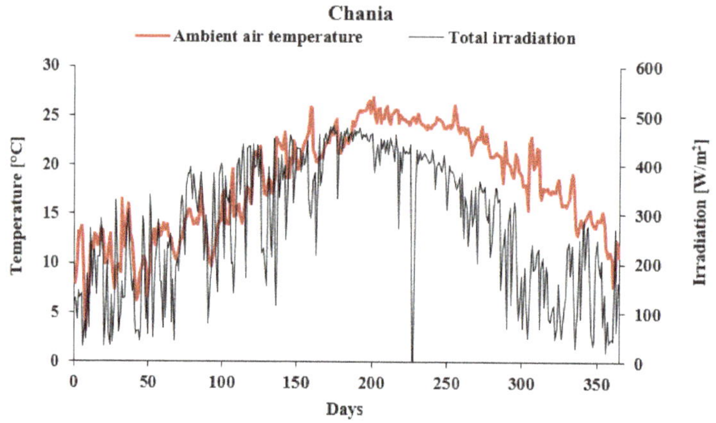

(**a**)

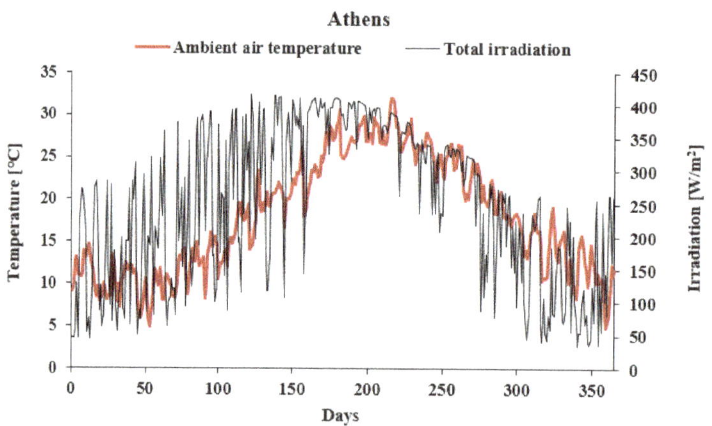

(**b**)

Figure A1. *Cont.*

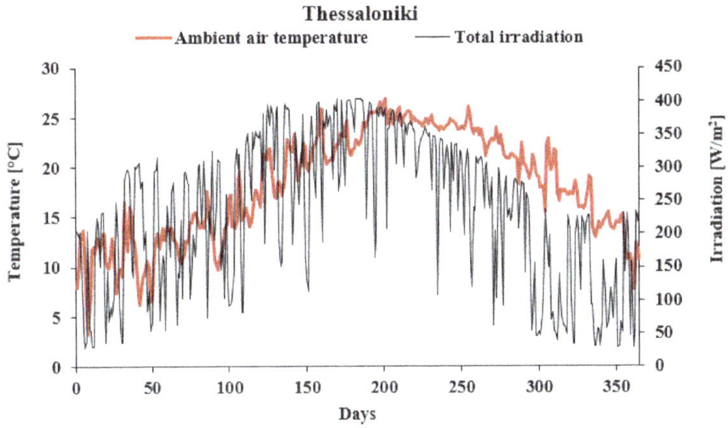

(c)

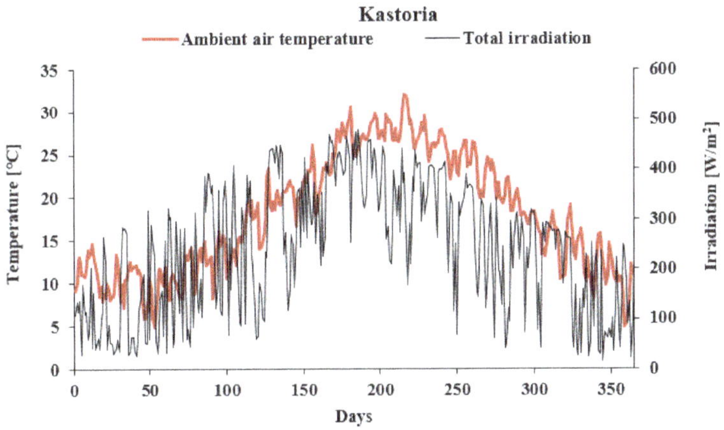

(d)

Figure A1. Average daily ambient air temperature and total solar irradiation for the cities of (**a**) Chania, (**b**) Athens, (**c**) Thessaloniki, and (**d**) Kastoria.

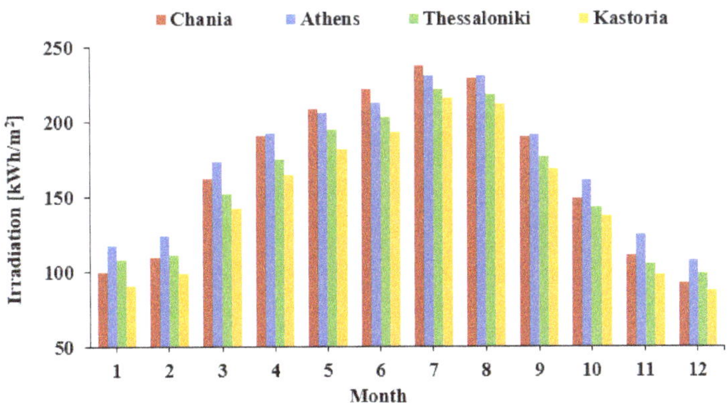

Figure A2. Monthly PV in-plane solar irradiation for the four examined cities.

Appendix B. Thermal Analysis Results

This appendix includes the calculation results of the thermal analysis simulations for every number of stories and for each of the four examined locations, given in Tables A1–A4. These results are also included in Figures 3–6.

Table A1. Thermal analysis calculations for the city of Chania.

Number of Stories	Specific Heating Energy [kWh/m^2]	Specific Cooling Energy [kWh/m^2]	Specific Electricity for Heating [kWh/m^2]	Specific Electricity for Cooling [kWh/m^2]
1	36.39	20.44	9.86	4.75
2	24.27	20.87	6.71	4.83
3	20.01	21.71	5.58	5.01
4	17.91	22.23	5.02	5.13
5	16.67	22.59	4.69	5.21
6	15.87	22.81	4.48	5.26
7	15.32	22.97	4.33	5.29

Table A2. Thermal analysis calculations for the city of Athens.

Number of Stories	Specific Heating Energy [kWh/m^2]	Specific Cooling Energy [kWh/m^2]	Specific Electricity for Heating [kWh/m^2]	Specific Electricity for Cooling [kWh/m^2]
1	41.73	37.09	11.58	8.92
2	28.22	35.75	8.00	8.61
3	21.32	35.39	6.10	8.51
4	21.22	35.04	6.11	8.45
5	19.87	34.86	5.74	8.41
6	18.99	34.71	5.50	8.37
7	18.69	34.41	5.42	8.30

Table A3. Thermal analysis calculations for the city of Thessaloniki.

Number of Stories	Specific Heating Energy [kWh/m^2]	Specific Cooling Energy [kWh/m^2]	Specific Electricity for Heating [kWh/m^2]	Specific Electricity for Cooling [kWh/m^2]
1	61.98	22.77	20.25	5.45
2	45.66	22.64	14.91	5.43
3	39.97	22.66	13.14	5.44
4	37.17	22.64	12.27	5.44
5	35.52	22.59	11.76	5.43
6	34.45	22.55	11.43	5.42
7	33.71	22.49	11.20	5.41

Table A4. Thermal analysis calculations for the city of Kastoria.

Number of Stories	Specific Heating Energy [kWh/m^2]	Specific Cooling Energy [kWh/m^2]	Specific Electricity for Heating [kWh/m^2]	Specific Electricity for Cooling [kWh/m^2]
1	99.28	7.87	51.83	1.93
2	76.48	9.07	37.93	2.21
3	68.62	9.56	33.46	2.32
4	64.73	9.82	31.23	2.39
5	62.45	9.93	30.04	2.41
6	60.97	10.02	29.19	2.43
7	59.94	10.07	28.60	2.44

References

1. Energy Statistics—An Overview. Available online: https://ec.europa.eu/eurostat/statistics-explained/index.php?title=Energy_statistics_-_an_overview (accessed on 29 March 2024).
2. ODYSSEE-MURE. Final Energy Consumption by Energy Sector in EU. Available online: https://www.odyssee-mure.eu/publications/efficiency-by-sector/overview/final-energy-consumption-by-sector.html (accessed on 29 March 2024).
3. National Energy Department. Statistical Data. Available online: https://ypen.gov.gr/ (accessed on 17 June 2024).
4. Statista. World Electricity Consumption Share by Sector. Available online: https://www.statista.com/statistics/859150/world-electricity-consumption-share-by-sector/ (accessed on 10 April 2024).
5. Kim, B.-J.; Kim, S.; Go, M.; Joo, H.-J.; Jeong, J.-W. Applicability performance evaluation of cascade heat pump for building electrification in winter. *J. Build. Eng.* **2024**, *82*, 108406. [CrossRef]
6. Share of Electricity Production from Renewables. Our World in Data. Available online: https://ourworldindata.org/grapher/share-electricity-renewables (accessed on 29 March 2024).
7. Hong, T.; Lee, S.H.; Zhang, W.; Sun, K.; Hooper, B.; Kim, J. Nexus of electrification and energy efficiency retrofit of commercial buildings at the district scale. *Sustain. Cities Soc.* **2023**, *95*, 104608. [CrossRef]
8. Bellos, E. Progress in beam-down solar concentrating systems. *Prog. Energy Combust. Sci.* **2023**, *97*, 101085. [CrossRef]
9. Li, Y.; Rosengarten, G.; Stanley, C.; Mojiri, A. Electrification of residential heating, cooling and hot water: Load smoothing using onsite photovoltaics, heat pump and thermal batteries. *J. Energy Storage* **2022**, *56*, 105873. [CrossRef]
10. Eurostat. Database. Available online: https://ec.europa.eu/eurostat/data/database (accessed on 4 April 2024).
11. Bódis, K.; Kougias, I.; Jäger-Waldau, A.; Taylor, N.; Szabó, S. A high-resolution geospatial assessment of the rooftop solar photovoltaic potential in the European Union. *Renew. Sustain. Energy Rev.* **2019**, *114*, 109309. [CrossRef]
12. O'Connor, J. Rooftop Solar PV Country Comparison Report. CAN Europe 2022. Available online: https://caneurope.org/rooftop-solar-pv-comparison-report/ (accessed on 3 April 2024).
13. Alipour, M.; Irannezhad, E.; Stewart, R.A.; Sahin, O. Exploring residential solar PV and battery energy storage adoption motivations and barriers in a mature PV market. *Renew. Energy* **2022**, *190*, 684–698. [CrossRef]
14. Sagani, A.; Mihelis, J.; Dedoussis, V. Techno-economic analysis and life-cycle environmental impacts of small-scale building-integrated PV systems in Greece. *Energy Build.* **2017**, *139*, 277–290. [CrossRef]
15. Arcos-Vargas, A.; Cansino, J.M.; Román-Collado, R. Economic and environmental analysis of a residential PV system: A profitable contribution to the Paris agreement. *Renew. Sustain. Energy Rev.* **2018**, *94*, 1024–1035. [CrossRef]
16. Rae Website. Available online: https://www.rae.gr/ (accessed on 24 May 2024).
17. Bruck, A.; Diaz Ruano, S.; Auer, H. Values and implications of building envelope retrofitting for residential Positive Energy Districts. *Energy Build.* **2022**, *275*, 112493. [CrossRef]
18. HELAPCO-Hellenic Association of Photovoltaic Companies. HELAPCO-Hellenic Association of Photovoltaic Companies 2024. Available online: https://helapco.gr (accessed on 28 June 2024).
19. Bastos, J.; Monforti-Ferrario, F.; Melica, G. *GHG Emission Factors for Electricity Consumption*; European Commission: Luxembourg, 2024.
20. Statista. Greece: Solar PV Electricity Output 2013–2023. Available online: https://www.statista.com/statistics/497570/electricity-production-from-solar-in-greece/ (accessed on 28 June 2024).
21. Bellos, E.; Iliadis, P.; Papalexis, C.; Rotas, R.; Mamounakis, I.; Sougkakis, V.; Nikolopoulos, N.; Kosmatopoulos, E. Holistic renovation of a multi-family building in Greece based on dynamic simulation analysis. *J. Clean. Prod.* **2022**, *381*, 135202. [CrossRef]
22. Dermentzis, G.; Ochs, F.; Franzoi, N. Four years monitoring of heat pump, solar thermal and PV system in two net-zero energy multi-family buildings. *J. Build. Eng.* **2021**, *43*, 103199. [CrossRef]
23. Thebault, M.; Gaillard, L. Optimization of the integration of photovoltaic systems on buildings for self-consumption–Case study in France. *City Environ. Interact.* **2021**, *10*, 100057. [CrossRef]
24. Feng, X.; Ma, T.; Yamaguchi, Y.; Peng, J.; Dai, Y.; Ji, D. Potential of residential building integrated photovoltaic systems in different regions of China. *Energy Sustain. Dev.* **2023**, *72*, 19–32. [CrossRef]
25. Roumpakias, E.; Zogou, O.; Stamatellou, A.-M. Optimization of Electrical and Thermal Storage in a High School Building in Central Greece. *Energies* **2024**, *17*, 1966. [CrossRef]
26. Stavrakakis, G.M.; Bakirtzis, D.; Drakaki, K.K.; Yfanti, S.; Katsaprakakis, D.A.; Braimakis, K.; Langouranis, P.; Terzis, K.; Zervas, P.L. Application of the Typology Approach for Energy Renovation Planning of Public Buildings' Stocks at the Local Level: A Case Study in Greece. *Energies* **2024**, *17*, 689. [CrossRef]
27. Sougkakis, V.; Lymperopoulos, K.; Nikolopoulos, N.; Margaritis, N.; Giourka, P.; Angelakoglou, K. An Investigation on the Feasibility of Near-Zero and Positive Energy Communities in the Greek Context. *Smart Cities* **2020**, *3*, 362–384. [CrossRef]
28. Vögele, S.; Poganietz, W.-R.; Kleinebrahm, M.; Weimer-Jehle, W.; Bernhard, J.; Kuckshinrichs, W.; Weiss, A. Dissemination of PV-Battery systems in the German residential sector up to 2050: Technological diffusion from multidisciplinary perspectives. *Energy* **2022**, *248*, 123477. [CrossRef]
29. Christiansen, J. European Market Outlook for Residential Battery Storage. Available online: https://resource-platform.eu/wp-content/uploads/files/statements/2820-SPE-EU-Residential-Market-Outlook-07-mr.pdf (accessed on 13 April 2024).
30. Orth, N.; Munzke, N.; Weniger, J.; Messner, C.; Schreier, R.; Mast, M.; Meissner, L.; Quaschnig, V. Efficiency characterization of 26 residential photovoltaic battery storage systems. *J. Energy Storage* **2023**, *65*, 107299. [CrossRef]

31. Liu, J.; Ma, T.; Wu, H.; Yang, H. Study on optimum energy fuel mix for urban cities integrated with pumped hydro storage and green vehicles. *Appl. Energy* **2023**, *331*, 120399. [CrossRef]
32. Forrousso, S.; Kaitouni, S.I.; Mana, A.; Wakil, M.; Jamil, A.; Brigui, J.; Azzouzi, H. Optimal sizing of off-grid microgrid Building-Integrated-Photovoltaic system with battery for a Net Zero Energy Residential Building in different climates of Morocco. *Results Eng.* **2024**, *22*, 102288. [CrossRef]
33. Chreim, B.; Esseghir, M.; Merghem-Boulahia, L. Recent sizing, placement, and management techniques for individual and shared battery energy storage systems in residential areas: A review. *Energy Rep.* **2024**, *11*, 250–260. [CrossRef]
34. DesignBuilder Software Ltd. *DesignBuilder*, Ver. 2024; DesignBuilder Ltd.: Stroud, UK, 2024. Available online: https://designbuilder.co.uk/ (accessed on 17 January 2023).
35. European Commission. JRC Photovoltaic Geographical Information System (PVGIS). Available online: https://re.jrc.ec.europa.eu/pvg_tools/en/#api_5.1 (accessed on 1 May 2024).
36. Census Building 2011—ELSTAT. Available online: https://www.statistics.gr/census-buildings-2011 (accessed on 4 April 2024).
37. TEE. Technical Guidelines of Technical Chamber of Greece. 2020. Available online: https://web.tee.gr/ (accessed on 9 January 2024).
38. Kitsopoulou, A.; Bellos, E.; Sammoutos, C.; Lykas, P.; Vrachopoulos, M.G.; Tzivanidis, C. A detailed investigation of thermochromic dye-based roof coatings for Greek climatic conditions. *J. Build. Eng.* **2024**, *84*, 108570. [CrossRef]
39. Kitsopoulou, A.; Bellos, E.; Lykas, P.; Vrachopoulos, M.G.; Tzivanidis, C. Multi-objective evaluation of different retrofitting scenarios for a typical Greek building. *Sustain. Energy Technol. Assess.* **2023**, *57*, 103156. [CrossRef]
40. Sharp-Solar Panel. Available online: https://www.sharp.eu/solar-energy/find-a-solar-panel (accessed on 21 March 2024).
41. *ISO 6946:2017*; Building Components and Building Elements—Thermal Resistance and Thermal Transmittance—Calculation Methods. ISO: Geneva, Switzerland, 2017. Available online: https://www.iso.org/standard/65708.html (accessed on 5 May 2024).
42. Karteris, M.; Theodoridou, I.; Mallinis, G.; Papadopoulos, A.M. Façade photovoltaic systems on multifamily buildings: An urban scale evaluation analysis using geographical information systems. *Renew. Sustain. Energy Rev.* **2014**, *39*, 912–933. [CrossRef]
43. Sassenou, L.-N.; Olivieri, L.; Olivieri, F. Challenges for positive energy districts deployment: A systematic review. *Renew. Sustain. Energy Rev.* **2024**, *191*, 114152. [CrossRef]
44. Sassenou, L.-N.; Olivieri, F.; Civiero, P.; Olivieri, L. Methodologies for the design of Positive Energy Districts: A scoping literature review and a proposal for a new approach (PlanPED). *Build. Environ.* **2024**, *260*, 111667. [CrossRef]
45. Zhang, X.; Penaka, S.R.; Giriraj, S.; Sánchez, M.N.; Civiero, P.; Vandevyvere, H. Characterizing Positive Energy District (PED) through a Preliminary Review of 60 Existing Projects in Europe. *Buildings* **2021**, *11*, 318. [CrossRef]
46. CORDIS; European Commission. Rotterdam, Umea and Glasgow: Generating Exemplar Districts In Sustainable Energy Deployment | Ruggedised Project | Fact Sheet | H2020. Available online: https://cordis.europa.eu/project/id/731198 (accessed on 30 May 2024).
47. POCITYF-POCITYF. Homepage. Available online: https://pocityf.eu/ (accessed on 30 May 2024).
48. Energy Cities. Smart Together. Available online: https://energy-cities.eu/project/smarter-together/ (accessed on 30 May 2024).
49. Ascione, F.; Bianco, N.; Mauro, G.M.; Napolitano, D.F. Retrofit of villas on Mediterranean coastlines: Pareto optimization with a view to energy-efficiency and cost-effectiveness. *Appl. Energy* **2019**, *254*, 113705. [CrossRef]
50. Kitsopoulou, A.; Bellos, E.; Lykas, P.; Sammoutos, C.; Vrachopoulos, M.G.; Tzivanidis, C. A Systematic Analysis of Phase Change Material and Optically Advanced Roof Coatings Integration for Athenian Climatic Conditions. *Energies* **2023**, *16*, 7521. [CrossRef]
51. Braulio-Gonzalo, M.; Bovea, M.D.; Ruá, M.J.; Juan, P. A methodology for predicting the energy performance and indoor thermal comfort of residential stocks on the neighbourhood and city scales: A Case Study Spain. *J. Clean. Prod.* **2016**, *139*, 646–665. [CrossRef]
52. Ministry of Environment and Energy. *Energy Inspections of Buildings, Statistical Analysis for the Year 2021 and the Period 2011–2019 06/20220*; Ministry of Environment and Energy: Athens, Greece, 2016. Available online: https://bpes.ypeka.gr/wp-content/uploads/TRANSLATION.11.04.2016.pdf (accessed on 1 April 2024).
53. Greece 2.0-National Recovery and Resilience Plan. Available online: https://greece20.gov.gr/ (accessed on 28 June 2024).
54. European Commission. €1 billion Greek State Aid. Available online: https://ec.europa.eu/commission/presscorner/detail/en/ip_24_1765 (accessed on 28 June 2024).
55. Inforegio—Connecting Greece's Cyclades Islands to the Mainland's Power Grid. Available online: https://ec.europa.eu/regional_policy/en/projects/greece/connecting-greece-s-cyclades-islands-to-the-mainland-s-power-grid (accessed on 28 June 2024).
56. Makrygiorgou, J.J.; Karavas, C.-S.; Dikaiakos, C.; Moraitis, I.P. The Electricity Market in Greece: Current Status, Identified Challenges, and Arranged Reforms. *Sustainability* **2023**, *15*, 3767. [CrossRef]

MDPI AG
Grosspeteranlage 5
4052 Basel
Switzerland
Tel.: +41 61 683 77 34

Designs Editorial Office
E-mail: designs@mdpi.com
www.mdpi.com/journal/designs

www.ingramcontent.com/pod-product-compliance
Lightning Source LLC
Chambersburg PA
CBHW040759250926
51850CB00007BA/1037
* 9 7 8 3 7 2 5 8 6 3 7 2 3 *